T0271878

Cutting-edge Technologies in Biological Sensing and Analysis

RIVER PUBLISHERS SERIES IN BIOMEDICAL ENGINEERING

Series Editors:

Dinesh Kant Kumar
RMIT University, Australia

The "River Publishers Series in Biomedical Engineering" is a series of comprehensive academic and professional books which focus on the engineering and mathematics in medicine and biology. The series presents innovative experimental science and technological development in the biomedical field as well as clinical application of new developments.

Books published in the series include research monographs, edited volumes, handbooks and textbooks. The books provide professionals, researchers, educators, and advanced students in the field with an invaluable insight into the latest research and developments.

Topics covered in the series include, but are by no means restricted to the following:

- Biomedical engineering
- Biomedical physics and applied biophysics
- Bio-informatics
- Bio-metrics
- Bio-signals
- Medical Imaging

For a list of other books in this series, visit www.riverpublishers.com

Cutting-edge Technologies in Biological Sensing and Analysis

Editors

Anh Hung Nguyen

Ph.D., Research Scientist, UC Irvine, USA

Sang Jun Sim

Ph.D., Professor of Chemical and
Biological Engineering, Korea University

Hung Cao

Associate Professor of Electrical and Biomedical Engineering and
Computer Science, UC Irvine, USA

River Publishers

Routledge
Taylor & Francis Group

NEW YORK AND LONDON

Published 2023 by River Publishers
River Publishers
Alsbjergvej 10, 9260 Gistrup, Denmark
www.riverpublishers.com

Distributed exclusively by Routledge
605 Third Avenue, New York, NY 10017, USA
4 Park Square, Milton Park, Abingdon, Oxon OX14 4RN

Cutting-edge Technologies in Biological Sensing and Analysis / by Anh Hung Nguyen, Sang Jun Sim, Hung Cao.

Routledge is an imprint of the Taylor & Francis Group, an informa business

ISBN 978-87-7022-379-9 (hardback)
ISBN 978-87-7004-051-8 (paperback)
ISBN 978-10-0381-062-9 (online)
ISBN 978-10-3262-943-8 (master ebook)

While every effort is made to provide dependable information, the publisher, authors, and editors cannot be held responsible for any errors or omissions.

Contents

3 Animal Models and Techniques Used in Cardiovascular Research   **95**

Jimmy Zhang, Samantha Lauren Laboy-Segarra, Anh H. Nguyen, Juhyun Lee, and Hung Cao

4 Phage-based Biosensors   **145**

Aarcha Shanmugha Mary, Vinodhini Krishnakumar, and Kaushik Rajaram

Preface

In the last few decades, advanced technologies have transformed the ways we conduct biological studies and deliver healthcare. Micro- and nanofabrication provided smaller sensors and systems with better performance; while novel materials enabled biocompatible implants and even degradable ones. Telecommunications and machine learning-based approaches have further brought those biomedical systems to the next level, with incredible computing capacity and real-time analysis.

The editors, with their complementary expertise in bioengineering, biological sciences, and electronics, are life-time collaborators. Therefore, in this book, we would like to cover a broad field of novel technologies used in biological assessment and analysis for humans, animal models and in-vitro platforms, in both health monitoring and biological studies. Biological scientists, bioengineers, and even entrepreneurs, at all levels, would be expected audience. We also hope this book will motivate and inspire the next generation and those who want to gain knowledge or enter related fields.

List of Figures

List of Tables

List of Contributors

Briand, Josephine, *School of Medicine, Oregon Health & Science University, USA*

Cao, Hung, *Department of Electrical Engineering and Computer Science, UC Irvine, USA; Department of Biomedical Engineering, UC Irvine, USA; Department of Computer Science, UC Irvine, Irvine, CA, USA*

Chiao, J.-C., *Department of Electrical and Computer Engineering, Southern Methodist University, USA*

Daniel Jilani, *Department of Electrical Engineering and Computer Science, UC Irvine, USA*

Dominguez, Cynthia, *Department of Bioengineering, University of Texas at Arlington, USA*

Jones, Carolyn, *University of Neurology, Oregon Health & Science University, USA*

Krishnakumar, Vinodhini, *Department of Lifescience, Central University of Tamil Nadu, India*

Laboy-Segarra, Samantha Lauren, *Department of Bioengineering, University of Texas, USA*

Lee, Jong Uk, *Department of Chemical and Biological Engineering, Korea University, Korea*

Lee, Juhyun, *Department of Bioengineering, University of Texas, USA*

Lim, Miranda M., *Department of Neurology, Oregon Health & Science University, USA*

Mary, Aarcha Shanmugha, *Department of Microbiology, Central University of Tamil Nadu, India*

Naderi, Amir Mohammad, *Department of Electrical Engineering and Computer Science, UC Irvine, USA*

Ngo, Thuy T. M., *School of Medicine, Oregon Health & Science University, USA*

Nguyen, Anh H., *Department of Electrical and Computer Engineering, UC Irvine, USA; Sensoriis Inc, Edmonds, USA*

Rajaram, Kaushik, *Department of Microbiology, Central University of Tamil Nadu, India*

Sim, Sang Jun, *Department of Chemical and Biological Engineering, Korea University, Korea*

Teranikar, Tanveer, *Department of Bioengineering, University of Texas at Arlington, USA*

Tran, Donna H., *Department of Biomedical Engineering, UC Irvine, USA*

Vishwanath, Manoj, *Department of Computer Science, UC Irvine, USA*

Wagner, Josiah T., *School of Medicine, Oregon Health & Science University, USA*

Yu, Shuai, *Rockley Photonics, USA; Department of Bioengineering, The University of Texas at Arlington, USA*

Zhang, Jimmy, *Department of Biomedical Engineering, UC Irvine, USA*

List of Abbreviations

2D	Two-dimensional
3D	Three-dimensional
AASM	American Academy of Sleep Medicine
AC	Alternative current
AFM	Atomic force microscopy
AFP	α-fetoprotein
Ag@SiO$_2$	SiO$_2$ shell-isolated silver nanoparticles
AHA	American Heart Association
AHE	Adaptive Histogram Equalization
ANN	Artificial neural network
ANOVA	Analysis of variance
ARANP	Ag-Au alloy shell Raman reporter
ATP	Adenosine triphosphate
AUC	Area under ROC
Au	Gold nanoparticle
AuNR/Ag	Silver-coated gold nanorods
BB-PEG	Final tissue clearing medium (refractive index (RI) 1.543), which is composed of 75% benzylbenzoate (BB), 22% PEGMMA, and 5% Quadrol
CA125	Cancer antigen
CART	Classification and regression tree
CB	Circulating biomarkers
CCD	Charged-coupled device
cDNA	complementary DNA
CDR	Complementarity determining regions
CD44	Cell surface adhesion receptor
CEA	Carcinoembryonic antigen
CFU	Colony forming units
celB	gene encoding cellulase enzymes
CLAHE	Contrast Limited Adaptive Histogram Equalization

CLARITY	Clear Lipid-exchanged Acrylamide-hybridized Rigid Imaging/Immunostaining/In situ hybridization-compatible Tissue-hydrogel
CNNs	Convolutional neural networks
CNS	Central neuron system
CO	Cardiac output
CT	Computed tomography
CTC	Circulating tumor cells
CUBIC	Clear, Unobstructed Brain Imaging Cocktails and computational analysis
CVDs	Cardiovascular diseases
CW	Continuous wave
DAC	Data-acquisition card
DAPI	4,6-diamidino-2-phenylindole
DC	direct current
DCM	Dilated cardiomyopathy
Dd	Diameter diastole
DENV NS1	Dengue virus non-structural protein 1
DISCO	Dimensional imaging of solvent-cleared organs
DL	Diameter Long
DNA	Deoxyribonucleic acid
DPF	Days post fertilization
Ds	Diameter systole
DS	Diameter Short
DT	Decision trees
ED	End-diastole
EDA	End-diastole area
EDV	End-diastolic volume
EEG	electroencephalogram
EF	Ejection fraction
EGF	Epidermal growth factor
EGFR	EGF receptor
EIS	Electrochemical impedance spectroscopy
ELISA	Enzyme-linked immunosorbent assay
EM	Electromagnetic
EOT	Extraordinary optical transmission
EpCAM	Targeted epithelial cells adhesion molecules
ES	End-systole
ESA	End-systole area

ESV	End-systolic volume
EV	Extracellular vesicle
Fab	Fragment antigen binding
FAC	Fractional area change
FC	Crystallisable fragment
FDA	Food and Drug Administration
FDOT	Fluorescence diffuse optical tomography
FISH	Fluorescence in situ hybridization
FOV	Field of view
FPS	Frames per second
FS	Fractional Shortening
FT	Fourier transform
FU	Focused ultrasound
Fvs	Variable fragment
GCS	Glasgow coma scale
GFP	Green fluorescent protein
GMM	Gaussian mixture model
hdf	Hours post fertilization
HDF5	Hierarchical data format version 5
HE4	Human epididymis protein
HR	Heart rate
ICCD	Intensified charged-coupled device
ICG	Indocyanine green
IFT	Inverse Fourier transform
IGF	Insulin-like growth factor
IGF-1	Insulin-like growth factor
IgG	Immunoglobulin G
inaW	ice nucleation gene
IoC	Intersection over union
IoV	Intersection over Union
ITNAA	Isothermal nucleic acid amplification
IV	Intravenous
KLB	Keller Lab Block
lacZ	Gene encoding beta-galactosidase
LAD	left anterior descending
LAMP	Loop-mediated isothermal amplification
LASSO	Least absolute shrinkage and selection operator
LCST	Lower critical solution temperature
LOD	Limit of detection

LPS	Liposaccharides
LSP	Localized surface plasmons
LSPR	Localized surface plasmon resonance
lux genes	luminescent reporter genes
mCpG	Anti-methyl-cytosine
ME	Magneto-elastic
miRNAs	microRNAs
ML	Normal
MLWA	Modified long wavelength approximation
MOF	Metal-organic framework
MRI	Magnetic resonance imaging
mRNA	messenger RNA
MRSA	Methicillin-resistant *Staphylococcus aureus*
mTBI	Mild TBI
NIR	Near-infrared
NSCLC	Non-small cell lung cancer
OP	Optoacoustic
Ori	Origin of replication
PAC	Phase-amplitude coupling
PACT	Photoacoustic computed tomography
PAM	Photoacoustic microscopy
PC-1 and PC-2	Two principal components
PCA	Principal component analysis
PCR	Polymerase chain reaction
PD	Phage display
PDF	Probability density function
PEDF	Rich pigment epithelium-derived factors
PEGMMA	Polyethylene glycol monomethyl ether monomethacrylate
PET	Positron emission tomography
PLV	Phase locking value
PNA	Peptide nucleic acid
PSA	prostate-specific antigen
PSD	Power spectral density
PSF	Point spread function
PSP	Propagating surface plasmon
PSS	Polystyrene sulfonate
QCM-D	Quartz crystal microbalance
QD	Quantum dots

QEEG	quantitative electroencephalogram
qPCR	quantitative PCR
RBC	Red blood cells
RBF	Radial basis function
RBP	Receptor binding proteins
RF	Radio-frequency
RFE	Recursive feature elimination
RI	Refractive index
RNA	Ribonucleic acid
ROC	Receiver operating characteristic
ROI	Region of interest
RP	Reporter phage
RT	Room temperature
SA chips	Streptavidin functionalized chips
SAM	Self-assembled monolayer
SBP	Streptavidin binding peptide
SBR	Signal-to-background ratio
scFv	Single chain variable fragment
SdAb	Single domain antibodies
SEM	Scanning electronic microscopy
Sen	Sensitivity
SERS	Surface-enhanced Raman scattering
SiO4	Silicate
SNR	Signal-to-noise ratio
SP	Surface plasmon
SPE	Screen-printed electrode
Spec	Specificity
SPECT	Single-photon emission computerized tomography
SPION	Superparamagnetic iron oxide nanoparticles
SPPs	Surface plasmon polaritons
SPR	Surface plasmon resonance
SV	Stroke volume
SVM	Support vector machines
SWNTs	Single-walled carbon nanotubes
SYBR Green	dsDNA binding dye
TA	thermoacoustic
TB	Tuberculosis
tB	Tert-butanol

TBI	Traumatic brain injury
TRUE	Time-reversed ultrasonically encoded
TTNtv	Titin truncated variants
UMF	Ultrasound-modulated fluorescence
USF	Ultrasound switchable fluorescence
UV	Ultraviolet
WHO	World Health Organization
XFCT	X-ray fluorescence computed tomography
XGB	XGBoost
XLCT	X-ray luminescence computed tomography
ZACAF	Zebrafish Automatic Cardiovascular Assessment Framework
zf	zebrafish

1

Plasmonic Sensors for Applications in Liquid Biopsies

Jong Uk Lee and Sang Jun Sim

Department of Chemical and Biological Engineering,
Korea University, Korea

Abstract

A liquid biopsy is a non-invasive means of gaining insights into the dynamics of disease using a patient's body fluids of patients. It has recently attracted much interest in cancer diagnosis and prognosis. Unlike a conventional tissue biopsy, a liquid biopsy provides a snapshot of the disease from the primary diagnosis along with recognizing distant tumor locations. Additionally, liquid biopsies can be used for repeated sampling of tumor markers and can be adjusted depending on the patient's response to personalized treatment. Among the various techniques developed for liquid biopsies, the nanoplasmonic biosensor presents a promising approach due to its high sensitivity and selectivity, as well as multiplexing capability for simultaneous target detection. This chapter will focus on emerging technologies for the detection of various disease markers using nanoplasmonic biosensors.

Keywords: Nanostructures, Plasmonics, Biosensors, Liquid biopsy, Biomarkers

1.1 Introduction

According to the World Health Organization (WHO), cancer is the second leading cause of death worldwide and was responsible for an estimated 9.6 million deaths in 2018. In addition, cancer metastasis poses a significant threat to patient recovery and positive outcomes. Thus, continuous monitoring of patients needs to be developed. Conventional tissue biopsies and imaging techniques only provide a glimpse into the heterogeneity of tumors, and the sampling bias associated with these methods can be severe. Moreover, repeated sample collection may not be feasible in many instances [1]. In light of these limitations of conventional techniques, liquid biopsy has attracted attention as a novel alternative for non-invasive diagnosis.

Liquid biopsy is a simple, rapid, non-invasive method for diagnosing and prognosing incurable and chronic diseases based on the analysis of biomarkers (CBs) in the biological fluids of patients. This new type of biopsy is likely to revolutionize disease management by allowing for the continuous monitoring of patients, the selection of high-precision treatments, and the rapid assessment of patient response [2]. CBs are biomolecules with excellent diagnostic, prognostic and therapeutic potential. They are found in biofluids, including blood, urine, tears, and saliva. The four most common CBs are protein, circulating tumor cells (CTCs), circulating nucleic acids (e.g., mRNA, circulating tumor DNA, and micro RNA), and exosomes [3]. Using liquid biopsy to detect the CBs in clinical samples, physicians can easily establish the disease condition, prognosis, and the effects of treatment in their patients. Moreover, this method allows clinicians to provide their patients with diagnostic results in a comparatively shorter time and more importantly, without inflicting pain.

Most conventional techniques for the detection of CBs require a long testing time, various steps, high costs, and an amplification process [4]. For these reasons, CBs-based liquid biopsies are partially limited for clinical usage. Hence, it is essential to develop a low-cost, rapid, real-time, ultra-sensitive, and selective platform for the detection of CBs in liquid biopsy. Recently, the nanoplasmonic biosensor has become a promising platform for detecting CBs in liquid biopsies. A nanoplasmonic biosensor comprises metal nanoparticles or substrate via optical scattering, including surface plasmon resonance (SPR), localized surface plasmon resonance (LSPR), and surface-enhanced Raman scattering (SERS). Typically, a signal readout is generated in the form of optical spectral shifts or fingerprint spectra [5]. The nanoplasmonic biosensor is usually a non-label, highly sensitive, and selective tool for a simple experimental process designed to allow for multiple rounds of detection using smaller sample volumes. Since nanoplasmonic biosensors essentially work as optical sensors, the evaluated parameter reflects the amount of the dry mass of the target biomolecule. Therefore, the nanoplasmonic biosensor is less likely to suffer interference from the solution [6]. Hence, the nanoplasmonic biosensors are a promising platform for point-of-care testing using liquid biopsies in the near future.

In this chapter, we will describe the basic theory of plasmonic phenomena, such as SPR, LSPR, and SERS, and discuss various nanoplasmonic biosensors for use in liquid biopsies according to the types of CBs, including proteins, circulating tumor cells, circulating nucleic acids and exosome. Finally, we will conclude with the future prospects of a nanoplasmonic biosensor in liquid biopsies.

1.2 Nanoplasmonic Biosensor

Surface plasmons (SPs) are free electrons that oscillate between the boundaries of metal and dielectric interfaces and are able to excite conductive electrons when coupled with photons [7, 8]. The propagation of SPs between metal and dielectric interfaces results in the formation of surface plasmon polaritons (SPPs) [8]. which propagate along the surfaces until their energy dissipates either through heat loss or radiation, into free space [9]. This phenomenon can be categorized into two classes: propagating surface plasmons (PSPs), which can be excited on metallic films, and localized surface plasmons (LSPs), which can be excited on metallic nanoparticles [9]. Both PSPs and LSPs can enhance the electromagnetic fields (EM) in the near-field region, and therefore, have been widely used in plasmon resonance sensors, surface-enhanced Raman scattering (SERS), fluorescence enhancement, refractive index (RI) and other detections devices [10, 11]. In this section, nanoplasmonic biosensors are discussed in the context of LSPR and SERS phenomena.

1.2.1 Localized surface plasmon resonance (LSPR) biosensor

LSPR occurs when plasmons are excited by incident light and the resulting electron oscillations are confined between metallic structures (Figure 1.1) [12]. As such, only materials with either negative real or small positive imaginary dielectric constants are capable of supporting surface plasmons. Noble metal nanoparticles, such as gold and silver, are generally used due to their strong optical extinction at visible spectra, near-infrared wavelengths, and the generative ability of the LSPR phenomenon which is sensitive to the surrounding medium. However, other metals, such as copper and aluminum, have also been used for exhibiting plasmon resonance [7].

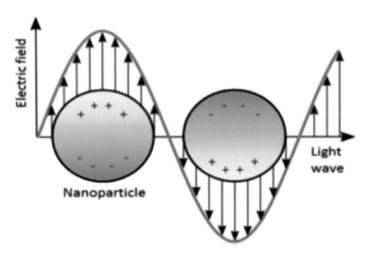

Figure 1.1   Schematic diagram illustrating the localized surface plasmon on a nanoparticle surface.

To satisfy the resonance conditions, the incident electromagnetic field has to match that of the oscillating electrons at the nanoparticle surface [13]. This is followed by polarization due to the polarity of light since electromagnetic waves behave differently [13]. Mie theory is often used to understand the factors that contribute to large increases in absorption and scattering during resonance [14]. Although Maxwell's equation is used to characterize the optical properties of LSPR for spherical metal nanoparticles isolated in a continuous surrounding medium, Mie theory represents the analytical solution to this equation in which the spherical boundary conditions are used to describe the extinction spectra for a given nanoparticle. To calculate the dielectric constants at various wavelengths and apply this theory to other convoluted shapes, the modified long wavelength approximation (MLWA) of Mie theory is applied [15]:

$$E\left(\lambda\right) = \frac{24\pi^2 N_a^3 \varepsilon_{\mathrm{D}}^{3/2} N}{\lambda \ln(10)} \left[\frac{\varepsilon_i}{\left(\varepsilon_r + \xi\varepsilon_{\mathrm{D}}\right)^2 + \varepsilon_i^2}\right],$$

where N_{a} is the radius of the particle, λ is the wavelength of the incident light, $\varepsilon = \varepsilon_r + i\varepsilon_i$ is the complex dielectric constant of the bulk metal, ε_{D} is the dielectric constant of the surrounding medium, N is the electron density, and ξ accounts for the shape of the particle.

This equation indicates that the LSPR of noble metal nanoparticles, appearing as strong absorption peaks in the optical spectra, are influenced by factors such as nanoparticle shape, incident light wavelength, composition, and the surrounding environment. Particularly, the effects of the surrounding dielectric constant on the extinction spectra of the plasmonic nanoparticle are vital for the study of LSPR biosensing, especially for LSPR extinction and scattering wavelength maximum, as they are sensitive to the dielectric constant. Since biological analyte binds to the surface of nanoparticles, alterations in the refractive index occur on the nanoparticle surface, resulting in LSPR peak frequency shifts. The shift of the LSPR frequency upon adsorbate coupling is explained by the following formula [12, 14]:

$$\Delta = m(\Delta n) \left[1 - \exp\left(\frac{-2d}{I_d} \right) \right],$$

where m is the refractive index sensitivity, Δn is the change in refractive index due to adsorbate binding, d is the effective adsorbate layer thickness, and I_d is the electromagnetic field decay length (approximated as an exponential decay). The main variables that determine the quantum of LSPR shifts are the difference in the refractive index of the absorbate relative to the solution (Δn) and the size (d) of the analyte binding to the nanoparticle's surface. According to the Drude model of the electronic structure of metals, the refractive index sensitivity (m) is the slope of the plotted line of LSPR frequency versus the refractive index, which is assumed to be linear (within a relatively small range of refraction index) [16]. Another variable that affects LSPR shifts is the electromagnetic field decay length, which has been previously shown to be sensitive to the nanoparticle shape [17].

Asymmetrical nanoparticles such as rod-shaped and bridge-shaped nanoparticles are more sensitive to changes in the localized refractive index to spherical nanoparticles, which is due to the shift in the LSPR frequency from blue to red as refractive index sensitivity increases linearly [18, 19]. For instance, a large rod-shaped nanoparticle with an aspect ratio of 3.6 was reported to show more sensitive change of localized refractive index spherical nanoparticle of 50 nm in diameter [20]. In terms of applying this to LSPR biosensors, biosensors that can detect cancer biomarkers at the attomolar ($\sim$ 1 aM) level are currently being fabricated [20–22].

LSPR biosensors are capable of detecting even infinitesimal amounts of markers of incurable diseases (e.g., Alzheimer's disease and cancer) in biological samples and are being applied for the ultra-sensitive detection of such biomarkers [23–26]. Relatively, nanoparticles with a diameter of 30–100 nm have an electromagnetic field decay length similar to that of other chemical and biological molecules, thereby allowing for the detection of the analyte binding due to their inherent sensitivity to the metal nanoparticles [27]. Furthermore, since LSPR biosensors can analyze changes in the RI of metal nanoparticle surfaces in real time, they are used as a platform to identify incurable diseases by monitoring biological phenomena, such as RNA splicing, DNA methylation, and immune system signal pathway in real-time [28–30].

1.2.2 SERS biosensor

Raman scattering is a type of inelastic light scattering that results from the gain or loss of energy in incident light via the scattering of molecules. Changes in the energy of scattered light, represented by the frequency shifts, are the result of photon-molecule interactions [31]. Due to the vibrational and rotational states of chemical bonding, each molecule has a distinctive Raman spectrum, making this "fingerprint" spectrum a valuable tool for analytic science. However, before the discovery of the SERS effect, and on extremely rare occasions, inelastic scattering hindered the practical application of Raman scattering [32].

SERS occurs when analyte molecules are adsorbed on the surface of nanoscale metal substrates. In particular, metals with high plasmon resonance, such as gold and silver, amplify the SERS signal. The signal amplification of SERS can generally be attributed to the combination of two mechanisms: electromagnetic enhancement and chemical enhancement [31]. The mechanism of electromagnetic enhancement involves the interaction between incident light and scattered light with electrons on the metal surfaces, resulting in the formation of SPPs. This excitation of SPPs results in a strong local enhancement of the electromagnetic field, which, when combined with the intensity of incident light, results in the formation of so-called electromagnetic hotspots on the metal nanostructure. The scattered photons undergo a similar interaction, thereby resulting in further enhancement. This overall electromagnetic enhancement effect has been extensively studied by Zeman and Schatz [33], and is defined by the following equation:

$$E = |E(\omega)|^2 |E(\omega')|^2,$$

where $E(\omega)$ and $E(\omega')$ denote local electromagnetic field enhancement at the incident light frequency ω and Raman-scattered frequency (ω'). In most cases, the difference between ω and (ω') is minute, which provides the simplified correlation between the SERS enhancement factor and electromagnetic field enhancement: $E=|E(\omega')|^4$ [34]. Electromagnetic enhancement is not only a major mechanism in SERS signal amplification, but is also unique to substrates and does not depend on the molecules. In comparison, chemical enhancement involves the formation of electrochemical bonds between the molecules and metal substrates [35]. As a consequence of the complex charge transfers, such as interaction, additional SERS signal enhancements are produced. On the contrary, chemical enhancement is considered a minor enhancement mechanism, as it depends on the type of analyte molecules present and the type of interaction with the metal.

The SERS effect amplifies the signal from the normal Raman spectrum by over 10^{10}-fold, thereby allowing the detection of single molecule analytes on the surface of metal substrates [36]. This superb sensitivity, in combination with its capability for specific fingerprinting and multiplexing, makes SERS an attractive technique in medical applications that aim to identify specific biomolecules dissolved in a complex mixture of body fluids. Disease development is often initiated via changes in cellular processes, which can be identified due to the abnormal expression levels of several biomarker molecules. Using relatively simple procedures, SERS biosensors, which are enabled with significantly powerful quantitative measurement of biomolecule panels, are able to guarantee the highly accurate diagnosis of various diseases [37]. Currently, biomarkers from various diseases are detected using SERS-based sensing techniques. For example, several microRNAs (miRNAs) detected from three types of breast cancer were detected and identified substrate using nanostructured SERS gold substrate [38, 39]. SERS has also been utilized to detect multiple proteins specifically related to diseases such as cancer and infectious diseases [29, 30]. Thus, SERS–based detection methods can also be applied as an analytical tool in the field of molecular pathologies, such as in the study of epigenetic modifications in DNA and RNA splicing [40, 41].

1.3 Detection of Liquid Biopsy Biomarkers

1.3.1 Proteins

Proteins are well known for their diverse roles in living organisms, including cell structure, metabolism, and the regulation of cell cycle. Due to the vital role of proteins in every cell, the anomalous expression or mutated protein expression is often a distinctive hallmark of various diseases, thereby making proteins significant biomarkers [42]. Although protein biomarkers are mainly obtained from the blood, they can also be obtained from body fluids such as urine or saliva. Some widely accepted methods for the analysis of protein biomarkers include enzyme-linked immunosorbent assay (ELISA), western blotting, and immunofluorescence. However, the use of protein biomarkers for the detection of disease also has disadvantages. The majority of protein biomarkers exist in trace amounts in bodily fluids, and cannot be amplified in the same manner as nucleic acids using polymerase chain reaction (PCR). Moreover, some highly abundant proteins, such as albumins in blood samples, can result in background noise, which may interfere with the detection result, thus further complicating the analysis. Therefore, for protein biomarkers to be accepted as standard diagnostic molecules, methods with high sensitivity and excellent reliability are required [43]. Further requirements for protein detection include simpler procedures with shortened time spans, lower reagents cost, and in many cases, the capacity for multiplex detection.

Plasmonics-based detection methods have proven to be excellent tools for protein detection. Combined with the complementary interaction of biomolecules, the clinical diagnosis of protein biomarkers can be achieved using nanoplasmonic detection systems [44] since this method is enabled with sensitivity and selectivity of single molecule detection. A large number of methods have been developed based on the plasmonic properties of metals, and thus, methods development is not limited to just LSPR or SERS.

For the plasmonic-based detection of proteins in body fluids, single nanoparticle LSPR-based biosensors are among the most popular methods. Single nanoparticle LSPR achieves sub-picomolar, at times even sub-femtomolar detection limits, by using individual nanoparticles as ultra-small biosensors. For example, Jun et al. developed a gold nanoparticle-based LSPR sensor for the detection of a mutated p53 protein [45]. Mutated p53 protein is often found in cancer cells and is known to induce abnormal gene expression due to a lowered binding affinity to the target promoter sequence. To detect a mutated p53 protein, the DNA promoter sequence GADD45 was immobilized on the gold nanoparticles. As the p53 protein was bound on GADD45, the LSPR peak shift was measured and compared with peak shift from non-mutated p53 protein (Figure 1.2 A and B) since the peak shift varies according to the binding affinity between GADD45 and p53. Thus, p53 proteins from various breast cancer and lung cancer cell lines were examined using this LSPR system, and by detecting the peak shift, it was possible to distinguish cell lines with mutated p53 proteins from those without the

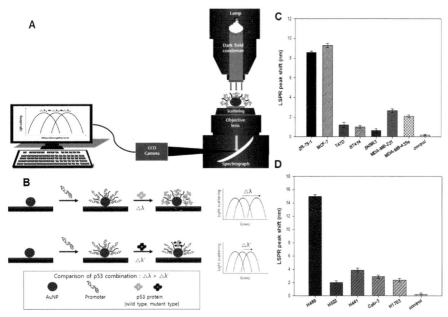

Figure 1.2 (A) Schematic diagram depicting dark-field microscope-integrated nanoplasmonic system measuring LSPR signal. (B) Schematics diagrams for identifying mutant p53 protein by its differential binding affinity against the GADD45 promoter. (C) LSPR peak shift according to the binding affinity of p53 from various breast cancer cell lines. (D) LSPR peak shift according to the binding affinity of p53 from various lung cancer cell lines. (C, D), SP6 RNA polymerase was used as a control. Reprinted from Jun et al. (2014) (copyright Wiley-VCH Verlag GmbH & Co.).

mutation (Figure 1.2 C and D). This approach is not only useful for diagnosing cancer but also for making personalized therapeutic choices suitable for individual patients. Similarly, Ma et al. detected telomerase in various cancer samples with high sensitivity and specificity by utilizing its specific binding to telomere sequences [46] Moreover, Lee et al. achieved multiplex detection of PSA, carcinoembryonic antigen (CEA), and α-fetoprotein (AFP) with a single nanoparticle LSPR sensor, with detection limits below 100 fM [26].

SERS is another method for protein detection. Distinctive fingerprints of each molecule facilitate the multiplexed detection of proteins, which is advantageous for more precise disease diagnosis. With the aid of multiple SERS tags targeting different proteins, SERS-based protein detection enables the simultaneous detection of various proteins on a single chip. For example,

Nguyen et al. detected four different protein biomarkers of breast cancer, namely, cancer antigen (CA125), HER2, human epididymis protein (HE4), and eotaxin-1, using a microfluidics-integrated SERS system (Figure 1.3 A) [29]. The SERS signal was maximized by the incorporation of the Fab fragment of an antibody, which shortened the distance between the nanoparticles, and the amplification of secondary signals by the deposition of silver atoms (Figure 1.3 B). In this work, intense signal enhancement provided exceptionally high sensitivity with a detection limit of $\leq$ 20 fM for each biomarker (Figure 1.3 C). SERS-based detection was applied to detect other cancer-related proteins. For instance, an Au-Ag alloy nanobox was utilized for the simultaneous detection of three independent proteins with clinically relevant sensitivity [47].

Many nanoplasmonic sensors have been developed for the detection of protein biomarkers. In general, the plasmonics-based detection of protein

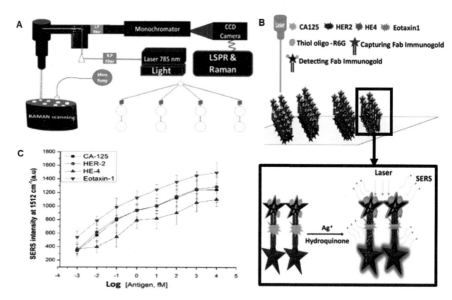

Figure 1.3   (A) Schematic diagram of microfluidics-integrated SERS system for multiplex detection of protein biomarkers. (B) Schematic diagram for the detection of proteins using the SERS system, and (below) secondary signal enhancement using silver atom deposition. (C) Correlation between biomarker concentration and SERS signal intensity. Reprinted with permission from Nguyen et al. (2015) (copyright 2015, Elsevier B.V.).

biomarkers is ultrasensitive and highly specific, with the capability of introducing multiplexing features. Due to these reasons, plasmonic-based detection has many advantages over more commonly available protein sensing methods, such as ELISA or fluorescent-based detection. Therefore, protein detection using nanoplasmonic sensors can be used to diagnose a diverse range of diseases, which, although not covered in this chapter, are equally significant [25, 48].

1.3.2 Circulating tumor cells

Circulating tumor cells (CTCs) are cancerous cells that have detached from malignant tumors and are able to circulate in the blood. CTCs in the bloodstream can be transported and deposited into distant organs, leading to cancer metastasis [49]. The relationship between CTCs and cancer, and the potential of CTCs as cancer biomarkers, was demonstrated in the 1990s in a study that correlated changes in the number of CTCs with post-therapy prognosis in cancer patients [50]. However, the analysis of CTCs is hindered by their low numbers in the blood (<100 CTCs/mL of whole blood), in contrast to the several millions of leukocytes and several billions of erythrocytes found in the same volume [51].

Currently, the only method for CTC analysis approved by the US Food and Drug Administration (FDA) is CellSearch® (Veridex, LLC), which involves the immunomagnetic separation and enrichment of CTCs. However, this method involves a complex series of preparation steps, low purity due to leukocyte contamination, and high error rates [52], thereby hindering the clinical application of CTC analysis in cancer diagnosis. Therefore, alternative methods capable of directly detecting CTCs with ultra-high sensitivity and specificity are highly desired, and many studies have focused on the development of such methods for the highly sensitive detection of CTCs.

SERS is one of the technologies capable of directly analyzing CTCs in the blood. The sharp fingerprint spectra of SERS-active molecules allow them to detect even the low numbers of CTCs in the complex mixtures of whole blood cells and thus removing the need for costly purification steps [53]. Research on SERS-based CTC detection has mainly focused on the development of SERS-active tags, by combining plasmonic nanoparticles with Raman-active molecules and selective ligands for CTC surface receptors. For example, Wang et al. designed a SERS reporter using gold nanoparticles decorated with QSY-21 Raman dye and epidermal growth factor (EGF) for the detection and identification of CTCs from EGF receptor (EGFR) overexpressing squamous

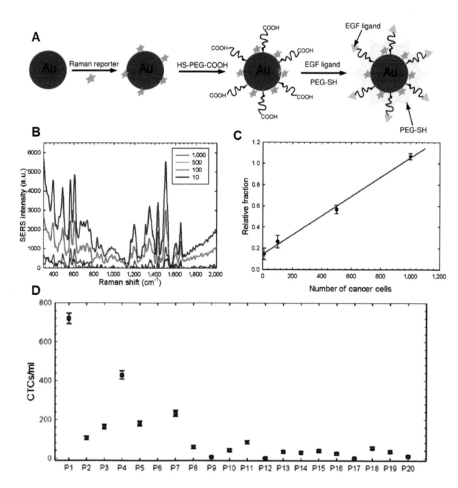

Figure 1.4 (A) Schematic diagram of the preparation of Raman-tagged, EGFR-targeting gold SERS probe. (B) SERS spectra from 2 mL mouse blood (containing $\sim 10^7$ leukocytes) spiked with 10, 100, 500, and 1,000 Tu212 cancer cells. (C) Correlation between relative SERS signal intensity and the number of Tu212 cells. (D) CTC counts from blood samples of 20 squamous cell carcinoma patients using a gold SERS probe. Reprinted with permission from Wang et al. (2011) (copyright 2011, American Association for Cancer Research).

cell carcinoma (Figure 1.4 A) [53]. The performance of e nanoparticles was examined using a cancer cell-spiked blood sample, resulting in a lower detection limit of 5–50 cells/mL blood (Figure 1.4 B and C). The sensitivity of SERS probe was further verified using samples from squamous cell carcinoma patients at various disease stages undergoing different treatments

(Figure 1.4 D). Similarly, Wu et al. targeted overexpressed folic acid receptors for the detection of CTCs from various types of cancer. Gold nanostar-based SERS-active probes had a more sensitive detection limit of 1 cell/mL blood without any need for optimization steps [54]. Moreover, to enhance the performance of the SERS-active probe, Xue et al. used a nanoparticle assembly of superparamagnetic iron oxide nanoparticles (SPION) and gold nanoparticles for the detection of CTCs from cervical cancer patients. The system could detect low concentrations of CTCs from two early-stage cancer patients, with a detection limit of 1 cell/mL blood [55].

This strategy can be expanded to multiplex detection and in vitro imaging of surface biomarkers of CTC. Nima et al. developed a multicolor SERS imaging agent for CTCs from breast cancer using silver-coated gold nanorods (AuNR/Ag), which targeted epithelial cells adhesion molecules (EpCAM), CD44 antigen, keratin-18, and insulin-like growth factor (IGF-1), respectively [56]. Each SERS tag is decorated with different Raman dyes, and representative Raman peaks were chosen for imaging CTCs in the whole blood sample (Figure 1.5 A–C). Au/Ag bimetallic nanomaterials significantly enhanced the SERS signal compared to the simple gold nanorods, leading to a shorter detection time and a higher sensitivity. The results confirmed the specific targeting of MCF-7 cells in the whole blood and separation of white blood cell samples within 30 minutes of incubation time without any background signal from blood cells (Figure 1.5 D) and with a detection limit of 1 CTCs in $\sim$7 million white blood cells. As a result, a highly specific analysis method for CTCs was established.

Alternatively, the SERS spectrum from the cell itself can be utilized as different cell types have a specific fingerprint. The cellular components vary among cells depending on their type and are represented by the relative intensity of specific SERS peaks. For example, erythrocytes are comprised of globin and porphyrin, which are not found in CTCs. Niciński et al. measured the spectral differences between different types of CTCs and normal blood cells to improve the detection of cancer cells [57]. In combination with a microfluidic separation device and SiO_2 shell-isolated silver nanoparticles ($Ag@SiO_2$), the label-free detection and identification of cancer cell types was achieved in a sensitive and reproducible manner via SERS (Figure 1.6 A). As shown in Figure 1.6 B, the separation of CTCs from blood cells was performed using inertial lift force and was based on the size difference. Isolated CTCs mixed with $Ag@SiO_2$ were allowed to flow into an optical chamber, where they were analyzed by SERS. The silver substrate in $Ag@SiO_2$ nanoparticles amplified the spectrum and enhanced the sensitivity

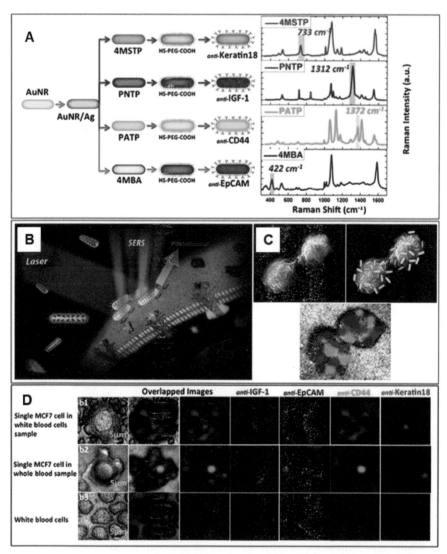

Figure 1.5   (A) Schematic representation of the preparation of four AuNR/Ag SERS probes for CTC surface biomarker analysis. Non-overlapping peaks from each probe were chosen and represented as different false colors in microscopic images. (B) Schematic diagram of CTC surface biomarker targeting and SERS imaging. (C) Schematic diagram of 2D multicolor imaging of CTC surface biomarkers using AuNR/Ag probe. (D) Microscopic SERS imaging of MCF-7 CTCs in white blood cells sample (first row), a whole blood sample (second low), and white blood cells without CTC (third row). Reprinted with permission from Nima et al. (2015) (copyright 2015, Springer Nature).

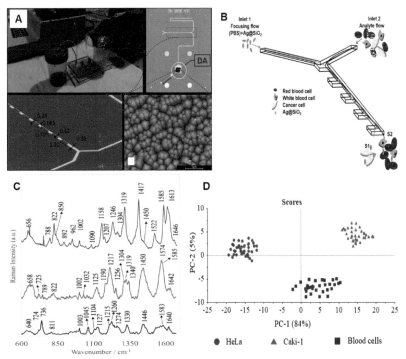

Figure 1.6 (A) Set-up of microfluidic devices for the separation of CTC from whole blood samples. Lower right panel: SEM image of SERS-active Ag substrate in the detection area (DA). (B) Schematic of inertial isolation of CTCs from blood cells. (C) The fingerprint SERS spectrum of blood cells (red), Caki-1 (yellow), and HeLa cells (blue) were measured using a microfluidics-integrated SERS system. (D) Principal component analysis (PCA) plot calculated from the SERS spectrum of normal blood cells, HeLa, and Caki-1 cells. Reprinted with permission from Niciński et al. (2019) (copyright 2019, Springer Nature).

of CTC detection. Figure 1.6 C shows the different spectra from normal blood cells to Caki-1 (renal carcinoma) and HeLa (cervical carcinoma) cells, with multiple peaks of different intensities, indicating different cellular components. Principal component analysis (PCA) was performed to categorize the similarities and differences among spectra. The results showed that the data could be clearly distinguished using only two principal components (PC-1 and PC-2) (Figure 1.6 D). As a result, a highly reproducible analysis with high accuracy for CTC-type identification was established.

To summarize, CTCs are an important index of cancer metastasis, and their use as biomarkers for distinguishing between different cancer stages is promising. However, the low number of CTCs in the blood samples of patients is a significant drawback for their use in clinical practice. This issue can be addressed using a plasmonics-based CTC detection system, which allows for the detection and identification of a minuscule number of CTCs in blood samples, with multiple benefits over alternative methods, including simple pre-treatment steps, ultrahigh sensitivity, and great specificity. As such, a plasmonic-based CTC detection system represents a major breakthrough for CTC-based cancer diagnosis.

1.3.3 Nucleic acids

Nucleic acids (mRNA, miRNA, and DNA) are promising biomarkers for disease diagnosis. Following apoptosis and necrosis of cancer cells, circulating nucleic acids are released into biofluids by the apoptosis and necrosis of cancer cells and actively secreted in the tumor microenvironment [58]. Since these molecules contain important information related to diseased cells, the analysis of circulating nucleic acids represents an alternative to tumor-tissue biopsies in certain diagnostic applications. Furthermore, circulating nucleic acids contain both sequence mutations and DNA methylation due to their origin, and their expression levels are elevated in the biofluids of patients compared to healthy controls. However, the concentrations of circulating nucleic acids in biofluid samples are extremely low, further, inconsistencies in their expression levels among patients are a major drawback to their use as biomarkers [59, 60]. Thus, the development of a sensitive and specific method for the detection of tumor-correlated sequences for clinical diagnosis, prognosis, and the therapeutic response has recently become the focus of many studies in cancer research.

The most commonly used circulating nucleic acid detection methods are based on polymerase chain reaction (PCR) or sequencing techniques [61, 62]. These techniques necessitate an amplification process to obtain clinically significant results from the patient samples due to low concentration of nucleic acids [63]. Moreover, they are not suitable for routine patient care due to their complexity and the large amounts of patient tissue or biofluid sample required [64]. For this reason, the development of advanced detection methods for circulating nucleic acid is essential for the diagnosis and prognosis of diseases using liquid biopsies.

The nanoplasmonic-based biosensors are one of the most advanced detection methods and can detect circulating nucleic acids in a liquid biopsy. This novel method requires a small sample size, has a sample detection process, and does not require an amplification step of clinical specimens, thereby considerably reducing its cost and turnaround time. Moreover, this technology features high sensitivity, a broad dynamic detection range, and label-free, and multiplex detection [65].

In this section, we will briefly describe recent advances in the development of a nanoplasmonic biosensor for the detection of circulating nucleic acids in a liquid biopsy. Surface plasmon resonance (SPR) is a plasmonic-related principle that is frequently used in sensing applications due to its sensitivity to environmental refractive index shifts generated by the changes in analyte concentration [44, 66]. SPR sensors based on the two-dimensional nanomaterial antimonene have been previously used for the detection of cancer-related biomarkers miRNA-21 and miRNA-155.67 This biosensor demonstrated a high sensitivity with a limit of detection (LOD) of 10 aM (Figure 1.7 A). Moreover, a label-free biosensing technique was previously developed for the detection of regional DNA methylation in MCF-7 cancer cells using SPR coupled with a molecular inversion probe (MIP). This method showed a linear correlation between SPR signal and concentration and was capable of detecting methylated DNA concentrations as low as 100 nM (Figure 1.7 B) [68].

Localized surface plasmon resonance (LSPR) is an optical phenomenon in which the metal nanoparticles and metal islands generate strong light absorption, resulting in the peak displacement of LSPR during coincidental incident photons with the frequency of electrons conduction [69]. Using the plasmon coupling mode of LSPR and gold nanoparticles, an advanced nanoplasmonic biosensor for dual detection of circulating tumor DNA (ctDNA), mutated E542K, E545K, and methylated DNA, was developed. In this study, peptide nucleic acid (PNA)-AuNP probes recognized and bound to 69bp PIK3CA ctDNA. Subsequently, PNA-AuNP-target genes were revealed at 200 fM sensitivity, resulting in an LSPR peak shift of 4.3 nm. For the simultaneous detection of methylation, immunogold colloids were fabricated to detect secondary responses to plasmon coupling and amplifying resonance signals. This amplification was approximately 107% and led to a four-fold increase in sensitivity (Figure 1.7 C) [30].

Although normal Raman scattering is generally considered unsuitable for the detection of ctDNA, the development of SERS with its intense signal enhancement has made it possible to overcome the limitations of the low scattering efficiency of Raman spectroscopy. For example, Zhou et al. developed a SERS-based ctDNA detection method, with the aid of single-walled carbon nanotubes (SWNTs) and RNase HII-assisted signal

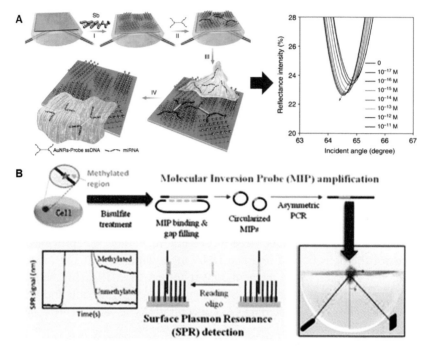

Figure 1.7 (A) Schematic representation of SPR-based miRNA sensor with antimonene (left); and change in SPR spectra with varying miRNA-21 concentration from 10–17 M to 10–11 M in the platform (right). (B) Schematic representation of SPR-based sensor for the detection of the regional methylation in DNA via signal amplification using a molecular inversion probe (MIP). (A) Reprinted with permission from Xue et al. (2019) (copyright 2019, Springer Nature). (B) Reprinted with permission from Carrascosa et al. (2014) (copyright 2014, Royal Society of Chemistry).

amplification. The triple-helix molecular switch, consisting of capture probe DNA and T-rich single-strand DNA, could recognize mutated ctDNA, KRAS G12DM, with a sensitivity of 0.3 fM (Figure 1.8 A) [70]. In another study, anti-methyl-cytosine (mCpG) antibody-conjugated gold nanoparticles were utilized to detect and analyze the global methylation of ctDNAs. By introducing positively-charged silver nanowires for the electrostatic absorption of DNAs that enhanced the SERS signal via the plasmon coupling effect between immunogold nanoparticles and silver nanowires, the system could detect methylated DNA at a concentration as low as 18 fg/mL (Figure 1.8 B) [40]. A subsequent study achieved the SERS spectrum fingerprint-based detection of cancer-specific circulating miRNAs in blood samples. A head-flocked nanopillar substrate, with high SERS enhancement factors between

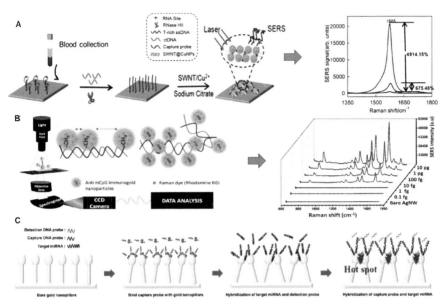

Figure 1.8 (A) Schematic representation of the detection of ctDNA via single-walled carbon nanotube-based SERS with the assistance of RNase HII (left); and resulting SERS intensity before the addition of target ctDNA (black), before (blue) and after (red) the addition of RNase HII (right). (B) Schematic representation of SERS-based detection of the global methylation state of ctDNA (left); and SERS intensity according to ctDNA concentration (right). (C) Schematic representation of head-flocked gold nanopillar structure-based detection of miRNAs in blood. (A) Reprinted with permission from Zhou et al. (2016) (copyright 2016, American Chemical Society). (B) Reprinted with permission from Nguyen et al. (2015) (copyright 2015, IOP Publishing). (C) Reprinted with permission from Kim et al. (2019) (copyright 2019, Royal Society of Chemistry).

gold head structures, was fabricated and modified to capture three circulating miRNAs, namely miR-10b, miR-21, and miR-373. Each miRNA produced a unique SERS signal, thereby allowing for the multiplex SERS detection of miRNAs with several femtomolar-level sensitivities (Figure 1.8 C) [38].

To summarize, circulating nucleic acids that are genetically and epi-genetically altered in cancer, can act as biomarkers for the detection and identification of cancer subtypes. However, the low expression levels of circulating nucleic acids, as well as the instability of free nucleic acids in biofluids, result in low concentrations of nucleic acids in the biofluids. Thus, their detection in patients' samples requires highly sensitive and specific detection methods. Nanoplasmonic sensors address this issue by pushing the limits of detection to the femtomolar range, thereby allowing for the molecular detection and identification of circulating nucleic acids in clinical samples.

1.3.4 Exosomes

Exosomes are small (~100 nm), cell-derived vesicles that are released from cells into the body fluids [71]. First identified in 1987, exosomes were initially considered to be tasked with the elimination of cellular waste [72]. However, further research revealed that exosomes perform far more complex functions, including cell-to-cell communication and immune response mediation [73]. For these reasons, exosomes are comprised of various cellular components, including lipids, proteins, and RNAs, which are specific to their originating cell, and are protected against proteases and other enzymes in body fluids. The potential use of exosomes as biomarkers for disease lies in the specificity of their components. For example, prostate cancer-derived urine exosomes contain two prostate cancer-specific mRNAs, PCA-3 and TMPRSS2:ERG, with distinct expression levels according to the cancer stage [74].

The molecular profiling of aberrantly-expressed biomolecules comprising circulating exosomes, which are stable in biofluids, holds immense potential in clinical applications. However, the analysis of exosomal contents in clinical settings using conventional methods is hindered by several drawbacks, including requirement for large sample volume, long processing times, high costs, and lack of sensitivity. To address these limitations, nanoplasmonic-based sensing techniques have been recently applied in several studies, in particular for the analysis of exosome contents.

Exosomes contain many tumor-related proteins on the membrane surface as well as inside vesicles. The utilization of membrane proteins for exosome-specific capturing and analysis is promising since the procedure for the enrichment, lysis, and biomarker purification of exosomes can be simplified. For example, Liang et al. utilized nanoplasmon-enhanced scattering for the detection of pancreatic cancer-derived exosomes [73, 75]. In this experiment, extracellular vesicle (EV) membrane protein CD81 was used to capture exosomes from blood plasma and loaded onto a sensor chip, with membrane proteins CD63 and CD9 selected as target biomarkers using CD63-targeting gold nanospheres and CD9-targeting gold nanorods, which scatter green and red light, respectively. When these two nanoparticles are in the vicinity, their scattered light is coupled, thereby increasing the scattering intensity (Figure 1.9 A–D). This method was found to be far more efficient than ELISA, with a detection limit of 0.23 ng EV/μl (vs. >10 ng EV/μl for

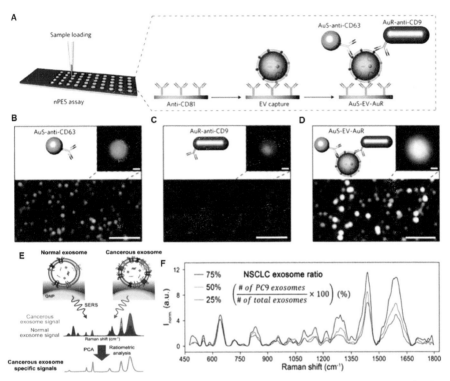

Figure 1.9 (A) Schematic representation of nanoplasmon-enhanced scattering assay for the detection of EVs. (B) Dark-field image of 50-nm gold nanosphere conjugated with anti-CD63 antibody. (C) Dark-field image of 25 × 60-nm gold nanorod conjugated with anti-CD9 antibody. (D) Dark-field image of nanosphere and nanorod complex. (A)-(D) were reprinted with permission from Liang et al. (2017) (copyright 2017, Springer Nature). (E) and (F) were reprinted with permission from Shin et al. (2018) (copyright 2018, American Chemical Society).

ELISA). For clinical testing, CD63 was replaced by a more relevant protein, ephrin type-A receptor 2 (EphA2), to examine the samples of individual patients at different stages of disease progression. The results indicated that the system could distinguish pancreatic cancer patients from normal subjects. Alternatively, SERS fingerprint spectra can be utilized for the analysis of exosomal protein biomarkers [76]. The differences in the contents of exosomal proteins results in unique SERS profiles between tumor-derived and normal cell-derived exosome (Figure 1.9 E and F). The unique SERS peaks from non-small cell lung cancer (NSCLC) cell-derived exosomes were identified to be several proteins, including CD9, CD81, epithelial cell adhesion molecules

(EpCAM), and epidermal growth factor receptors (EGFR). In another study, nanohole-based extraordinary optical transmission (EOT) was utilized for the multiplex profiling of CD24 and EpCAM on the exosome surface of ovarian cancer cells [77]. Similarly, surface plasmon resonance (SPR) was utilized for the profiling of EGFR and programmed death-ligand 1 (PD-L1) on the exosome surface of NSCLCs [78].

Nucleic acids in exosomes include mRNAs and miRNAs. These RNAs are known to be transferred to exosome-receiving cells and to regulate cell function [77, 79]. Therefore, the nucleic acids in exosomes have a more direct correlation with cancer compared to exosomal proteins. As a result, exosome-derived nucleic acids have a high potential for use as target biomarkers for cancer detection. In 2015, Joshi et al. were the first to report on the plasmonics-based detection of exosomal miRNA-10b (miR-10b) derived from pancreatic cancer. A gold nanoprism was selected as a single-particle sensor with ultra-high sensitivity [80]. Their results showed an attomolar detection limit that was at least 1000-fold more sensitive than RT-PCR, microarray, or fluorescence-based exosomal miRNA detection, with the ability to identify the one-nucleotide difference between miR-10b and miR-10a (Figure 1.10 A and B). Another study utilized a SERS-based platform for the detection of multiplex miRNAs in exosomes derived from breast cancer cells. Lee et al. introduced a head-flocked gold nanopillar substrate with dense plasmonic hot spots for the detection of low-abundance exosomal miRNAs. Using this method, the miRNAs (miR-21, 222, and 200c) in serum samples could be detected at attomolar level. Furthermore, this SERS sensor allowed to discriminate between breast cancer subtypes (luminal, HER2-positive, and triple-negative) based on the exosomal miRNA expression patterns (Figure 1.10 C and D).39 Mir-21 in NSCLC cell-derived exosomes was also detected with the aid of a Au core/R6G/Ag-Au alloy shell Raman reporter (ARANP) and a duplex specific nuclease. The probe DNA was bound to ARANP and silicon microbead at each end, and when miR-21 was hybridized with the probe DNA, duplex-specific nuclease cleaved the miRNA-DNA heteroduplex, thereby releasing ARANP and generating the SERS signal [81].

The nanoplasmonics-based detection and analysis of the exosome contents shows great potential for the identification of tumor cells and opens up new avenues for cancer therapy and prognosis by allowing the use of minimally- to non-invasive assays. The high sensitivity and specificity of the nanoplasmonic sensors provides a breakthrough for the difficulties encountered in exosome-based diagnosis, including the scarcity of exosomes

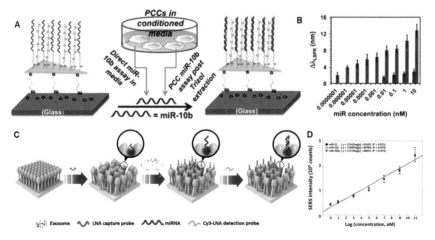

Figure 1.10 Schematic representation of miR-10b detection via LSPR using a complementary probe DNA-modified nanoprism. (B) Correlation of LSPR peak shift with miR-10b (red) and miR-10a (blue) shows highly sensitive and selective detection of exosomal miRNA. (C) Schematic detection of head-flocked nanopillar SERS substrate for the detection of exosomal miRNAs. (D) Correlation of SERS signal intensity with miR-21, miR-222, and miR-200c concentration. (A) and (B) were reprinted with permission from Joshi et al. (2015) (copyright 2015, American Chemical Society). (C) and (D) were reprinted with permission from Lee et al. (2019) (copyright 2019, Wiley-VCH Verlag GmbH & Co.).

in body fluids, which often result in tedious, low-efficiency isolation steps. Although the precise role and clinical significance of the exosome contents in various cancers are yet to be elucidated [82], nanoplasmonics-based exosome analysis techniques are promising tools in the diagnosis and analysis of cancer development.

1.4 Conclusions and Future Perspectives

Nanoplasmonic biosensors are emerging as promising tools for use in liquid biopsies due to their rapid turnaround time, high sensitivity, and label-free detection. By harnessing subtle biological interactions, biomolecules that exist in infinitesimal amounts in body fluids can be detected in real-time. Moreover, the simplicity of the surface modification involved in the attachment of bioconjugates to probe elements allows for the detection of almost any type of biomolecule.

However, a number of hurdles must be overcome before the nanoplasmonic biosensor-based liquid biopsy can find broader use clinically.

Until recently, the majority of research groups working on this technique have focused on improving its detection performance. However, the main issues and limitations are related to the complexity of the detection processes, including sample pre-treatment and analysis. Since biosensor-based clinical diagnosis instruments are often used by clinical staff or patients rather than highly skilled experts, they must be simple to use, rapid, and automated. Therefore, improving the applicability of nanoplasmonic biosensors as integrated devices should be explored alongside increasing their sensing performance. Furthermore, recent studies have only focused on increasing the detection sensitivity and selectivity of nanoplasmonic biosensors to singular biomarkers. However, variations in the relative concentration of biomarkers will need to be monitored and compared to identify diseases, rather than determining the absolute concentration of a single biomarker.

Overall, nanoplasmonic biosensors have proven to be useful liquid biopsy tools with an intrinsic, optical-based technology. Extensive efforts to integrate detection instruments by combining various technologies (e.g., microfluidics) and the simultaneous detection of relevant biomarkers have made it possible for nanoplasmonic biosensors to meet the strict demands of clinical diagnostics towards their application in point-of-care testing.

References

[1] Siravegna, G.; Marsoni, S.; Siena, S.; Bardelli, A., Integrating liquid biopsies into the management of cancer. *Nat Rev Clin Oncol* 2017, 14 (9), 531-548.

[2] Reimers, N.; Pantel, K., Liquid biopsy: novel technologies and clinical applications. *Clin Chem Lab Med* 2019, 57 (3), 312-316.

[3] Soda, N.; Rehm, B. H. A.; Sonar, P.; Nguyen, N. T.; Shiddiky, M. J. A., Advanced liquid biopsy technologies for circulating biomarker detection. *J Mater Chem B* 2019, 7 (43), 6670-6704.

[4] Leary, R. J.; Kinde, I.; Diehl, F.; Schmidt, K.; Clouser, C.; Duncan, C.; Antipova, A.; Lee, C.; McKernan, K.; De la Vega, F. M.; Kinzler, K. W.; Vogelstein, B.; Diaz, L. A.; Velculescu, V. E., Development of Personalized Tumor Biomarkers Using Massively Parallel Sequencing. *Sci Transl Med* 2010, 2 (20).

[5] Willets, K. A.; Van Duyne, R. P., Localized surface plasmon resonance spectroscopy and sensing. *Annu Rev Phys Chem* 2007, 58, 267-297.

[6] Ferhan, A. R.; Jackman, J. A.; Park, J. H.; Cho, N. J.; Kim, D. H., Nanoplasmonic sensors for detecting circulating cancer biomarkers. *Adv Drug Deliver Rev* 2018, 125, 48-77.

[7] Chen, Y.; Ming, H., Review of surface plasmon resonance and localized surface plasmon resonance sensor. *Photonic Sensors* 2012, 2, 13.

[8] Brockman, J. M.; Nelson, B. P.; Corn, R. M., Surface plasmon resonance imaging measurements of ultrathin organic films. *Annu Rev Phys Chem* 2000, 51, 41-63.

[9] Maier, J. M., *Plasmonics: Fundamentals and Applications*. Springer-Verlag: New York, 2007; p 14.

[10] Sun, M. T.; Zhang, Z. L.; Wang, P. J.; Li, Q.; Ma, F. C.; Xu, H. X., Remotely excited Raman optical activity using chiral plasmon propagation in Ag nanowires. *Light-Sci Appl* 2013, 2.

[11] Fang, Y. R.; Zhang, Z. L.; Chen, L.; Sun, M. T., Near field plasmonic gradient effects on high vacuum tip-enhanced Raman spectroscopy. *Phys Chem Chem Phys* 2015, 17 (2), 783-794.

[12] Haes, A. J.; Van Duyne, R. P., A nanoscale optical biosensor: Sensitivity and selectivity of an approach based on the localized surface plasmon resonance spectroscopy of triangular silver nanoparticles. *J Am Chem Soc* 2002, 124 (35), 10596-10604.

[13] Sagle, L. B.; Ruvuna, L. K.; Ruemmele, J. A.; Van Duyne, R. P., Advances in localized surface plasmon resonance spectroscopy biosensing. *Nanomedicine-Uk* 2011, 6 (8), 1447-1462.

[14] Mie, G., Articles on the optical characteristics of turbid tubes, especially colloidal metal solutions. *Ann Phys-Berlin* 1908, 25 (3), 377-445.

[15] Yang, W. H.; Schatz, G. C.; Vanduyne, R. P., Discrete Dipole Approximation for Calculating Extinction and Raman Intensities for Small Particles with Arbitrary Shapes. *J Chem Phys* 1995, 103 (3), 869-875.

[16] Mayer, K. M.; Hafner, J. H., Localized Surface Plasmon Resonance Sensors. *Chem Rev* 2011, 111 (6), 3828-3857.

[17] Zhang, Y.; Hong, H.; Cai, W. B., Imaging with Raman Spectroscopy. *Curr Pharm Biotechno* 2010, 11 (6), 654-661.

[18] Miller, M. M.; Lazarides, A. A., Sensitivity of metal nanoparticle plasmon resonance band position to the dielectric environment as observed in scattering. *J Opt a-Pure Appl Op* 2006, 8 (4), S239-S249.

[19] Sepulveda, B.; Angelome, P. C.; Lechuga, L. M.; Liz-Marzan, L. M., LSPR-based nanobiosensors. *Nano Today* 2009, 4 (3), 244-251.

[20] Truong, P. L.; Ma, X.; Sim, S. J., Resonant Rayleigh light scattering of single Au nanoparticles with different sizes and shapes. *Nanoscale* 2014, 6 (4), 2307-2315.

[21] Truong, P. L.; Cao, C.; Park, S.; Kim, M.; Sim, S. J., A new method for non-labeling attomolar detection of diseases based on an individual gold nanorod immunosensor. *Lab Chip* 2011, 11 (15), 2591-2597.

[22] Truong, P. L.; Kim, B. W.; Sim, S. J., Rational aspect ratio and suitable antibody coverage of gold nanorod for ultra-sensitive detection of a cancer biomarker. *Lab Chip* 2012, 12 (6), 1102-1109.

[23] Kang, M. K.; Lee, J.; Nguyen, A. H.; Sim, S. J., Label-free detection of ApoE4-mediated beta-amyloid aggregation on single nanoparticle uncovering Alzheimer's disease. *Biosens Bioelectron* 2015, 72, 197-204.

[24] Kim, H.; Lee, J. U.; Song, S.; Kim, S.; Sim, S. J., A shape-code nanoplasmonic biosensor for multiplex detection of Alzheimer's disease biomarkers. *Biosens Bioelectron* 2018, 101, 96-102.

[25] Kim, H.; Lee, J. U.; Kim, S.; Song, S.; Sim, S. J., A Nanoplasmonic Biosensor for Ultrasensitive Detection of Alzheimer's Disease Biomarker Using a Chaotropic Agent. *Acs Sensors* 2019, 4 (3), 595-602.

[26] Lee, J. U.; Nguyen, A. H.; Sim, S. J., A nanoplasmonic biosensor for label-free multiplex detection of cancer biomarkers. *Biosens Bioelectron* 2015, 74, 341-346.

[27] Myroshnychenko, V.; Rodriguez-Fernandez, J.; Pastoriza-Santos, I.; Funston, A. M.; Novo, C.; Mulvaney, P.; Liz-Marzan, L. M.; de Abajo, F. J. G., Modelling the optical response of gold nanoparticles. *Chem Soc Rev* 2008, 37 (9), 1792-1805.

[28] Song, S.; Nguyen, A. H.; Lee, J. U.; Cha, M.; Sim, S. J., Tracking of STAT3 signaling for anticancer drug-discovery based on localized surface plasmon resonance. *Analyst* 2016, 141 (8), 2493-2501.

[29] Nguyen, A. H.; Lee, J.; Choi, H. I.; Kwak, H. S.; Sim, S. J., Fabrication of plasmon length-based surface enhanced Raman scattering for multiplex detection on microfluidic device. *Biosens Bioelectron* 2015, 70, 358-365.

[30] Nguyen, A. H.; Sim, S. J., Nanoplasmonic biosensor: Detection and amplification of dual bio-signatures of circulating tumor DNA. *Biosens Bioelectron* 2015, 67, 443-449.

[31] Pilot, R.; Signorini, R.; Durante, C.; Orian, L.; Bhamidipati, M.; Fabris, L., A Review on Surface-Enhanced Raman Scattering. *Biosensors-Basel* 2019, 9 (2).

[32] Unser, S.; Bruzas, I.; He, J.; Sagle, L., Localized Surface Plasmon Resonance Biosensing: Current Challenges and Approaches. *Sensors-Basel* 2015, 15 (7), 15684-15716.

[33] Zeman, E. J.; Schatz, G. C., An Accurate Electromagnetic Theory Study of Surface Enhancement Factors for Ag, Au, Cu, Li, Na, Al, Ga, in, Zn, and Cd. *J Phys Chem-Us* 1987, 91 (3), 634-643.

[34] Schlucker, S., Surface-Enhanced Raman Spectroscopy: Concepts and Chemical Applications. *Angew Chem Int Edit* 2014, 53 (19), 4756-4795.

[35] Lombardi, J. R.; Birke, R. L.; Lu, T. H.; Xu, J., Charge-Transfer Theory of Surface Enhanced Raman-Spectroscopy - Herzberg-Teller Contributions. *J Chem Phys* 1986, 84 (8), 4174-4180.

[36] Le Ru, E. C.; Etchegoin, P. G., Single-Molecule Surface-Enhanced Raman Spectroscopy. *Annual Review of Physical Chemistry, Vol 63* 2012, 63, 65-87.

[37] Laing, S.; Gracie, K.; Faulds, K., Multiplex in vitro detection using SERS. *Chem Soc Rev* 2016, 45 (7), 1901-1918.

[38] Kim, W. H.; Lee, J. U.; Song, S.; Kim, S.; Choi, Y. J.; Sim, S. J., A label-free, ultra-highly sensitive and multiplexed SERS nanoplasmonic biosensor for miRNA detection using a head-flocked gold nanopillar. *Analyst* 2019, 144 (5), 1768-1776.

[39] Lee, J. U.; Kim, W. H.; Lee, H. S.; Park, K. H.; Sim, S. J., Quantitative and Specific Detection of Exosomal miRNAs for Accurate Diagnosis of Breast Cancer Using a Surface-Enhanced Raman Scattering Sensor Based on Plasmonic Head-Flocked Gold Nanopillars. *Small* 2019, 15 (17).

[40] Nguyen, A. H.; Lee, J. U.; Sim, S. J., Plasmonic coupling-dependent SERS of gold nanoparticles anchored on methylated DNA and detection of global DNA methylation in SERS-based platforms. *J Optics-Uk* 2015, 17 (11).

[41] Nguyen, A. H.; Lee, J. U.; Sim, S. J., Nanoplasmonic probes of RNA folding and assembly during pre-mRNA splicing. *Nanoscale* 2016, 8 (8), 4599-4607.

[42] Lee, H. J.; Wark, A. W.; Corn, R. M., Microarray methods for protein biomarker detection. *Analyst* 2008, 133 (8), 975-983.

[43] Wu, L.; Qu, X. G., Cancer biomarker detection: recent achievements and challenges. *Chem Soc Rev* 2015, 44 (10), 2963-2997.

[44] Jackman, J. A.; Ferhan, A. R.; Cho, N. J., Nanoplasmonic sensors for biointerfacial science. *Chem Soc Rev* 2017, 46 (12), 3615-3660.

[45] Jun, H. J.; Nguyen, A. H.; Kim, Y. H.; Park, K. H.; Kim, D.; Kim, K. K.; Sim, S. J., Distinct Rayleigh Scattering from Hot Spot Mutant p53 Proteins Reveals Cancer Cells. *Small* 2014, 10 (14), 2954-2962.

[46] Ma, X.; Truong, P. L.; Anh, N. H.; Sim, S. J., Single gold nanoplasmonic sensor for clinical cancer diagnosis based on specific interaction between nucleic acids and protein. *Biosens Bioelectron* 2015, 67, 59-65.

[47] Li, J. R.; Wang, J.; Grewal, Y. S.; Howard, C. B.; Raftery, L. J.; Mahler, S.; Wang, Y. L.; Trau, M., Multiplexed SERS Detection of Soluble Cancer Protein Biomarkers with Gold-Silver Alloy Nanoboxes and Nanoyeast Single-Chain Variable Fragments. *Anal Chem* 2018, 90 (17), 10377-10384.

[48] Nguyen, A. H.; Shin, Y.; Sim, S. J., Development of SERS substrate using phage-based magnetic template for triplex assay in sepsis diagnosis. *Biosens Bioelectron* 2016, 85, 522-528.

[49] Aceto, N.; Bardia, A.; Miyamoto, D. T.; Donaldson, M. C.; Wittner, B. S.; Spencer, J. A.; Yu, M.; Pely, A.; Engstrom, A.; Zhu, H. L.; Brannigan, B. W.; Kapur, R.; Stott, S. L.; Shioda, T.; Ramaswamy, S.; Ting, D. T.; Lin, C. P.; Toner, M.; Haber, D. A.; Maheswaran, S., Circulating Tumor Cell Clusters Are Oligoclonal Precursors of Breast Cancer Metastasis. *Cell* 2014, 158 (5), 1110-1122.

[50] Racila, E.; Euhus, D.; Weiss, A. J.; Rao, C.; McConnell, J.; Terstappen, L. W. M. M.; Uhr, J. W., Detection and characterization of carcinoma cells in the blood. *P Natl Acad Sci USA* 1998, 95 (8), 4589-4594.

[51] Baccelli, I.; Schneeweiss, A.; Riethdorf, S.; Stenzinger, A.; Schillert, A.; Vogel, V.; Klein, C.; Saini, M.; Bauerle, T.; Wallwiener, M.; Holland-Letz, T.; Hofner, T.; Sprick, M.; Scharpff, M.; Marme, F.; Sinn, H. P.; Pantel, K.; Weichert, W.; Trumpp, A., Identification of a population of blood circulating tumor cells from breast cancer patients that initiates metastasis in a xenograft assay. *Nat Biotechnol* 2013, 31 (6), 539-U143.

[52] Shen, Z. Y.; Wu, A. G.; Chen, X. Y., Current detection technologies for circulating tumor cells. *Chem Soc Rev* 2017, 46 (8), 2038-2056.

[53] Wang, X.; Qian, X. M.; Beitler, J. J.; Chen, Z. G.; Khuri, F. R.; Lewis, M. M.; Shin, H. J. C.; Nie, S. M.; Shin, D. M., Detection of Circulating Tumor Cells in Human Peripheral Blood Using Surface-Enhanced Raman Scattering Nanoparticles. *Cancer Res* 2011, 71 (5), 1526-1532.

[54] Wu, X. X.; Xia, Y. Z.; Huang, Y. J.; Li, J.; Ruan, H. M.; Chen, T. X.; Luo, L. Q.; Shen, Z. Y.; Wu, A. G., Improved SERS-Active Nanoparticles with Various Shapes for CTC Detection without Enrichment Process

with Supersensitivity and High Specificity. *Acs Appl Mater Inter* 2016, 8 (31), 19928-19938.

[55] Xue, T.; Wang, S. Q.; Ou, G. Y.; Li, Y.; Ruan, H. M.; Li, Z. H.; Ma, Y. Y.; Zou, R. F.; Qiu, J. Y.; Shen, Z. Y.; Wu, A. G., Detection of circulating tumor cells based on improved SERS-active magnetic nanoparticles. *Anal Methods-Uk* 2019, 11 (22), 2918-2928.

[56] Nima, Z. A.; Mahmood, M.; Xu, Y.; Mustafa, T.; Watanabe, F.; Nedosekin, D. A.; Juratlin, M. A.; Fahmi, T.; Galanzha, E. I.; Nolan, J. P.; Basnakinan, A. G.; Zharov, V. P.; Biris, A. S., Circulating tumor cell identification by functionalized silver-gold nanorods with multicolor, super-enhanced SERS and photothermal resonances. *Sci Rep* 2015, 4, 4752.

[57] Nicinski, K.; Krajczewski, J.; Kudelski, A.; Witkowska, E.; Trzcinska-Danielewicz, J.; Girstun, A.; Kaminska, A., Detection of circulating tumor cells in blood by shell-isolated nanoparticle - enhanced Raman spectroscopy (SHINERS) in microfluidic device. *Sci Rep-Uk* 2019, 9.

[58] Das, J.; Kelley, S. O., High-Performance Nucleic Acid Sensors for Liquid Biopsy Applications. *Angew Chem Int Ed Engl* 2020, 59 (7), 2554-2564.

[59] Schwarzenbach, H.; Hoon, D. S. B.; Pantel, K., Cell-free nucleic acids as biomarkers in cancer patients. *Nat Rev Cancer* 2011, 11 (6), 426-437.

[60] Finotti, A.; Allegretti, M.; Gasparello, J.; Giacomini, P.; Spandidos, D. A.; Spoto, G.; Gambari, R., Liquid biopsy and PCR-free ultrasensitive detection systems in oncology (Review). *Int J Oncol* 2018, 53 (4), 1395-1434.

[61] Pantel, K.; Alix-Panabieres, C., Circulating tumour cells and cell-free DNA in gastrointestinal cancer. *Nat Rev Gastro Hepat* 2017, 14 (2), 73-74.

[62] Austin, L. K.; Avery, T.; Jaslow, R.; Fortina, P.; Sebisanovic, D.; Siew, L.; Zapanta, A.; Talasaz, A.; Cristofanilli, M., Concordance of circulating tumor DNA (ctDNA) and next-generation sequencing (NGS) as molecular monitoring tools in metastatic breast cancer (MBC). *Cancer Res* 2015, 75.

[63] Campuzano, S.; Pedrero, M.; Pingarron, J. M., Electrochemical genosensors for the detection of cancer-related miRNAs. *Anal Bioanal Chem* 2014, 406 (1), 27-33.

[64] Bellassai, N.; Spoto, G., Biosensors for liquid biopsy: circulating nucleic acids to diagnose and treat cancer. *Anal Bioanal Chem* 2016, 408 (26), 7255-7264.

[65] Soler, M.; Huertas, C. S.; Lechuga, L. M., Label-free plasmonic biosensors for point-of-care diagnostics: a review. *Expert Rev Mol Diagn* 2019, 19 (1), 71-81.

[66] Nguyen, H. H.; Park, J.; Kang, S.; Kim, M., Surface Plasmon Resonance: A Versatile Technique for Biosensor Applications. *Sensors-Basel* 2015, 15 (5), 10481-10510.

[67] Xue, T. Y.; Liang, W. Y.; Li, Y. W.; Sun, Y. H.; Xiang, Y. J.; Zhang, Y. P.; Dai, Z. G.; Duo, Y. H.; Wu, L. M.; Qi, K.; Shiyananju, B. N.; Zhang, L. J.; Cui, X. Q.; Zhang, H.; Bao, Q. L., Ultrasensitive detection of miRNA with an antimonene-based surface plasmon resonance sensor. *Nat Commun* 2019, 10.

[68] Carrascosa, L. G.; Ibn Sina, A.; Palanisamy, R.; Sepulveda, B.; Otte, M. A.; Rauf, S.; Shiddiky, M. J. A.; Trau, M., Molecular inversion probe-based SPR biosensing for specific, label-free and real-time detection of regional DNA methylation. *Chem Commun* 2014, 50 (27), 3585-3588.

[69] Li, X. Y.; Ye, M. S.; Zhang, W. Y.; Tan, D.; Jaffrezic-Renault, N.; Yang, X.; Guo, Z. Z., Liquid biopsy of circulating tumor DNA and biosensor applications. *Biosens Bioelectron* 2019, 126, 596-607.

[70] Zhou, Q. F.; Zheng, J.; Qing, Z. H.; Zheng, M. J.; Yang, J. F.; Yang, S.; Ying, L.; Yang, R. H., Detection of Circulating Tumor DNA in Human Blood via DNA-Mediated Surface-Enhanced Raman Spectroscopy of Single-Walled Carbon Nanotubes. *Anal Chem* 2016, 88 (9), 4759-4765.

[71] Zhang, Y.; Liu, Y. F.; Liu, H. Y.; Tang, W. H., Exosomes: biogenesis, biologic function and clinical potential. *Cell Biosci* 2019, 9.

[72] Johnstone, R. M.; Adam, M.; Hammond, J. R.; Orr, L.; Turbide, C., Vesicle Formation during Reticulocyte Maturation - Association of Plasma-Membrane Activities with Released Vesicles (Exosomes). *J Biol Chem* 1987, 262 (19), 9412-9420.

[73] Properzi, F.; Logozzi, M.; Fais, S., Exosomes: the future of biomarkers in medicine. *Biomark Med* 2013, 7 (5), 769-778.

[74] Nilsson, J.; Skog, J.; Nordstrand, A.; Baranov, V.; Mincheva-Nilsson, L.; Breakefield, X. O.; Widmark, A., Prostate cancer-derived urine exosomes: a novel approach to biomarkers for prostate cancer. *Brit J Cancer* 2009, 100 (10), 1603-1607.

[75] Liang, K.; Liu, F.; Fan, J.; Sun, D. L.; Liu, C.; Lyon, C. J.; Bernard, D. W.; Li, Y.; Yokoi, K.; Katz, M. H.; Koay, E. J.; Zhao, Z.; Hu, Y., Nanoplasmonic quantification of tumour-derived extracellular vesicles in plasma microsamples for diagnosis and treatment monitoring. *Nat Biomed Eng* 2017, 1 (4).

[76] Shin, H.; Jeong, H.; Park, J.; Hong, S.; Choi, Y., Correlation between Cancerous Exosomes and Protein Markers Based on Surface-Enhanced Raman Spectroscopy (SERS) and Principal Component Analysis (PCA). *Acs Sensors* 2018, 3 (12), 2637-2643.

[77] Im, H.; Shao, H. L.; Park, Y. I.; Peterson, V. M.; Castro, C. M.; Weissleder, R.; Lee, H., Label-free detection and molecular profiling of exosomes with a nano-plasmonic sensor. *Nat Biotechnol* 2014, 32 (5), 490-U219.

[78] Liu, C.; Zeng, X.; An, Z. J.; Yang, Y. C.; Eisenbaum, M.; Gu, X. D.; Jornet, J. M.; Dy, G. K.; Reid, M. E.; Gan, Q. Q.; Wu, Y., Sensitive Detection of Exosomal Proteins via a Compact Surface Plasmon Resonance Biosensor for Cancer Diagnosis. *Acs Sensors* 2018, 3 (8), 1471-1479.

[79] Valadi, H.; Ekstrom, K.; Bossios, A.; Sjostrand, M.; Lee, J. J.; Lotvall, J. O., Exosome-mediated transfer of mRNAs and microRNAs is a novel mechanism of genetic exchange between cells. *Nat Cell Biol* 2007, 9 (6), 654-U72.

[80] Joshi, G. K.; Deitz-McElyea, S.; Liyanage, T.; Lawrence, K.; Mali, S.; Sardar, R.; Korc, M., Label-Free Nanoplasmonic-Based Short Noncoding RNA Sensing at Attomolar Concentrations Allows for Quantitative and Highly Specific Assay of MicroRNA-10b in Biological Fluids and Circulating Exosomes. *Acs Nano* 2015, 9 (11), 11075-11089.

[81] Ma, D. D.; Huang, C. X.; Zheng, J.; Tang, J. R.; Li, J. S.; Yang, J. F.; Yang, R. H., Quantitative detection of exosomal microRNA extracted from human blood based on surface-enhanced Raman scattering. *Biosens Bioelectron* 2018, 101, 167-173.

[82] Kowal, J.; Arras, G.; Colombo, M.; Jouve, M.; Morath, J. P.; Primdal-Bengtson, B.; Dingli, F.; Loew, D.; Tkach, M.; Thery, C., Proteomic comparison defines novel markers to characterize heterogeneous populations of extracellular vesicle subtypes. *P Natl Acad Sci USA* 2016, 113 (8), E968-E977.

2

Analysis Methods and Clinical Applications of Circulating Cell-free DNA and RNA in Human Blood

Josiah T. Wagner, Josephine Briand, and Thuy T. M. Ngo

Knight Cancer Institute, Oregon Health & Health Science University (OHSU)
Email: ngth@ohsu.edu

Abstract

Circulating cell-free DNA (cfDNA) and cell-free RNA (cfRNA) in bodily fluids can provide a rich source of information with clinical value in diagnosis, prognosis, and treatment monitoring of various diseases. For the last three-decades, the development of affordable massively parallel sequencing and highly sensitive PCR-based techniques has allowed us to produce a remarkable amount of cfDNA and cfRNA data, advancing our understanding of basic cfDNA and cfRNA biology and elucidating its clinical utilities. In this book chapter, we will discuss the key technologies for the analysis of cfDNA and cfRNA and their clinical applications. Methods to purify cfDNA and cfRNA from human plasma and to analyze their aberrations are detailed. We will summarize an overview of clinical applications of cfDNA and cfRNA in prenatal diagnostic testing, oncology, transplantation, microbial analysis, autoimmune disorders, and Alzheimer's disease.

Keywords: Circulating cell-free DNA (cfDNA), Cell-free RNA (cfRNA), Clinical diagnosis, Prognosis, Deep sequencing

2.1 Introduction

For many diseases, treatment decisions rely on the precise molecular analyses of suspected tissues and organs. The most common medical procedure that provides direct access to an internal organ, the needle biopsy, can be problematic for several reasons, including excessive invasiveness to access a suspected region, inability to capture disease heterogeneity, initiation of disease progression, and potential accompaniment of morbidity. To overcome the challenges associated with tissue biopsies, many liquid biopsy applications are being developed with the goal of replacing invasive diagnostic or treatment monitoring procedures. Several blood-based analytes have been gaining interest for liquid biopsy development, including circulating cells (CTCs, CHCs, and TAMs), circulating nucleic acids, platelets, extracellular vesicles, and protein panels [1–10]. In this chapter, we will focus on the key technologies for analyzing circulating nucleic acids and strategies for their use in the clinic. A summary of the sources and analysis methods of circulating cell-free DNA and RNA in human blood is depicted in Figure 2.1.

Extracellular DNA, now more commonly referred to as cell-free DNA (cfDNA), was first detected in blood by Mandel and Metais in 1948 [11]. However, the phenomenon was underappreciated for the next half century until the observation of fetal-derived and tumor-derived cfDNA was reported in the 1990s [12–14]. The origin of circulating cfDNA remains poorly understood. The two main mechanisms proposed to be involved in cfDNA generation are cellular breakdown (e.g., necrosis, apoptosis, pyroptosis, autophagy, and mitotic catastrophe) and active secretion (e.g., EVs,

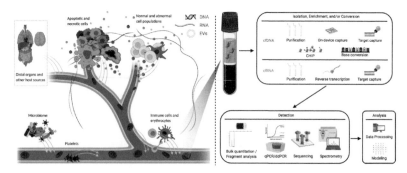

Figure 2.1   Sources and analysis methods of circulating cell-free DNA and RNA in human blood.

virtosomes, and Argonaute 2) [15]. Circulating cfDNA has a short half-life in biological fluids, suggesting an ongoing cfDNA release process followed by rapid degradation and filtration [16, 17]. In healthy individuals, hematopoietic cell death is the major source of cfDNA circulating in the blood. Circulating cfDNA is also known to increase during vigorous exercise in healthy individuals, potentially due to mechanical and metabolic muscular damage. Although the release of cfDNA into circulation appears to be a normal physiological process, additional biological events may cause additional release of cfDNA. Phagocytes, for example, digest cells and discharge their DNA fragments and this process may increase in frequency during infection [15]. During periods of inflammation, leukocyte inflammation response can lead to cfDNA release [15]. The normal progression of aging may also contribute to cfDNA abundance, due to increasing cellular senescence and death as well as decreased cfDNA clearance. In transplant patients, injury of a donated organ due to host immune response leads to the increase of donor-specific cfDNA [18, 19]. During pregnancy, both fetal DNA and maternal DNA levels increase likely by apoptosis due to the short length of the DNA fragments [20–22]. The concentration of cfDNA in cancer patients is generally higher than in healthy subjects. The sources of circulating tumor DNA (ctDNA) are attributed to apoptosis, necrosis, and active secretion of cancer cells as well as the destruction of circulating tumor cells and micrometastases under stress responses [15]. These examples of cfDNA utility suggest circulating cfDNA could be used to detect or monitor a wide variety of diseases and processes.

Circulating cfDNA in bodily fluids can provide a rich source of information with clinical value, including somatic mutations, copy number alterations, gene fusions, and DNA fragmentation patterns [23–33]. More recently, profiling of epigenetic circulating cfDNA alterations such as methylation status and nucleosome positioning has also provided exciting insights for liquid biopsy technology [34–46]. In disease processes, affected tissue types contribute a portion of their DNA to the circulating cfDNA pool. Polymerase chain reaction (PCR)-based assays, while highly sensitive, have a limited ability to measure a large number of genes simultaneously. In the last decade, the increasing affordability of massively parallel next-generation sequencing (NGS) has allowed us to produce a remarkable amount of cfDNA data, advancing our understanding of basic cfDNA biology and elucidating its clinical utility [23–46]. Like traditional PCR-based assays, high-throughput sequencing of cfDNA can reveal copy number alterations, somatic variants, gene fusion, and epigenetic alterations but at a much higher scale. In 2008, Fan et al. and Chiu et al. demonstrated the utility of cfDNA in maternal blood for prenatal diagnosis of Down syndrome by the detection of the extra copy of the 21 chromosomes [47, 48]. Somatic variants and gene fusions in circulating cfDNA contributed from fetuses in pregnant women, donor organs in transplant patients, and tumors of cancer patients have been detected using ultradeep sequencing of target gene panels and digital PCR (18, 19, 23–33, 49). Epigenetic status of circulating cfDNA, such as maps of modifications on cytosine and nucleosome footprints, may reflect the alterations under pathology conditions [34–46]. Snyder et al. generated maps of nucleosome occupancy in cfDNA and showed its correlation with nuclear architecture, gene structure, and expression. Their results suggest that nucleosome footprints inform the cfDNA tissues-of-origin [44]. Importantly, the majority of epigenetic modifications found in human disease are associated with DNA methylation at regions enriched with cytosine and guanine dinucleotide (CpG) islands. Gene promoters enriched with CpG regions are often hypomethylated or hypermethylated in tumor cells. Global hypomethylation which affects gene stability and integrity was observed in cancer as well. Many groups have demonstrated the ability to detect the methylation status of cfDNA in blood for sensitive detection of cancer originating from more than 20 organ types [34–46].

While cfDNA has been the predominate circulating nucleotide investigated for biological and clinical importance, recent studies have reported the presence of extracellular RNA or cell-free RNA (cfRNA) in a diverse array of body fluids, including cerebrospinal fluid (CSF), saliva, serum, plasma, and urine [50–57]. The mechanism of cfRNA biogenesis, carriers, and biotypes composition is poorly understood [54]. Circulating cfRNA is found to be associated with multiple subclasses of carriers such as EVs, ribonucleoproteins, and lipoprotein complexes [51, 52]. Types and proportions of RNA cargo carried by different carriers may depend on the fluid type, pre-analytical factors, and physiology or pathology conditions [52]. Different RNA isolation methods preferentially extract RNA from specific subclasses of carriers and the analysis methods may bias the selection of RNA biotypes [52]. Therefore, the heterogeneity of cfRNA carriers adds challenges to efficient and reproducible methods for RNA isolation. The most common RNA biotypes reported are small noncoding RNAs (ncRNAs) including microRNAs (miRNAs), transfer RNA (tRNA) fragments, and piwi-interacting RNAs (piRNAs) [53]. Messenger RNAs (mRNAs) and long noncoding RNAs (lncRNAs) have also been reported as cfRNA in biofluids [58–62]. A recent study that used small ncRNA sequencing across five biofluids to evaluate 10 different common methods for cfRNA isolation showed that the majority of cfRNA in plasma is carried by EVs and AGO2 complexes [51, 52, 63]. Physiology and pathology conditions that alter EV generation and AGO2 release may be reflected in RNA cargo profiles. Therefore, these recent discoveries have sparked interest in investigating the potential utility of cfRNA as a diagnostic, prognostic, and therapeutic biomarker.

2.2 Method of Analysis and Detection

2.2.1 Purification of cell-free DNA

Variation in donor demographic, disease status, drug use, recent exercise, and other biological factors can affect circulating cfDNA yield [64]. In addition, the choice of cfDNA isolation, enrichment, and detection method will affect both cfDNA yield and genomic targets detected. When planning a circulating cfDNA experiment, the investigator will need to consider the amount of sample available, composition and quality of sample, cost versus sensitivity required to detect target(s) of interest, and whether the genetic targets are known or unknown.

2.2.2 Circulating cfDNA isolation techniques: commercial kits

Prior to downstream quantitative or qualitative analysis, circulating cfDNA usually must be made physically accessible to intercalating dyes or enzymes. Contaminating components in the sample, such as lipids, free proteins, DNA-associated proteins, or circulating RNA, must also be removed in order to reduce background noise and inhibitors. Lo et al. [65] first demonstrated the potential clinical utility of plasma and serum cfDNA using a rapid-boiling method. However, this method results in a crude DNA extract that can retain enzymatic inhibitors, and since then several commercialized kits have been developed to improve cfDNA extraction purity and reproducibility. Commercial circulating cfDNA kits generally follow one of three enrichment methods: silica column-based enrichment, polymer-mediated enrichment (PME), or magnetic bead enrichment. In the first method, cfDNA is freed from other components by using detergent and proteinase K digestion. Then, cfDNA is bound to a silica membrane in the presence of chaotropic salts. Bound cfDNA is then washed and selectively eluted using low salt conditions. In the second method, cfDNA is bound to a polymer and the complex is pelleted by centrifugation. The cfDNA–polymer complex is then lysed and free cfDNA is bound to a filter column before being washed and eluted. In the final method, samples are lysed using a detergent and proteinase K digestion.

Freed cfDNA is bound to selective paramagnetic beads and the cfDNA-magnetic bead complex is concentrated using an externally applied magnetic force and washed. Then, the cfDNA is eluted from the cfDNA-magnetic bead complex and the beads are removed.

The wide selection of in-house and commercially available kits for cfDNA extraction has prompted head-to-head comparisons of extraction capabilities. Mauger et al. [66] evaluated five published protocols and six commercial kits for extracting circulating cfDNA from plasma. Sample input ranged from 200 – 1000 μL depending on the minimum input indicated by respective protocols and the extracted cfDNA was quantified by qPCR and Qubit dsDNA HS Assay. The authors reported similar extraction cfDNA yields and integrity between the Norgen Plasma/Serum Circulating DNA Purification MiniKit and the Qiagen QIAamp Circulating Nucleic Acid Kit, both of which involve binding circulating cfDNA to a column-embedded membrane following lysis. However, a slight increase in cfDNA yield and better reproducibility gave an edge to the Norgen Plasma/Serum Circulating DNA Purification MiniKit.

To test commercially available kits in a clinically relevant scenario, recent work by Sorber et al. [67] compared the efficiency of commercially available silica column-based (Qiagen QIAamp circulating nucleic acid kit), PME (AnalytikJena PME free-circulating DNA Extraction Kit, and magnetic bead (Promega Maxwell RSC ccfDNA Plasma Kit; EpiGentek EpiQuick Circulating Cell-Free DNA Isolation Kit; SanBio NEXTprep-Mag cfDNA Isolation Kit versions 1 and 2) kits for the extraction of circulating cfDNA from normal and cancer patients. The cohort included patients with pancreatic ductal adenocarcinoma that had a known *KRAS* mutation. Circulating cfDNA isolation efficiency was determined using digital droplet PCR (ddPCR) to quantify total cfDNA and *KRAS* mutated ctDNA. Overall, the highest cfDNA yields were found using the QIAamp Circulating Nucleic Acid and Maxwell RSC ccfDNA Plasma kits. Notably, no cfDNA was detected in the EpiQuick Circulating Cell-Free DNA Isolation Kit or NEXTprep-Mag cfDNA Isolation Kit version 1. Patient *KRAS* mutations were also only consistently detected across samples when extracted using the QIAamp circulating nucleic acid or Maxwell RSC ccfDNA Plasma kits. Although the reason for the lack of cfDNA yield in some of the kits, the inclusion of sample lysis prior to extraction may have had a role in increasing cfDNA yield. Sorber et al. [67] conclude that the QIAamp Circulating Nucleic Acid and Maxwell RSC ccfDNA Plasma kits provided the highest cfDNA recovery rate and least PCR inhibitors in the eluant compared to the other kits tested. In a separate study the same year, van Ginkel et al. [68] compared cfDNA extraction efficiency and reproducibility of two silica column-based kits (QIAamp Circulating Nucleic Acid Kit and Zymo research Quick cfDNA serum and plasma kit), two magnetic bead-based kits (Qiagen QIAsymphony circulating DNA Kit and Roche MagNA pure LC DNA isolation kit – large volume) and one PME-based kit (AnalytikJena PME free-circulating DNA extraction kit). Focusing on wild-type or BRAF/EGFR-mutant ctDNA targets, van Ginkel et al. [68] found that the Zymo Research Quick cfDNA Serum and Plasma kit had the highest overall cfDNA yield, but the Qiagen QIAamp Circulating Nucleic Acid Kit had nearly equal yield and better reproducibility. Also the same year, Diefenbach et al. [69] compared several commercial kits for cfDNA extraction and found similar results to Mauger et al. [66], in that Norgen Plasma/Serum Circulating DNA Purification MiniKit and the QIAamp Circulating Nucleic Acid Kit had similar extraction yield, but in the end, gave a slight edge to the QIAamp circulating nucleic acid kit due to consistency.

While total cfDNA yield is an important indicator of extraction performance, an ideal circulating cfDNA extraction should efficiently capture the 185 to 200 bp length DNA fragments expected of typical cfDNA profiles [70], but also may need to capture larger cfDNA fragments. These larger cfDNA fragments are thought to be released under disease conditions, such as the case with ctDNA [71, 72]. Specific size classes of cfDNA fragments may contain the majority of targets of interest and retaining those fragments is crucial for success [27–74]. Kloten et al. [75] compared the silica column-based QIAamp Circulating Nucleic Acid and magnetic bead-based Maxwell RSC ccfDNA kits for extracting cfDNA from serum and plasma. Both methods were able to successfully extract cfDNA from serum and plasma at sufficient quantities for *KRAS* mutation identification. Although the silica column-based method and the bead-based method had similar yields for smaller molecular weight cfDNA fragments (<600 bp), the silica column-based method had significantly greater yields of high molecular weight cfDNA (>600 bp). This finding suggests that the silica-column-based method is superior for capturing higher molecular weight circulating cfDNA, and thus may be a preferential method when high molecular weight cfDNA fragments are expected in the sample (e.g., ctDNA). In a similar study performed a year later by Solassol et al. [76] targeting two *EGFR* mutations in lung cancer patient plasma samples, the authors compared three silica column-based kits (Zymo Research Quick cfDNA Serum and Plasma Kit, QIAamp Circulating Nucleic Acid Kit, Norgen Plasma/Serum Cell-Free Circulating DNA Purification Mini Kit) and two magnetic bead-based kits (Bioo Scientific NextPrep-Mag cfDNA Isolation Kit, Promega Maxwell RSC ccfDNA

Plasma Kit) using a consistent 1 mL of plasma for all kits. Solassol et al. [76] found that the Zymo Research Quick cfDNA Serum and Plasma Kit had the most consistent extraction efficiency and interestingly was also the only kit to consistently capture dinucleosome cfDNA fragments. In two samples with known low levels of an EGFR point mutation, the mutation was only detected in the Zymo and Norgen kits. The authors conclude that although the Zymo Research Quick cfDNA Serum and Plasma Kit did not always capture the most cfDNA, the kit likely had the least PCR inhibitors in the eluant which allowed for ultra-sensitive detection of the EGFR point mutation.

Overall, the choice of cfDNA extraction method will depend on the final analysis application. Head-to-head comparisons of cfDNA extraction kits have revealed that there are several commercial kits with nearly equivalent performance. Silica-based column extractions, specifically the Zymo Research Quick cfDNA Serum and Plasma Kit, QIAamp Circulating Nucleic Acid Kit, Norgen Plasma/Serum Cell-Free Circulating DNA Purification Mini Kit, appear to hold a slight advantage over bead-based and PME methods because of higher cfDNA yields and higher performance in sensitive enzymatic reactions required for mutation analysis. However, there is not an overall consensus on the top-performing extraction kit likely because of between-lab variation. Thus, it is likely researchers will need to continue performing in-house head-to-head comparisons to determine the best kit to suit their cfDNA extraction and analysis needs.

2.2.3 Circulating cfDNA isolation techniques: direct capture from sample

Commercial kits for extracting circulating cfDNA have the benefit of generally only requiring commonly found laboratory equipment and reagents for cfDNA isolation. However, microfluidic chips have recently been gaining popularity because of their potential for high-throughput screening, low sample input, and point-of-care clinical translation. Microfluidic chip-based technology for cfDNA isolation generally falls under either "solid phase" or "liquid phase" isolation strategies. In "solid-phase" isolation, the chips take advantage of the non-specific adsorption between DNA and silica similar to the strategy of silica column-based methods. Silica particles are immobilized to an on-chip matrix or combined with magnetic beads [77, 78]. Recent work has also integrated pressure and immiscibility-based extraction (PIBEX) with microfluidic technology to enhance the capture of cfDNA to a silica membrane [79]. The cfDNA can then be captured and washed on-chip before being eluted for downstream analysis. For "liquid phase" isolation methods, cfDNA is captured from a sample onto a microfluidic chip using chemicals or electric fields. Due to the possibility of less organic contaminants remaining after chemical extraction [80], microfluidic chips that use electric fields to capture cfDNA have largely gained traction over chemical-based methods.

2.2.4 Circulating cfDNA isolation techniques: increasing sample yield or concentration

For many analysis applications that have a limited sample input volume, it may be desirable to increase the final cfDNA concentration. At lower cfDNA concentrations, the addition of carrier RNA during extraction may be beneficial to increasing sample yield. Many commercial circulating cfDNA kits include carrier RNA as an optional additive during DNA extraction. However, at the typical sample input of ≥ 1 mL, the addition of carrier RNA may not affect cfDNA or ctDNA yield [75, 81]. In addition, carrier RNA left in the eluted DNA sample may also interfere with downstream quantitation or analysis [82]. Therefore, the addition of carrier RNA during cfDNA extraction appears to provide negligible sample yield improvement while also adding a confounding factor to sample analysis.

The final concentration of cfDNA can also be increased post-extraction. Traditional methods of concentrating DNA, such as vacuum (e.g., SpeedVac) and chemical (e.g., ethanol precipitation) concentration, risk sample loss because cfDNA is often found at low initial concentration. There are now several commercial kit options for concentrating cfDNA samples. Similar to cfDNA extraction, DNA concentrators are generally silica column (e.g., Zymo DNA Clean and Concentrator Kit series) or magnetic bead (e.g., Dynabeads DNA DIRECT Universal) binding methods. DNA concentrator kits vary based on total DNA binding capacity (amount of total input cfDNA that can be handled) and final elution volume; therefore, the choice of DNA concentrator will depend on the concentration of the input sample and the volume required by the final application.

2.2.5 Purification of cell-free RNA

2.2.5.1 Circulating cfRNA extraction

Like circulating cfDNA kits, circulating cfRNA extraction kits can differ based on silica-based or magnetic bead binding methods. The choice of circulating cfRNA extraction kit will depend on the size class of RNAs being targeted for downstream analysis. Depending on kit chemistry, specific size classes of RNAs can be enriched or excluded. Because the majority of circulating cfRNA research has focused on small ncRNAs such as miRNA and piRNAs, there are several kits tailored to maximizing the yield of these ~22 nt molecules. Tan et al. [83] evaluated the yield and reproducibility of five kits for extracting miRNA from plasma and serum: Qiagen miRNeasy Serum/Plasma kit, Exiqon miRCURY RNA Isolation Kit – Biofluids, ThermoFisher mirVana PARIS Kit, Macherey-Nagel NucleoSpin miRNA Plasma, and Norgen Biotek Plasma/Serum Circulating and Exosomal RNA Purification Kit. Evaluation of the kits was based on qPCR quantitation of spiked-in synthetic controls and plasma-derived circulating miRNAs. The authors found that the Exiqon miRCURY RNA Isolation Kit – Biofluids, ThermoFisher mirVana PARIS Kit, and Macherey-Nagel NucleoSpin miRNA Plasma kits had the highest average recoveries of the spike-in control. In addition, the NucleoSpin miRNA Plasma kit had the lowest between-extraction variability overall. In a similar study the same year, Brunet-Vega et al. [84] included five kits in their comparison like Tan et al. [83] but included the Zymo Direct-zol RNA MiniPrep instead of the ThermoFisher mirVana PARIS Kit and used a slightly different Norgen Biotek kit, the Norgen Biotek Plasma/Serum RNA Purification Mini Kit. The authors found that when measured by qPCR, the Exiqon miRCURY RNA Isolation Kit – Biofluids had the best recovery of synthetic miRNAs that were spiked-in at the RNA purification step while the Macherey-Nagel NucleoSpin miRNA Plasma kit had the worst. Overall, Brunet-Vega et al. found the five kits to have comparable performance when recovering endogenous miRNAs from plasma. The reason for the discrepancy between the spike-in and endogenous miRNA recovery results was unclear. However, Brunet-Vega et al. suggest the discrepancy may be due to biological differences, whereby endogenous miRNAs are associated with protein or vesicle complexes and spike-in miRNAs are not.

Both Tan et al. [83] and Brunet-Vega et al. [84] focused on silica column-based extraction kits for their comparisons. In a subsequent circulating cfRNA kit comparison by Al-Qatati et al. [85], the magnetic-bead-based ThermoFisher TaqMan miRNA ABC Purification Kit was compared to kits from Qiagen, Machery Nagel, and Analytik Jena. The authors reported the highest plasma miRNA yield from the ThermoFisher Kit, although it should be noted that the authors did not specifically state which kits were used by the other three companies. If specific miRNA targets are known prior to the extraction step, the TaqMan miRNA ABC Purification Kit can be used to enrich targets using hybrid-capture and thus potentially increase overall sensitivity compared to that generating complete cDNA libraries.

However, targets must be chosen carefully because if any are omitted from the hybrid-capture step, those targets would not be measurable during subsequent analysis. Using serum as input material, Guo et al. [86] sequenced purified circulating cfRNA from five commercial kits: the Qiagen Circulating Nucleic Acid Kit, ThermoFisher Scientific Invitrogen TRIzol LS Reagent, Qiagen miRNEasy Serum/Plasma Kit, QiaSymphony RNA extraction kit, and the Exiqon miRCURY RNA Isolation—BioFluids Kit. The QIAsymphony SP platform uses a magnetic rod system to capture and wash RNA, and when used with the QiaSymphony RNA extraction kit, can be automated for high-throughput isolation of circulating RNA samples. The authors found that the Qiagen Circulating Nucleic Acid Kit had the highest total yield of circulating cfRNA from 200 μL serum relative to the other kits, but the kit also had a strong outlier that caused it to have the highest range in technical variability.

2.2.6 cfDNA genetic analysis
2.2.6.1 cfDNA quantitation and quality control
Following circulating cfDNA extraction, most workflows will require quantitation and quality control assessment prior to final application. A successful circulating cfDNA extraction will have sufficient cfDNA quantity for analysis but also DNA that is not over-fragmented. Bulk cfDNA quantitation is the non-specific quantitation of total cfDNA and is usually the first step of quality control immediately following cfDNA purification. Total DNA concentration and protein contamination are traditionally determined using ultraviolet (UV) spectroscopy by measuring sample absorbance at 260 and 280 nm. However, the low concentration of cfDNA prevents accurate quantitation using UV spectroscopy and the method is prone to erroneous measurements due to residual chemical contaminants from the extraction process [87]. Thus, bulk quantitation of cfDNA is now commonly performed using fluorometric methods. Fluorometric methods for cfDNA quantitation rely on the selective binding of a fluorescent dye to double-stranded DNA (dsDNA). The fluorescence intensity is then measured using a plate reader or fluorometer and the intensity is directly proportional to the amount of intact dsDNA in the sample within the working range. Importantly, the dyes rely on the presence of dsDNA for binding so free nucleotides do not contribute to the overall signal. Popular commercial fluorometric kits for quantifying cfDNA include the Invitrogen Quant-iT dsDNA Assay Kit HS, Invitrogen Qubit dsDNA HS Assay Kit, and the Invitrogen Quant-iT PicoGreen dsDNA Assay Kit.

Characterization of cfDNA fragmentation patterns is also an important step in quality control. Generally, circulating cfDNA fragments have a dominant peak at ~167 bp, which is approximately the length expected of a DNA strand wrapped around a nucleosome and linker histone [88]. An excess of cfDNA fragments <167 bp may indicate atypical DNA degradation during extraction, while an excess of fragments >167 bp may indicate genomic DNA contamination. Fragment analysis can be performed using capillary electrophoresis such as the chip-based Agilent 2100 bioanalyzer or high-throughput Agilent Fragment Analyzer systems. Along with size distribution, fragment analyzers provide a quantitative assessment of concentration for DNA molecules that fall within the size range of detection.

For samples with extremely low concentrations of cfDNA, the most sensitive circulating cfDNA quantitation and fragmentation analysis methods are quantitative PCR (qPCR) and digital droplet PCR (ddPCR). By comparing to a standard curve (as with qPCR) or by direct counting (as with ddPCR), absolute concentrations of cfDNA can be determined in a sample. However, these methods rely on locus-specific probes that measure only region per primer pair. Genomic regions that are farther from the centromere may be exposed to enzymatic hydrolysis sooner than regions closer to the centromere. If this potential fragmentation pattern is not taken into consideration, the final concentration may be over- or underestimated. Therefore, choosing genomic targets carefully and measuring multiple reference genomic regions can provide a more reliable cfDNA quantitation [89]. Commercial kits with preoptimized primer pairs and DNA standards, such as the Roche KAPA hgDNA Quantification and QC Kit, can be used to quantify cfDNA and estimate the degree of fragmentation by targeting different size genomic regions. When possible, qPCR or ddPCR results should be compared to fluorometric or capillary electrophoresis results for concordance.

2.2.7 Circulating cfDNA detection by qPCR and ddPCR

The first step of cfDNA analysis following QC is often to target specific gene loci using PCR-based detection methods. Accurate and sensitive methods of detection are especially crucial for ctDNA applications because of the low-frequency nature of most ctDNA targets, especially at early tumor stages [90]. The gold standard for sensitive nucleic acid detection remains qPCR, but ddPCR has also gained traction in applications that require single-copy resolution. There are now many variations of qPCR, with all methods relying on the basic selective PCR amplification of an oligonucleotide region using targeted forward and reverse primers. Amplification is detected by the accumulation of fluorescence either due to a dsDNA-specific intercalating dye (e.g., SYBR green or Evagreen) or hydrolysis of a probe that releases dye from its quencher (e.g., TaqMan). Positive fluorescence signal is determined at the critical threshold cycle (Ct, also sometimes referred to as quantitation cycle or Cq) which is set automatically by the qPCR machine or manually. Because PCR methods rely on targeted primer pairs, all methods must have prior knowledge of the variant in question. Multiple targets can be measured simultaneously in a process known as multiplexing if the qPCR or ddPCR fluorescence detection platform supports the detection of multiple colors. Alternatively, microfluidics can be used to divide samples into microreactions for probing many primer pairs simultaneously, as is the strategy with Fluidigm BioMark HD qPCR system. QPCR and ddPCR methods have successfully been used to detect single-nucleotide polymorphisms (SNPs), copy-number variants (CNVs), as well as insertion/deletion events (INDELs) in cfDNA [91–93]. Variations in qPCR detection of cfDNA variants generally attempt to either increase the limit of detection or increase the number of targets that can be multiplexed. There are several commercial and free software packages available to analyze qPCR data, such as qBasePLUS, GenEX, TaqMan Genotyper, DataAssist, and ExpressionSuite. One of the simplest qPCR variations, allele-specific PCR (AS-PCR), was first described by Newton et al. [94] and relies on the lack of PCR amplification when an SNP causes a mismatch at the 3' end in of one of the primers. Although generally considered to be a semi-quantitative assay, AS-PCR has been used on circulating cfDNA samples to identify ctDNA mutation signatures [75, 95]. An extension of AS-PCR, "Allele-Specific Non-Extendable Primer Blocker PCR" (AS-NEPB-PCR), includes the addition of a non-extendable primer blocker that has an exact match to the WT DNA variant and prevents unintentional amplification of WT sequences [96]. Although AS-NEPB-PCR may permit a 10-fold increase in sensitivity of AS-PCR for detecting SNPs, as of yet AS-NEPB-PCR has not been applied to circulating cfDNA. In another extension of AS-PCR, "Pooled, Nested, WT-Blocking qPCR" (PNB-qPCR) is a two-reaction nested qPCR assay that has been applied to ctDNA [97]. Similar to the premise of AS-NEPB-PCR, the first round of PCR amplification in PNB-qPCR uses WT-blocking primers that promote the enrichment of mutant ctDNA fragments only. Unlike AS-NEPB-PCR, however, a second round of PCR using allele-specific primers to amplify the first round PCR products is performed and variants are detected by LNA probe hydrolysis. Another method that utilizes WT-blocking primers is the peptide nucleic acid-locked nucleic acid (PNA-LNA) PCR clamp [98]. In this method, a PNA clamp is used to block WT sequence amplification during PCR in a gene of interest. A mutation-specific LNA probe is used to detect mutant signals and the amount of mutation-specific probe signal is compared to total copies of the gene of interest measured by a second probe. The PNA-LNA PCR clamp method has been successfully used in detecting EGFR mutations in plasma-derived circulating cfDNA [99, 100]. Strategies that leverage differential melting temperatures of WT and mutant sequences have also been used to detect low levels of mutant circulating cfDNA. One method known as "coamplification at lower denaturation temperature PCR" (COLD-PCR) begins by binding target-specific primers at a melting temperature that favors primer binding to mutant sequences and thus increased amplification during PCR [101]. Following PCR, amplicons are melted and cooled such that mutant-WT heteroduplexes are allowed to form. Finally, high-resolution melting (HRM) detects the heteroduplexes and infers the presence of mutant sequences, as the mismatches formed in the heteroduplexes cause the dsDNA to melt at lower temperatures. COLD-PCR/HRM has been applied to detect *KRAS* mutations in plasma ctDNA from thoracic malignancy patients [102].

PCR-based assays that can detect CNVs are less straightforward than SNP or INDEL assays. Multiplex ligation-dependent probe amplification (MLPA) is based on the region-specific hybridization of probe pairs that can be PCR amplified using universal primers [103]. Prior to PCR amplification, a probe pairs are ligated such that only perfectly matching hybridization events with adjacent probe pairs permit ligation. Then, PCR is performed using universal primers with one of the primers fluorescently labeled and amplicons are separated by capillary electrophoresis. By knowing the exact length of the amplicon sequences, multiple regions can be deconvoluted using the resulting electropherogram and relative CNV levels can be estimated. MLPA has been described to detect CNVs in cfDNA of bladder cancer patients [93]. However, the use of MLPA for CNV analysis in cfDNA can still be hindered when there is an overabundance of normal WT copies of the region of interest.

As clinical liquid-biopsy research moves toward earlier detection of diseases, there has been an increased interest in nucleic acid detection platforms that can reliably measure single-copy events in genes of interest. Like conventional qPCR, ddPCR relies on the accumulation of fluorescent signals derived from intercalating dyes or probe hydrolysis as targets of interest are amplified. However, ddPCR reactions are first fractionated into $\sim$ 20,000 droplets of known volume using water-oil emersion technology. Each droplet contains PCR reaction components, primers, and probe (if probe-based assay), but may or may not contain the target sequence of interest. Following the end of PCR, droplets that initially contained template are counted based on the resulting fluorescent signal and the total concentration of template can be calculated using Poisson statistics [104]. The digital (i.e., binary) outcome of the ddPCR allows for the detection of fold changes <2, the theoretical limit of qPCR, as well as the detection of single-copy events. Using the leading platform in ddPCR technology, the QX200 Droplet Digital PCR System (Bio-Rad), low levels of *KRAS* and other gene SNPs have been detected in circulating cfDNA from cancer patients [105–109]. Additionally, ddPCR has been used for the detection of CNVs in circulating cfDNA [110, 111], although an excess of WT copies may still convolute results. Bio-Rad offers commercial kits for gene-specific SNP and CNV detection. Similar to ddPCR, BEAMing (beads, emulsions, amplification, and magnetics) relies on emulsion-based reaction compartmentalization and a digital readout but includes steps that covalently bind amplified cfDNA template targets to magnetic beads [112]. The beads are then counted by hybridizing fluorescent probes to the target sequences and measuring fluorescent signals using a high-resolution flow cytometer. BEAMing has been demonstrated to have accurately low levels of circulating cfDNA [113, 114]. Head-to-head comparisons of circulating cfDNA mutation detection using ddPCR and BEAMing have suggested good concordance between the methods, although BEAMing may have a higher overall annual cost [115, 116].

2.2.8 Circulating cfDNA detection by Raman spectroscopy or mass spectrometry

Because of the limited multiplexing ability of solely PCR-based detection methods, detection techniques using mass spectrometry have been developed to detect circulating cfDNA targets. Surface-Enhanced Raman Spectroscopy (SERS) has been shown to detect low-frequency mutations in circulating cfDNA when combined with PCR (SERS-PCR) [117, 118]. In SERS-PCR, labeled tags that hybridize to targets of interest are detected using their specific Raman scattering pattern. However, SERS-PCR has not been demonstrated to detect more than a few cfDNA variants at a single time and thus may have limited ability to be multiplexed. Platforms such as the UltraSEEK MassARRAY (Agena) can capture hundreds of targets of interest using biotinylated probes and detect variants based on the resulting mass spectrometry spectra of products. UltraSEEK MassARRAY has been demonstrated to have high concordance with ddPCR for detecting circulating cfDNA variants while having increased multiplexing ability [119].

2.2.9 Circulating cfDNA detection by sequencing

One limitation of qPCR and ddPCR methods for detecting cfDNA targets is their limited multiplexing scalability. Sequencing methods, on the other hand, are able to characterize tens to thousands of targets simultaneously. Next-generation sequencing (NGS), also sometimes referred to as "second-generation sequencing," is the current method of choice for high-throughput sequencing of circulating cfDNA. Millions of short DNA sequences, also known as reads, can be obtained in parallel with the potential for basepair resolution of genomic targets. While "third-generation sequencing" platforms such as the PacBio Sequel II (Pacific Biosciences) and Oxford Nanopore systems have emerged promising longer read lengths [120, 121], few published circulating cfDNA studies have taken advantage of this technology due to their high cost per base or high error rate. NGS workflows follow a general process of 1) preparing a sequencing library from isolated cfDNA, 2) optionally enriching for target sequences, 3) obtaining sequencing reads by sequencing the prepared libraries, 4) computational alignment of reads to a reference human genome, and 5) inferring SNPs, INDELs, CNVs, and fragment size using bioinformatics. Commercial kits for NGS library preparation are offered for the specific sequencing instrument to be used. Illumina sequencing remains the most commonly used NGS technology for circulating cfDNA sequencing because of low cost per sequenced base. The Illumina HiSeq series and Illumina MiSeq series use the process of "bridge amplification" and reversible terminator chemistry to generate millions of reads in parallel [122]. The closest NGS competitor to Illumina is the Ion Torrent (Life Technology) series of sequencing platforms which are based on semiconductor chip technology, rather than optics as with Illumina. Ion Torrent GeneStudio S5, Ion Torrent Genexus, Ion Torrent Personal Genome Machine, and Ion Torrent Proton generate DNA sequences by recognizing the release of a hydrogen ion when a correct nucleotide is incorporated during DNA synthesis [123].

An important consideration of all NGS platforms is the internal error rate caused by wrongly incorporated bases during the sequencing process. Higher error rates require higher sequencing depth (also known as sequencing "coverage") at sites of interest in order to reliably detect variants. Average error rate estimates range from 0.1 to 1% depending on the error type (e.g., single nucleotide substitution or deletion) and platform [124, 125]. Because reads from circulating cfDNA variants of interest are almost always found at much lower frequencies than WT reads, a strategy during library preparation is to increase the sequencing depth at the region of interest using target enrichment. Kits such as Illumina TruSight Oncology 500 ctDNA and Ion Torrent Oncomine cfDNA assays use targeted, predesigned gene panels to enrich informative SNVs, indels, and CNVs in circulating cfDNA. Alternatively, researchers can customize their own gene panels for enrichment prior to sequencing using the Illumina TruSeq Custom Amplicon v1.5 or Ion AmpliSeq HD Panels. The use of molecular barcoding during library preparation has also gained popularity because of its potential to reduce error introduction from DNA polymerase base misincorporation during PCR. Molecular barcoding adds unique molecular indexes (UMIs) to each DNA template strand during library preparation and prior to PCR such that each DNA template can be later identified using bioinformatic software [126–128]. The Illumina TruSeq UMI system claims to reduce sequencing error rates in samples to $\leq 0.007\%$ and many other NGS workflows that require ultra-sensitive detection of variants now routinely use UMIs in some form. Error suppression also be performed post-sequencing using sophisticated bioinformatics software packages such as Tri-Nucleotide Error Reducer (TNER) [129] and PCR Error Correction (PER) [130].

Although NGS target enrichment coupled with sequencing error suppression allows for highly sensitive at basepair resolution, requirement of prior knowledge of regions of interest limits the ability for de novo variant discovery. Whole exome sequencing (WES) targets all protein-coding regions in the genome and has the potential to discover circulating cfDNA variants that fall within these regions with high sensitivity [131, 132]. Whole genome sequencing (WGS) has the potential to characterize both novel and known variants in a circulating cfDNA sample because it sequences all DNA template indiscriminately [133]. Recent work by Mauger et al. [134] compared five WGS library preparation methods for analyzing circulating cfDNA: ThruPLEX Plasma-seq, QIAseq cfDNA All-in-One, NEXTFLEX Cell-Free DNA-seq, Accel-NGS 2 S PCR FREE DNA and Accel-NGS 2 S PLUS DNA. Using plasma-derived circulating cfDNA as input, the authors found that detected SNV, CNV, and indel counts were mostly shared between the five kits. For low abundance somatic SNVs ($<5\%$ frequency), the Thru-PLEX Plasma-seq kit was able to identify the most variants. While WGS is appealing for its ability to profile whole genomes, a limiting factor of WGS is the sample cfDNA input requirement is often higher than cfDNA sequencing workflows that have target enrichment. Whole genome amplification using rolling circle amplification may provide increased variant sensitivity for WGS while limiting PCR bias [135].

Following sequencing, reads are computationally aligned, or "mapped," to their appropriate reference genome. Reads can optionally be trimmed and filtered for quality and adapter sequences using software such as Trimmomatic [136] or fastp [137]. Commonly used and open-source short read mappers include bowtie2 [138], BWA [139], and BWA-SW [140]. Alignment files can then be examined directly or passed to variant identification software depending on the desired end analysis, for example, SNPs: ABEMUS [141] and muTect [142]; CNVs: GSTIC2 [143]; and FACETS [144]; genomic rearrangements: PARE [145], fetal cfDNa fraction: FetalQuant. Many commercial sequencing kits such as the Illumina TruSight and Ion Torrent Ion AmpliSeq include comprehensive online bioinformatics analysis portals when using their kits.

2.2.10 cfRNA analysis

2.2.10.1 Special considerations for circulating cfRNA analysis: quality control

Like circulating cfDNA, purified circulating cfRNA is normally too low in concentration to be quantitated and quality-checked using spectroscopy. Circulating cfRNA analysis can be performed by fluorometric, capillary electrophoresis, and PCR methods using systems similar to cfDNA with some additional considerations. Using fluorometric methods, the Quant-iT^{TM} RiboGreen^{TM} RNA Assay Kit (ThermoFisher) and QuantiFluor RNA System (Promega) can quantitate RNA from ≥ 100 pg/μL and ≥ 250 pg/μL, respectively. Fluorometric methods are specific to RNA and contaminating cfDNA should have minimal impact on signal. The Agilent 2100 Bioanalyzer and Agilent Fragment Analyzer both offer RNA kits that can quantitate and reliably size RNA from ≥ 50 pg using high-resolution capillary electrophoresis. Both the Agilent 2100 Bioanalyzer and Agilent Fragment Analyzer additionally produce an RNA quality score, RNA Index Number (RIN), or RNA Quality Number (RQN), respectively. However, since the RNA quality scores measure a 18S to 28S rRNA peak ratio and circulating cfRNA is normally highly degraded, the scores may not accurately reflect the quality of RNA transcripts. Circulating cfRNA integrity can instead be determined using a qPCR-based 5'/3' ratio mRNA integrity assay [146]. The 5'/3' ratio mRNA integrity assay relies on the assumption that as the protective 3' polyA tail shortens on a transcript, mRNA degrades in a 3'/5' direction. This assay is performed by measuring the relative abundance of a 5' exon versus a 3' exon in a particular gene, for example, *ACTB* exon 1–2/ACTB exon 3–4. Because qPCR is measured on a log2 scale, intact mRNAs would reveal a

5'/3' exon ratio of approximately 0 ([5' exon Ct] − [3' exon Ct] = 5'/3' exon ratio). Therefore, excessive degradation at 3' exons would have 5'/3' exon ratios >0 and samples with high ratios may be unsuitable for downstream analysis.

2.2.10.2 Special considerations for cfRNA analysis: cDNA synthesis and qPCR

The inherent instability of RNA necessitates the conversion of circulating cfRNA to complementary DNA (cDNA) prior to most analysis applications. Because circulating cfRNA is generally fragmented, cDNA synthesis that included both random hexamer and oligo-dT primers may provide the most comprehensive cDNA libraries. To increase the sensitivity of PCR-based applications, the resulting cDNA can be pre-amplified using gene-specific primers. Synthesis of cDNA and pre-amplification can be performed in a single reaction, such as with the SuperScript^{TM} IV One-Step RT-PCR System (ThermoFisher), or cDNA synthesis and pre-amplification can be performed separately. Small RNA classes such as miRNA cannot be converted to cDNA with typical cDNA synthesis kits. Using kits such as the miScript II RT Kit (Qiagen) or TaqMan MicroRNA Reverse Transcription Kit (ThermoFisher) a 5' adapter and 3' polyA tail are added to the miRNAs to extend the templates prior to cDNA synthesis. Following cDNA synthesis and optional pre-amplification, cDNA can be analyzed with the same qPCR and ddPCR procedures as circulating cfDNA. If using the TaqMan system, miRNA qPCR profiling can also be performed using preconfigured microfluidic array cards that can support up to eight samples, 380 assay targets, and four replicates per card. Many companies offer complete workflows that include reagents for cDNA synthesis, preamplification, and qPCR quantitation of targets. A head-to-head comparison of miRNA cDNA synthesis and qPCR workflows by Tan et al. [83] suggested that the TaqMan miRNA PCR system (ThermoFisher) had higher sensitivity and reproducibility than the miScript miRNA PCR system (Qiagen). Similar workflows exist from ThermoFisher and Qiagen for performing cDNA synthesis and qPCR quantitation of larger RNA targets such as mRNAs.

2.2.10.3 Special considerations for circulating cfRNA analysis: sequencing

Unlike circulating cfDNA sequencing workflows, which generally have less emphasis on sequencing specific fragment lengths, the fragment size of circulating cfRNA in a sequencing library will greatly affect the RNA species type that is sequenced. Sequencing kits for RNA are divided into two main categories: small RNA (smRNA) kits that target RNA species <200 nt and RNA kits that target RNA species approximately >200 nt. Small RNA kits such as the Illumina TruSeq Small RNA Library Preparation kit and Ion Total RNA-Seq Kit v2 are sequenced on their respective platforms and can identify a wide variety of small RNAs, such as miRNAs, piRNAs, tRNA fragments, and snoRNAs. For analyses that require miRNA-specific quantitation, miRNA kits like the Qiagen QIAseq miRNA Library Kit are available that target the ~22 nt size class expected of miRNAs. For larger RNAs, such as mRNAs, long noncoding RNAs, and circular RNAs, total RNA sequencing kits with low input RNA requirements are used. The SMARTer Stranded Total RNA-Seq Kit has successfully been used in circulating cfRNA applications from plasma [147], and there are several other kits that provide low-input requirement RNA sequencing [148]. In addition, Bio-Rad has recently introduced a low-input requirement RNA sequencing kit, the SEQuoia Complete Stranded RNA Library Prep Kit, that can sequence both small and long RNA species. However, the SEQuoia Complete Stranded RNA Library Prep Kit has yet to comprehensively be tested on circulating cfRNA samples.

Following circulating cfRNA sequencing, resulting reads are quality controlled and mapped to their respective references similar to cfDNA. However, the choice of mapping software and reference will depend on the type of RNAseq performed. For miRNA, common mappers include STAR aligner [149] and bowtie1 [150]. The choice of mapping software can impact numbers and types of miRNAs identified [151]. Typical miRNA analyses that seek to identify and quantify known miRNAs and piRNAs use the miRBase database as a reference [152]. Many miRNA kits also include comprehensive online bioinformatics support, such as the QIAseq miRNA Library Kit, which can analyze UMIs and perform differential expression from raw data. For longer RNA species, such as mRNAs and long noncoding RNAs, reads are mapped using a short-read mapper (e.g., STAR aligner) that allows for reads to be split across an appropriately annotated reference genome. Following read counting, differential expression can be performed using RNAseq statistical tools such as DEseq2 [153] and edgeR [154].

Retention or loss of specific circulating cfRNA classes during the extraction step can be especially apparent following profiling applications such as sequencing. Guo et al. [86] used the circulating cfRNA from their four extraction kit head-to-head comparisons as input to a smRNA sequencing pipeline. For all extracted cfRNA samples, the TruSeq Small RNA sample preparation kit was used and samples were sequenced identically such that any observed variation could be attributed principally to the RNA extraction methods. Although the Qiagen Circulating Nucleic Acid Kit was determined to have the highest overall yield of RNA material, it also had the lowest rate of smRNAs mapping to the genome, and suggested a high abundance of material present that was not related to smRNAs. Interestingly, the four extraction kits did not identify the same smRNA profiles and it is possible that the kits were demonstrating some smRNA selection bias. Nonetheless, the authors conclude that the Invitrogen TRIzol method had the overall highest mapping rate and deemed it the best of the four kits compared.

2.2.11 cfDNA epigenetics analysis

Epigenetics has been defined as changes to DNA that impact gene activity or expression without changing the DNA sequence itself. DNA methylation and histone modifications are the two major mechanisms in epigenetics [155, 156]. Importantly, aberrant DNA methylation and chromatin structure is a hallmark of several diseases, including cancer [155–157]. While genetic analysis of circulating cfDNA can often be challenging due to an underrepresentation of mutated targets compared to high WT background, epigenetic aberration may offer an alternative or supplemental solution thanks to its wide spreading across the genome [34–46 157]. In this section, we review the methods which have been used to profile DNA methylation patterns in circulating cfDNA (Figure 2.2).

2.2.12 DNA methylation

DNA methylation results from the addition of a methyl group (CH3) to cytosine in a CpG dinucleotide. This addition takes place through the action of an enzyme family called DNA methyltransferases (DNMT). The demethylation processes mediated by a family of enzymes called a ten-eleven translocation (TET) are able to oxidize 5-methylcytosine into 5-hydroxymethylcytosine (5hmC), then into 5-formylcytosine (5fC) and finally 5-carboxylcytosine (5caC). When DNA is methylated, it can induce changes in gene expression, primarily by inhibiting transcription through reduced transcription factor fixation to promoters and other DNA physical properties [158, 159]. Methods for profiling cfDNA methylation are divided into three main categories: base conversion-based, methylation-sensitive restriction enzyme-based, and enrichment-based methods.

Bisulfite conversion is a base conversion technique described for the first time in 1992 by Frommer [160]. Bisulfitation is a chemical modification

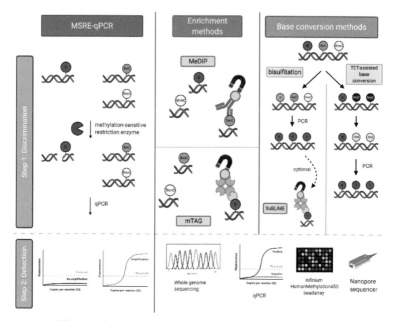

Figure 2.2   Technologies for cfDNA methylation analysis.

that consists in transforming non-methylated cytosines into uracil. It is composed of three steps: sulfonation, deamination, and desulfonation [161]. Methylation protects cytosines from bisulfite conversion while methylated cytosines are converted. The advantages of this technique are that it permits to have an overview of every cytosine in the genome after whole genome sequencing, and bioinformatics analysis to determine which one is significant. The bisulfite conversion protocol has been standardized and is fast. However, the aggressive bisulfite treatment leads to severe DNA degradation. Moreover, it is an indirect study of cytosine methylation, because only non-methylated cytosines are modified, and the bisulfite conversion reaction can be incomplete [162]. Another problem is that bisulfite conversion will induce the reading of 5hmC as 5mC after sequencing, so this technique is not able to discriminate between 5mC and 5hmC. To solve that problem, a method called Tet-assisted oxidative bisulfite conversion has been developed. It is based on the use of β-glucosyltransferase to protect 5hmC from deoxydation by TET, resulting in a conversion of 5mC into 5caC (read as a T) and then 5hmC will be read as C. By comparing Tet-assisted oxidative conversation with regular bisulfite conversion it is possible to discriminate non-methylated cytosines, 5mC and 5hmC.

Because the harshness of bisulfite treatment usually causes significant sample loss, there has been interest in developing bisulfite-free base conversion techniques to profile DNA methylation. Recently, Song et al. developed a novel conversion technique to discriminate methylated from unmethylated cytosines without bisulfite treatment. In this method, TET-assisted pyridine borane conversion permits to transform 5mC and 5hmC into 5caC, followed by chemically modified dihydrouracil (DHU) [163]. Subsequence PCR converts DHU to into thymine, enabling a C-to-T transition of 5mC and 5hmC. Aligning of sequences before and after conversion infers methylated and hydroxymethylated cytosines. As previously described for bisulfite conversion, a first step with β-glucosyltransferase allows to discriminate 5mC from 5hmC. TET-assisted pyridine borane conversion is less destructive than the bisulfite treatment. The method directly reports on 5mC and 5hmC modifications. Like standard bisulfite treatment, the main challenge of this technique is incomplete base conversion, and the functional consequences of the modified cytosines are not trivial to infer.

2.2.13 cfDNA methylation detection methods: sequencing, array, and qPCR

After conversion, discrimination of methylated and/or hydroxymethylated from unmethylated cytosines can be achieved by sequencing, methylation array, and methylation-specific qPCR. Using base-converted cfDNA as input, WGS followed by computational alignment to a reference genome can report a comprehensive map of the methylome for every cytosine, including CpG islands and non-CpG sites. For low-input cfDNA, the end repair and adapter ligation are usually performed before bisulfite treatment to increase amplification efficiency. Due to the coverage required to sequence across the entire genome, WGS of base-converted cfDNA can be costly. This is because the samples can contain a mixture of methylated and unmethylated cytosines at a given genome location, requiring sufficient coverage of both types in order to make accurate statistical inferences. Target enrichment using primer sets to amplify selected target regions or probe hybridization is more clinically applicable because it is more cost-effective and allows a higher sequencing depth. Both whole genome and targeted methylation sequencing of cfDNA have been demonstrated as promising technologies for cancer detection in multiple cancer types [164, 165]. The Infinium HumanMethylation450 bead array is a method commonly used to study DNA methylation for more than 450 000 CpG across the genome (out of 28 million, or 1.6%). These CpG sites have been chosen in order to cover the different regions of the genome (promoter region, 5'-UTR, first exon, gene body, and 3'-UTR) to provide an overview of the methylation state. This compromise between WGS and targeted sequencing has made it possible to identify a DNA demethylation signature of two genes that are a potential biomarker for spontaneous preterm labor [166]. Traditional qPCR with specific primers for the target region can also be used. Using probes that will hybridize only if DNA is methylated (or, contrariwise, demethylated) following base conversion, the methylation status of a genome region can be determined without sequencing [167, 168].

2.2.14 Enrichment-based cfDNA methylation detection

Specific antibodies have been developed to recognize methylated cytosine and therefore allow enrichment of methylated DNA fragments before downstream detected methods. Methylated DNA Immunoprecipitation sequencing (MeDIP-seq) was successfully developed to capture methylated cfDNA followed by sequencing [169]. This method gives a global map of methylation status for a DNA fragment but does not allow a base-resolution identification. By adding 'filler DNA' before immunoprecipitation, the antibody/DNA ratio is kept and allows the use of 1–10 ng of DNA instead of 100 ng minimum [40] and then can be used with cell-free DNA and enable to detect and classify tumors [40, 169].

Another methylation enrichment method is based on the conversion of modified cytosine to biotin. This method, also called streptavidin bisulfite ligand methylation enrichment (SuBLiME) is based on bisulfite conversion, but with a supplementary step that consists in adding biotin to bisulfite-converted non-methylated cytosines [170]. After an affinity capture with streptavidin-coupled magnetic beads, a deep-sequencing can be performed. SuBLiME can be seen as a combination of enrichment methods and bisulfite sequencing. Instead of bisulfite conversion, biotin can be attached to non-methylated cytosines with an engineered SssI DNA methyltransferase and Ado-6-amine, a cofactor analog, and finally N-hydroxy succinimidyl ester [171]. This method is known as methyltransferase-directed transfer of activated groups (mTAG), and because it requires only 100–300 ng of DNA, it can be adapted to cfDNA applications.

2.2.15 Enzyme-based cfDNA methylation detection

Methylation-sensitive restriction enzymes (MSRE) cleave DNA at a specific nucleotide sequence. Methylated cytosines at the restriction site either protect or promote the DNA from cleavage. After digestion, the fraction of methylated compared to unmethylated DNA fragments is quantified by digestion efficiency measurement using several methods: an electrophoretic mobility shift assay, ligation assay, qPCR with primers flanking on two sides of the restriction digestion site, and sequencing [172–175]. Conventionally, an MSRE assay has been conducted clinical for analyses of DNA methylation changes in human diagnosis. For cfDNA applications, MSRE digestion followed by multiplexed qPCR or ddPCR may allow the detection of less than 10 copies (0.1–1%) of specific methylated targets in unmethylated DNA background for a panel of up to 48–96 targets [176].

2.3 Clinical Applications

In this section, we will discuss the clinical applications of nucleic acids in prenatal diagnosis testing, oncology, transplantation, microbial analysis, diabetes, autoimmune disorders, and Alzheimer's disease. A list of notable cfDNA and cfRNA clinical trials is summarized in Table 2.1.

2.3.1 Non-invasive prenatal diagnosis testing

Conventionally, echography was used routinely to monitor fetal growth and detect abnormalities such as Down's syndrome or fetal malformation. When pathology is suspected, it may require invasive diagnostic confirmation procedures such as amniocentesis. During the last two decades, non-invasive prenatal testing (NIPT) using circulating cfDNA in the mother's blood has emerged as a mainstream approach for fetal diagnosis. NIPT was first used to determine fetal sex for mothers carrying X-linked disorders, such as hemophilia or Duchenne muscular dystrophy, which can affect a male fetus [177]. NIPT was also applied for early detection of the development of external genitalia to prevent excessive virilization of a female fetus by treating the mother with dexamethasone [178]. More recently, NIPT is used for the determination of fetal sex, aneuploidies, fetal rhesus D genotyping, pre-eclampsia, paternally inherited monogenic disorders, and detection of maternally inherited mutations using modern WGS methods [179].

Circulating cfDNA of fetal origin is detectable from 4 weeks of pregnancy in maternal blood [180]. During pregnancy, fetal DNA represents between 0.39 and 11.4% of total cfDNA from maternal plasma and this rate is dynamic during pregnancy [13]. Compared to NIPT based on fetal cells in the mother's blood, cfDNA-based NIPT methods give a higher quantity of fetal fraction

Table 2.1　Circulating cell-free DNA studies in clinical trials.

Test or study name	Biomarker type	Target	Application	Phases 1 and 2	Phase 3	Phase 4	Clinical trial number
				Discovery and prediction, assay validation	Clinical validation	Clinical utility	
Non-invasive prenatal testing							
Vanadis® NIPT, IONA	cfDNA	Counting chromosomes number	NIPT for screening of trisomy 21, 18 and 13				NCT03559374 – Recruitment status unknown NCT01472523 – Recruitment completed
Oncology							
Concordance of Key Actionable Genomic Alterations as Assessed in Tumor Tissue and Plasma in Non-small Cell Lung Cancer	cfDNA	17 genes panel	Genomic variants detection in advanced non-small cell lung cancer				NCT02762877 – Recruitment terminated
Detect-A, Cancer seek	Proteins and cfDNA	Mutations in regions of 16 genes + 9 protein biomarkers	Early detection of colorectal, ovary, pancreas, breast, upper gastrointestinal tract, lung, and liver cancer				NCT04213326 – Recruiting
STRIVE study, CCGA, SUMMIT, Pathfinder	cfDNA/cfRNA	Cell-free nucleic acids profiles, including DNA methylation	Early detection of more than 50 cancer types				NCT04241796 – Recruiting NCT03934866 – Enrolling by invitation NCT02889978 – Active, not recruiting NCT03085888 – Active, not recruiting
SafeSEQ, OncoBEAM	cfDNA	Mutations in several genes	Genotyping for colorectal, breast, lung, melanoma, head and neck squamous cell carcinoma				NCT02751177 – Completed
Epi proLung	cfDNA	SHOX2 and PTGER4 methylation	Lung cancer detection for risk patients				
Epi proColon	cfDNA	SEPT9 methylation	Cancer detection for patients with an average risk of CRC				NCT00855348 – Completed
HCCBloodTest	cfDNA	SEPT9 methylation	Liver cancer detection for patients with cirrhosis				NCT03804593 – Completed
Therascreen MGMT Pyro	cfDNA	MGMT methylation	Predict response to alkylating agent in gliomas				
COLVERA	cfDNA	IKZF1, BCAT1	Detection of residual and recurrent disease in CRC patients				NCT03706235 – Completed NCT03706248 – Completed

Table 2.1 (*Continued*).

Test or study name	Biomarker type	Target	Application	Phases 1 & 2	Phase 3	Phase 4	Clinical trial number
Ivygene	cfDNA	DNA methylation	Breast, colon, liver, and lung cancer detection				NCT03694600 – Recruiting
cobas® EGFR Mutation Test v2	cfDNA	42 EGFR mutation	Identify EGFR mutations in non-small-cell lung cancer				NCT04035486 – Recruiting
Guardant360 CDx	cfDNA	Comprehensive genomic profiling	Identify patients with NSCLC who should be treated with osimertinib				NCT04497285 – Not yet recruiting
FoundationOne Liquid CDx	cfDNA	Comprehensive genomic profiling	Identify patients able to benefit from targeted therapy for NSCLC, prostate, ovary, and breast cancers				NCT04484636 – Recruiting
Transplantation							
AlloSure, Prospera	cfDNA	Quantification of donor-derived cfDNA	Acute rejection in kidney transplantation				NCT04566055 – Active, not recruiting NCT04091984 – Recruiting NCT03765203 – Study completed
Allosure Heart	cfDNA	Quantification of donor-derived cfDNA	Acute rejection in heart transplantation				NCT03695601 – Recruiting
AlloSure	cfDNA	Quantification of donor-derived cfDNA	Determine if Allosure can be used to identify acute rejection in lung transplantation				NCT04318587 – Recruiting
Prospera	cfDNA	Quantification of donor-derived cfDNA	Determine if Prospera can be extended to several other transplantation types				NCT03984747 – Recruiting
Cell-free DNA as a Biomarker After Lung Transplantation, Longitudinal Study of Cell-Free DNA in Lung Transplant (LoSt)	cfDNA	Quantification of donor-derived cfDNA	Acute rejection in lung transplantation				NCT04271267 – Study completed NCT04234919 – Enrolling by invitation
Non-invasive Test for Acute Rejection Identification in Heart Transplanted Patients (INNO-GRAFTRS001), A New Biomarker for the Non-invasive Diagnosis of Rejection After Heart Transplantation (BIODRAFT)	cfDNA	Quantification of donor-derived cfDNA	Acute rejection in heart transplantation				NCT03477383 – Recruiting NCT04274712 – Enrolling by invitation
Genome Transplant Dynamics	cfDNA	Quantification of donor-derived cfDNA	Rejection after heart and lung transplantation				NCT02423070– Recruiting
Quantitative Detection of Circulating Donor-Specific DNA in Organ Transplant Recipients (DTRT-Multi-Center Study) (DTRT)	cfDNA	Quantification of donor-derived cfDNA	Determine the range of ddcfDNA indicating that rejection in several transplantation types				NCT02109575– Active, not recruiting

Table 2.1   *(Continued).*

Test or study name	Biomarker type	Target	Application	Phases 1 & 2	Phase 3	Phase 4	Clinical trial number
Cf-DNA Assay During Treatment of Acute Rejection	cfDNA	Quantification of donor-derived cfDNA after anti-rejection treatment	Successful treatment of an acute rejection episode after kidney transplantation				NCT04019353 – Recruiting
Diabetes Atherothrombosis Markers in Diabetics (MADI)	cfDNA	Quantification	Determine if cfDNA can represent a trigger for atherothrombotic plaques in type 2 diabetic patients				NCT02898467 – Unknown status
				Discovery and prediction, assay validation	Clinical validation	Clinical utility	
Non-invasive prenatal testing							
Vanadis® NIPT, IONA	cfDNA	Counting chromosomes number	NIPT for screening of trisomy 21, 18 and 13				NCT03559374 – Recruitment status unknown NCT01472523 – Recruitment completed
Concordance of Key Actionable Genomic Alterations as Assessed in Tumor Tissue and Plasma in Non-small Cell Lung Cancer	cfDNA	17 genes panel	Genomic variants detection in advanced non-small cell lung cancer				NCT02762877 – Recruitment terminated
Detect-A, Cancer seek	Proteins and cfDNA	Mutations in regions of 16 genes + 9 protein biomarkers	Early detection of colorectal, ovary, pancreas, breast, upper gastrointestinal tract, lung, and liver cancer				NCT04213326 – Recruiting
STRIVE study, CCGA, SUMMIT, Pathfinder	cfDNA/cfRNA	Cell-free nucleic acids profiles, including DNA methylation	Early detection of more than 50 cancer types				NCT04241796 – Recruiting NCT03934866 – Enrolling by invitation NCT02889978 – Active, not recruiting NCT03085888 – Active, not recruiting
SafeSEQ, OncoBEAM	cfDNA	Mutations in several genes	Genotyping for colorectal, breast, lung, melanoma, and head and neck squamous cell carcinoma				NCT02751177 – Completed
Epi proLung	cfDNA	SHOX2 and PTGER4 methylation	Lung cancer detection for risk patients				
Epi proColon	cfDNA	SEPT9 methylation	Cancer detection for patients with an average risk of CRC				NCT00855348 – Completed
HCCBloodTest	cfDNA	SEPT9 methylation	Liver cancer detection for patients with cirrhosis				NCT03804593 – Completed
Therascreen MGMT Pyro	cfDNA	MGMT methylation	Predict response to alkylating agent in gliomas				
COLVERA	cfDNA	IKZF1, BCAT1	Detection of residual and recurrent disease in CRC patients				NCT03706235 – Completed NCT03706248 – Completed
Ivygene	cfDNA	DNA methylation	Breast, colon, liver, and lung cancer detection				NCT03694600 – Recruiting

Table 2.1 (*Continued*).

Test or study name	Biomarker type	Target	Application	Phases 1 & 2	Phase 3	Phase 4	Clinical trial number
cobas® EGFR Mutation Test v2	cfDNA	42 EGFR mutation	Identify EGFR mutations in non-small cell lung cancer				NCT04035486 – Recruiting
Guardant360 CDx	cfDNA	Comprehensive genomic profiling	Identify patients with NSCLC who should be treated with osimertinib				NCT04497285 – Not yet recruiting
FoundationOne Liquid CDx	cfDNA	Comprehensive genomic profiling	Identify patients able to benefit from targeted therapy for NSCLC, prostate, ovary, and breast cancers				NCT04484636 – Recruiting
AlloSure, Prospera	cfDNA	Quantification of donor-derived cfDNA	Acute rejection in kidney transplantation				NCT04566055 – Active, not recruiting NCT04091984 – Recruiting NCT03765203 – Study completed
Allosure Heart	cfDNA	Quantification of donor-derived cfDNA	Acute rejection in heart transplantation				NCT03695601 – Recruiting
AlloSure	cfDNA	Quantification of donor-derived cfDNA	Determine if Allosure can be used to identify acute rejection in lung transplantation				NCT04318587 – Recruiting
Prospera	cfDNA	Quantification of donor-derived cfDNA	Determine if Prospera can be extended to several other transplantation types				NCT03984747 – Recruiting
Cell-free DNA as a Biomarker After Lung Transplantation, Longitudinal Study of Cell-Free DNA in Lung Transplant (LoSt)	cfDNA	Quantification of donor-derived cfDNA	Acute rejection in lung transplantation				NCT04271267 – Study completed NCT04234919 – Enrolling by invitation
Non-invasive Test for Acute Rejection Identification in Heart Transplanted Patients (INNO-GRAFTRS001), A New Biomarker for the Non-invasive Diagnosis of Rejection After Heart Transplantation (BIODRAFT)	cfDNA	Quantification of donor-derived cfDNA	Acute rejection in heart transplantation				NCT03477383 – Recruiting NCT04274712 – Enrolling by invitation
Genome Transplant Dynamics	cfDNA	Quantification of donor-derived cfDNA	Rejection after heart and lung transplantation				NCT02423070 – Recruiting
Quantitative Detection of Circulating Donor-Specific DNA in Organ Transplant Recipients (DTRT-Multi-Center Study) (DTRT)	cfDNA	Quantification of donor-derived cfDNA	Determine the range of ddcfDNA indicating that rejection in several transplantation types				NCT02109575 – Active, not recruiting
Cf-DNA Assay During Treatment of Acute Rejection	cfDNA	Quantification of donor-derived cfDNA after anti-rejection treatment	Successful treatment of an acute rejection episode after kidney transplantation				NCT04019353 – Recruiting
Atherothrombosis Markers in Diabetics (MADI)	cfDNA	Quantification	Determine if cfDNA can represent a trigger for atherothrombotic plaques in type 2 diabetic patients				NCT02898467 – Unknown status

and are more reliable [181]. Circulating cfDNA from the fetus can come from several sources, such as apoptotis of fetal hematopoietic cells [182], transfer through placenta [183], and trophoblast destruction [184]. Several factors can influence the fetal cfDNA fraction, including the mother's body mass index, fetal aneuploidies, or twin pregnancy [185]. Circulating cfDNA methylation can also be used to discriminate maternal DNA from fetal DNA. For example, maspin gene promoter has been found as poorly methylated in placenta, whereas highly methylated in maternal blood cells [186].

The main goal of NIPT is to increase the detection rate for chromosomal abnormalities while reducing the rate of invasive testing [187]. The most common screening modality for fetal chromosomal abnormalities is the combined first-trimester screening (cFTS) after echography to identify fetus age, number of fetuses, and major structural abnormalities. Patients are then stratified into three groups: high, intermediate, and low risk. The high-risk group should undergo invasive testing and NIPT is offered to the intermediate group. If NIPT returns a high-risk score, the mother will go through invasive testing. An alternative option to prenatal screening is to have NIPT systematically performed in parallel to cFTS, thus maximizing the chances to detect chromosomal abnormalities. However, it is costly to perform cFTS and NIPT simultaneously. Most NIPT options provide testing for trisomies 21, 18, 13, and for aneuploidies of sex chromosomes. Some of them are also able to detect microdeletions with higher accuracy than chromosome aneuploidies. Another major pregnancy risk is miscarriage which could be predictable with new biomarkers indicating pregnancy involuntary interruption. For example, preeclampsia can be detected by an increase in fetal cfDNA [188] and the risk of preterm birth can be predicted up to 2 months prior to the labor by measuring the level of an mRNA panel in the mother's blood [59]. Moreover, unmethylated sequences of the maspin promoter are elevated in women with preeclampsia [189]. NIPT has been widely adopted worldwide over the past decade. However, prescriptions for NIPT should be weighted carefully by healthcare professionals since false-positive NIPT results may lead to an increase in invasive testing procedures.

2.3.2 Oncology

2.3.2.1 Cancer screening and diagnosis

Genomic alterations including somatic mutations, chromosomal instability (copy number alteration, inversion, insertions, and translocation), and epigenetic modifications are common features of cancer [23–46]. Analysis of these cancer-associated DNA signatures in circulation (e.g., DNA-derived directly from tumors, circulating ctDNA) has the potential to revolutionize cancer screening, diagnosis, and treatment selection [23–46]. Before the era of NGS, PCR-based assays were developed to detect common mutations on driver genes such as *KRAS* and *EGFR* [105, 190]. Modern methods such as massively parallel NGS have enabled us to survey a broad spectrum of genetic alterations in cancer (23–33). It has now been well-established that ctDNA often presents at very low variant allele frequencies (VAFs) of less than 0.01% at early stages [191–193]. Several targeted sequencing approaches using hybrid capture (CAPPSeq [32], digital sequencing [194], and exome sequencing [30]) or PCR amplicons (SafeSeqS [195], NG-TAS [196], and Tam-Seq [197]) have been demonstrated to effectively capture highly recurrent mutations for sequencing at high depth to sensitively detect ctDNA with low VAFs. These methods have shown promising applications in the detection of a wide range of cancers at almost all organ sites, for example, brain [198], head and neck [199], [200, 201], thyroid [202], lung [10, 32], breast [10, 203], kidney [204], stomach [10, 205], colon [206], ovary [10, 207], prostate [208], smooth muscle tissue [209] and hematologic malignancies [210]. However, due to the limited number of total cfDNA copies available, ctDNA mutations of a given target panel may not present in a practical blood draw volume (e.g., a few tubes of blood), especially when there is a low disease burden. A genome-wide mutational approach was proposed to increase sampling probability and demonstrated potential application in the detection of minimal residual disease in lung adenocarcinoma, colorectal cancer, and melanoma [211]. Improvement of detection sensitivity could also be achieved by combining mutations with aneuploidy and a protein panel [212, 213]. Another challenge for circulating ctDNA detection is differentiating the cancer-derived DNA from benign mutations increasing with age [214]. The majority of mutations detected in cfDNA are consistent with the mutation spectra of Clonal Hematopoiesis of Indeterminate Potential (CHIP) [214]. Therefore, before assigning mutations detected in circulating cfDNA as a positive signal of cancer, mutations related to CHIP must be taken into consideration.

Epigenetic aberrations, including DNA methylation and chromatin structural changes, are widespread across multiple cancers and are known to have specific patterns associated with the tissue or region of origin [34–46]. Indeed, the first FDA-approved circulating cfDNA test for screening colon cancer, Epi ProColon, is based on the detection of methylated Septin 9 in plasma cfDNA by PCR [215]. Deconvolution using methylation patterns revealed the composition of tissue contribution to circulating cfDNA in healthy donors and diseases including cancer of colon, breast, and prostate [216]. Bisulfite whole genome and targeted sequencing demonstrated that detection of methylation aberrance in cfDNA is more promising than mutation detection with organ specificity in more than 50 cancer types [38] and as early as four years before clinically diagnosed by the current standard of care [39]. To improve sensitivity compared to bisulfite sequencing, immunoprecipitation-based enrichment and selection method for methylation sequencing developed for low input cfDNA (cfMeDIP–seq), demonstrated potential application in the detection and classification of tumor subtypes in cancers of lung, breast, colon, pancreas, kidney, and brain [40–42]. In addition to the conventional methylcytonsine mark, as described, a recent identified epigenetic mark 5-Hydroxymethylcytosine (5hmC) was also shown diagnostic potential in cfDNA for pancreatic, lung, and liver cancers [43]. CfDNA is highly fragmented in circulating with its protected patterns as a proxy of nucleosome positioning and chromatin structure which varies between cell types of malignant cells [45]. Analysis of cfDNA nucleosome footprint and fragmentation inform the tissue of origin and detect the presence of breast, colorectal, lung, ovarian, pancreatic, gastric, or bile duct cancer [45].

Due to mutations or epigenetic changes in genomic promoter regions or gene bodies, expressed RNAs can be misregulated in cancer and function as tumor suppressors and oncogenes [217–219]. Accumulating evidence suggests the stability of circulating RNA, particularly miRNA, in blood as and exhibits their potential for personalized cancer diagnostics [220]. Additionally, since RNA profiles display tissue and lineage cell specificity, they have the potential to differentiate tumor subtypes. miRNA profiles are distinctive for cancers of breast [221], ovary [222], lung [223], esophagus [224], prostate [50, 225, 226], liver[227], colon [228], and brain [229]. Other non-coding RNAs, including lncRNA, siRNA, piRNA, snRNA, circular RNA, and others have also been detected at high levels in blood of cancer patients, with diagnostic potential [230–232].

2.3.3 Guiding treatment stratification and prognosis and monitoring

The selection of personalized treatment for cancer depends on molecular characteristics of the tumor reported from morphological pathology examinations, immunostainings, and genetic tests of the biopsy [233]. For example, tyrosine kinase inhibitor (TKI) therapies are effective when this signaling pathway is aberrant in cancer. In lung cancer, patients with activating mutations (L858R or exon 19 deletions) on the gene encoded for the epidermal growth factor receptor (EGFR) benefit from TKIs gefitinib and erlotinib treatment [234, 235]. However, the tumor location may not accessible for biopsy for many patients [236, 237]. Moreover, needle biopsy may not capture the heterogeneity of the tumor. Circulating nucleic acids have the potential to determine clinically-relevant tumor molecular class and mutational status, as with needle biopsies, but can sample a greater fraction of the body at once. Importantly, cfDNA may disclose mutations not detected in tissue biopsy [238, 239]. Mutations detected in cfRNA have shown the potential to guide treatment selection and to predict prognosis in prostate cancer [240], melanoma, thyroid [241] and breast cancers [242], and others [243]. Along with circulating cfDNA, circulating epigenetic and transcriptomic biomarkers have also been highlighted as predictors of cancer progression on treatment in some cancers such as lung cancers [244] and gastric cancers [245]. In cancers associated with chronic virus infection, detection of active viral genome in blood could help to guide treatment selection and predict response. For example, in metastatic cervical cancer, detection of human papillomavirus cfDNA could be used to select patients for HPV type-specific T-cell-based immunotherapies and to monitor treatment efficiency [246]. In pediatric B-non-Hodgkin lymphomas, the rate of reduction in Epstein-Barr virus DNA level in blood was associated with treatment response [247].

To achieve robust and tolerable administration of cancer treatments, monitoring of therapeutic efforts should be adaptive and guided by measurements that reflect evolution of tumors and host factors [248]. With current clinical practice, treatment monitoring mostly relies on radiological imaging, blood tests of conventional cancer markers, organ functions, and the performance status of the patients. In cancer types that allow accessible biopsies such as hematological malignancies and breast cancers, serial biopsies may be performed to monitor the response and molecular evolution of the tumor to direct the subsequent therapy [248, 249]. Cell-free DNA and RNA approaches emerged as promising tools for monitoring treatment response, clonal evolution, minimal residual disease, and early prediction of relapse [250–257].

2.3.4 Transplantation

Transplantation is a medical procedure to restore a failing organ or tissue with a functional replacement. Depending on the procedure and organ, the donor can be living or deceased. The most grafted organs are kidney, liver, heart, lung pancreas, and cornea of skin. Despite matching between the donor and the receiver to ensure compatibility, rejection is the major cause of graft failure. Therefore, it is crucial to monitor new organ health to detect a possible rejection after transplantation. Invasive biopsy of the transplant is the current standard clinical procedure for monitoring after transplantation. Circulating cfDNA in the receiver's blood has been leveraged as a new approach to monitor the transplanted organ since the detection of donor-derived cfDNA from kidney and liver transplant recipients was first published in 1998 [258].

To distinguish donor from recipient-derived DNA, two strategies have been employed: donor–recipient sex-mismatch and sequence-mismatches. For organ transplanted from a male donor to a female recipient, regions on Y chromosomes can be used to calculate donor-derived cfDNA levels in the recipient's blood [259]. While this method provides reliable quantification of cfDNA-derived from the transplanted organ, sex-mismatch between the donor and the recipient represents less than 25% of total transplantations [19]. Sequence-mismatches-based approaches rely on the genetic difference between the donor and the recipient at particular locations. Thousands of SNPs between the donor and the recipient identified by using high-throughput sequencing can be used to distinguish donor-derived cfDNA from recipient cfDNA and measure the level of DNA released from the transplanted organ [19]. This method has been used to monitor heart transplantation rejection [260] and can be adapted to other transplantations [261]. SNPs can also be measured by a droplet digital PCR assay and demonstrated potential applications in kidney, liver, and heart transplantation [262].

The kinetics of donor-derived cfDNA released into the recipient's circulation is organ-specific. For lung and heart transplants, cfDNA from donor releases at a high rate within one day after transplant and rapidly decreases for heart within a few days, assuming the heart is not rejected, whereas for lung the decrease is much slower [263, 264]. When acute rejection occurs, the level of donor-derived cfDNA increases for both heart and lung transplantations. This increase in donor-derived cfDNA comes from dying cells from transplants due to apoptosis of acute and chronic rejection. Transplanted patients are often required immunosuppressive therapies to avoid rejection leading to susceptibility to infections with hidden symptoms. De Vlaminck et al. demonstrated the utility of cfDNA sequencing for simultaneous monitoring of rejection and to detect viral DNA fragments after transplantation [265]. The most commonly detected virus is torque teno viruses (TTVs) from the Anelloviridae family that is ubiquitous in humans and asymptomatic. Importantly, TTV detection also allows for the stratification of patients at risk for rejection [266].

2.3.5 Microbial analysis

Until recently, human blood was widely considered to be sterile. There is accumulating evidence that there is a human-blood microbiome that may have important implications for disease [267]. Human microbiomes are considered temporally stable and are personalized to the host [268]. Nucleic acids derived from the complex human microbiome communities may be leaked into circulation and thus circulating cfDNA and cfRNA analysis may be able to provide a snapshot of these interactions. Due to the extremely low abundance of microbe-derived nucleic acids in circulation, detection of microbes is especially susceptible to contamination and proper negative controls must always be performed in parallel [269]. Recent work by Kowarsky et al. [270] revealed hundreds of previously uncharacterized microbes in circulating cfDNA samples. Analysis of microbial communities by circulating cfDNA has also revealed promising advances in cancer diagnosis [271–273] and infection detection [274, 275]. Similar microbial community analysis has also been performed in circulating cfRNA to detect infection [276]; however, the sensitivity of circulating cfRNA to identify disease-specific microbial communities (e.g., cancer) remains to be determined.

2.3.6 Chronic diseases

2.3.6.1 Diabetes

Chronic diabetes impacts more than 382 million people in 2013 and is expected to increase with the addition of 592 million new cases by 2035 [277, 78]. The abnormal insulin production characteristic of diabetes leads to an increase in blood glucose rate and can cause increased risks for other diseases, such as heart attack, neuropathy, retinopathy, and kidney failure. Diabetes requires periodic monitoring of the disease progression for management. Although glucose rate can be easily measured in blood, it does not reflect the development of pathology complications. For early stages of type 2 and gestational diabetes, several studies have demonstrated the difference in circulating miRNA expression profile, compared to healthy control patients [279–283]. The identified cfRNA biomarkers can also reveal potential underlying pathologies. For example, mir-25 is found in new diabetes type 1 children and is associated with residual beta-cell function and glycemic control during disease progression [284]. Interestingly, miR-192 and 193b are found only at an early stage of diabetes and can be modulated by physical activity [285]. This finding suggests that exercise may delay or prevent diabetes development.

2.3.6.2 Autoimmune diseases

Autoimmune diseases are a large group of disorders in which the immune system attacks functional body parts. Because autoimmune diseases progress in unpredictable phases, circulating nucleic acid biomarkers could provide important information about disease progression for patient management. Rheumatoid arthritis (RA) is prevalent in 0.24% of the worldwide population [286] with negative impacts on patients' quality of life and the death rate [287]. Its diagnosis is based on two protein dosages: rheumatoid factor and anti-citrullinated peptide/protein antibodies [288]. cfDNA level was found to be higher in RA patients than in healthy control with low sensitivity in early stages [289, 290]. An increase in cfDNA rate after administration of the treatment is correlated with positive outcomes [291]. Several circulating miRNA was shown to discriminate disease activities and treatment responses [292]. For example, miR-223 and miR-23 could be used as a predictor and a biomarker of response to treatment [293].

Another autoimmune disease, multiple sclerosis (MS), is a degenerative disorder without cure but can be managed to reduce symptoms. The major need for MS biomarkers is to detect the switch between the beginning of demyelination versus clinically definitive MS and to inform management plans for the prevention of disease progression. Currently, protein biomarkers in cerebrospinal fluid (CSF) are used for monitoring the disease [294–296]. Blood circulating biomarkers are emerging as an alternative method to obviate the need for invasive CSF sampling. miRNAs such as miR-30e, miR-93, miR-15579 [297], and an exosomal miRNA profile has shown as promising biomarkers for detection and monitoring response to treatment [298].

2.3.6.3 Alzheimer's diseases

Alzheimer's disease (AD) is a neuropathology leading to dementia and memory loss and is thought to be caused by aggregation of neurotoxic amyloid β-peptide (Aβ) protein in the brain. Early detection of AD could provide avenues to slow down disease progression and improve patient quality of life. Many circulating miRNAs have been highlighted as potential biomarkers for AD [299–303]. Additional isolation of circulating extracellular vesicles (EV) from brain using an antibody against L1CAM+ could increase sensitivity [304]. Prediction of AD progression can also be predicted by changes in serum level for a panel of miRNA [305]. Large-scale clinical studies to further integrate and validate these exploratory publications are needed to advance the application of circulating RNA in diagnosis and prognosis of AD.

2.4 Challenges and Outlooks

CfDNA has demonstrated clinical utility in NIPT and transplantation and has also shown great promise for precision oncology. Though cfDNA is stable in circulation and across the day [220], its low overall abundance is a major challenge. For example, detection of a certain panel of mutation in cancer at early stages may require infeasibility large volume of input blood [192, 193]. Major breakthroughs in methods for effective, robust, and convenient sampling, and isolation methods will advance the field of cfDNA-based fluid biopsy toward clinical application. For example, one could improve cfDNA sampling by placing a convenient medical device for in-vivo cfDNA collection instead of ex-vivo cfDNA isolation methods from drawn blood. The core principle for cfDNA analysis is the detection of non-self or mutated genomic materials such as fetal DNA in NIPT, DNA from donors in transplantation, and ctDNA originated from the tumor cells. The low copies of mutated or non-self-genomic materials are embedded in high background of germline DNA. Ultra-deep sequencing and ddPCR have been demonstrated in research labs to tackle this lesser allele fraction problem [30, 32, 191–195]. To lower the cost of analysis, we can improve the enrichment of mutated or non-self-DNA depletion in the germline DNA background. We also need sequencing strategies to diminish false positive calls from sequencing errors. A final challenge for applying circulating cfDNA in diseases such as cancer is the nature of diverse mutation spectra and the presence of mutations in benign conditions. Significant reductions in sequencing cost in the near future may allow robust coverage of the mutation spectrum and enough sequencing depth to detect low copies of these mutations compared to the germline background. To avoid false positive calls of mutations in benign conditions such as clonal hematopoiesis of indeterminate potential (CHIP), which increases exponentially with age, we need to employ safe-guard algorithms and additional measurements from orthogonal analytes. Analysis of epigenetic aberrances may allow the detection of rare DNA variants at the early stages of disease. Novel methods for base conversion of cytosine modifications that can also handle low-input DNA are especially highly desirable. Future DNA isolation methods that retain bound histone proteins will enable the utility of the epigenetic codes on histones for cfDNA analysis.

While the field of circulating cfDNA is beginning to reach maturity, circulating cfRNA has received relatively less attention. The major challenges for cell-free RNA analysis are the low abundance and degradation of RNA, discordance between extraction or analysis protocols due to the diversity of biotypes (e.g., mRNA vs miRNA) and carriers, and preanalytical variability. The complex distribution of RNA biotypes with different lengths, secondary structures, and association with carriers leads to bias and variability in isolation and detection. Circulating cfRNAs of different biotypes are sheltered and protected in a variety of carriers such as extracellular vesicles, including exosomes and microvesicles, as well as lipoprotein and AGO2 complexes. Isolation methods and sequencing library protocols may preferentially extract, recover and capture either small RNA or long RNA associated with certain classes of carriers while against others. Moreover, preanalytical variability dampens the reproducibility of cell-free RNA measurements. Understanding the components of RNA carriers in biofluids and the composition of RNA content is needed to standardize sample processing and analysis protocols for a certain class of RNA of interest. The delineation of RNA packaging within these carriers may also lead to an increase in contrast of biomarkers between pathological and physiological conditions.

Measuring cfDNA and cfRNA individually as single analytes may not be sufficient to achieve the sensitivity and specificity needed for routine clinical applications. In addition, it is important to consider individual-specific baselines due to significant interpersonal variation in normal physiology. A combination of circulating cfDNA and cfRNA with other analytes may offer a solution to increase the resolution and performance of liquid biopsy tests, permitting earlier detection of diseases and disorders. Future development of platforms allowing simultaneous measurements of multiple analytes longitudinally in a high-throughput manner, as well computational tools for multiomic data integration, will enable liquid-biopsy applications to be fully realized in the clinic.

References

[1] K. E. Sundling, A. C. Lowe, Circulating Tumor Cells: Overview and Opportunities in Cytology. *Advances in anatomic pathology* 26, 56-63 (2019).

[2] L. M. Millner, M. W. Linder, R. Valdes, Jr., Circulating tumor cells: a review of present methods and the need to identify heterogeneous phenotypes. *Annals of clinical and laboratory science* 43, 295-304 (2013).

[3] J. A. Thiele, K. Bethel, M. Králíčková, P. Kuhn, Circulating Tumor Cells: Fluid Surrogates of Solid Tumors. *Annual Review of Pathology: Mechanisms of Disease* 12, 419-447 (2017).

[4] Y. Liu, X. Cao, The origin and function of tumor-associated macrophages. *Cellular And Molecular Immunology* 12, 1 (2014).

[5] D. L. Adams et al., Circulating giant macrophages as a potential biomarker of solid tumors. *Proceedings of the National Academy of Sciences* 111, 3514 (2014).

[6] C. E. Gast et al., Cell fusion potentiates tumor heterogeneity and reveals circulating hybrid cells that correlate with stage and survival. *Science Advances* 4, eaat7828 (2018).

[7] M. G. Best et al., RNA-Seq of Tumor-Educated Platelets Enables Blood-Based Pan-Cancer, Multiclass, and Molecular Pathway Cancer Diagnostics. *Cancer cell* 28, 666-676 (2015).

[8] M. G. Best, P. Wesseling, T. Wurdinger, Tumor-Educated Platelets as a Noninvasive Biomarker Source for Cancer Detection and Progression Monitoring. *Cancer research* 78, 3407-3412 (2018).

[9] In, S. G. J. G. t Veld, T. Wurdinger, Tumor-educated platelets. *Blood*, blood-2018-2012-852830 (2019).

[10] J. D. Cohen et al., Detection and localization of surgically resectable cancers with a multi-analyte blood test. *Science* 359, 926 (2018).

[11] P. Mandel, P. Metais, [Nuclear Acids In Human Blood Plasma]. *C R Seances Soc Biol Fil* 142, 241-243 (1948).

[12] Y. M. Lo et al., Presence of fetal DNA in maternal plasma and serum. *Lancet (London, England)* 350, 485-487 (1997).

[13] Y. M. D. Lo et al., Quantitative Analysis of Fetal DNA in Maternal Plasma and Serum: Implications for Noninvasive Prenatal Diagnosis. *The American Journal of Human Genetics* 62, 768-775 (1998).

[14] X. Q. Chen et al., Microsatellite alterations in plasma DNA of small cell lung cancer patients. *Nature medicine* 2, 1033-1035 (1996).

[15] J. Aucamp, A. J. Bronkhorst, C. P. S. Badenhorst, P. J. Pretorius, The diverse origins of circulating cell-free DNA in the human body: a critical re-evaluation of the literature. *Biological reviews of the Cambridge Philosophical Society* 93, 1649-1683 (2018).

[16] V. J. Gauthier, L. N. Tyler, M. Mannik, Blood clearance kinetics and liver uptake of mononucleosomes in mice. *Journal of immunology (Baltimore, Md.: 1950)* 156, 1151-1156 (1996).

[17] A. Kustanovich, R. Schwartz, T. Peretz, A. Grinshpun, Life and death of circulating cell-free DNA. *Cancer biology & therapy* 20, 1057-1067 (2019).

[18] I. De Vlaminck et al., Circulating cell-free DNA enables noninvasive diagnosis of heart transplant rejection. *Science translational medicine* 6, 241ra277 (2014).

[19] S. R. Knight, A. Thorne, M. L. Lo Faro, Donor-specific Cell-free DNA as a Biomarker in Solid Organ Transplantation. A Systematic Review. *Transplantation* 103, 273-283 (2019).

[20] H. C. Fan, Y. J. Blumenfeld, U. Chitkara, L. Hudgins, S. R. Quake, Analysis of the size distributions of fetal and maternal cell-free DNA by paired-end sequencing. *Clin Chem* 56, 1279-1286 (2010).

[21] L. Qiao et al., Sequencing shorter cfDNA fragments improves the fetal DNA fraction in noninvasive prenatal testing. *American Journal of Obstetrics and Gynecology* 221, 345.e341-345.e311 (2019).

[22] P. Hu et al., An enrichment method to increase cell-free fetal DNA fraction and significantly reduce false negatives and test failures for noninvasive prenatal screening: a feasibility study. *Journal of Translational Medicine* 17, 124 (2019).

[23] R. J. Leary et al., Development of personalized tumor biomarkers using massively parallel sequencing. *Science translational medicine* 2, 20ra14 (2010).

[24] D. J. McBride et al., Use of cancer-specific genomic rearrangements to quantify disease burden in plasma from patients with solid tumors. *Genes, chromosomes & cancer* 49, 1062-1069 (2010).

[25] J. He et al., IgH gene rearrangements as plasma biomarkers in Non-Hodgkin's lymphoma patients. *Oncotarget* 2, 178-185 (2011).

[26] T. Forshew et al., Noninvasive identification and monitoring of cancer mutations by targeted deep sequencing of plasma DNA. *Science translational medicine* 4, 136ra168 (2012).

[27] R. J. Leary et al., Detection of chromosomal alterations in the circulation of cancer patients with whole-genome sequencing. *Science translational medicine* 4, 162ra154 (2012).

[28] A. Narayan et al., Ultrasensitive measurement of hotspot mutations in tumor DNA in blood using error-suppressed multiplexed deep sequencing. *Cancer research* 72, 3492-3498 (2012).

[29] S. J. Dawson et al., Analysis of circulating tumor DNA to monitor metastatic breast cancer. *The New England journal of medicine* 368, 1199-1209 (2013).

[30] M. Murtaza et al., Non-invasive analysis of acquired resistance to cancer therapy by sequencing of plasma DNA. *Nature* 497, 108-112 (2013).

[31] E. Crowley, F. Di Nicolantonio, F. Loupakis, A. Bardelli, Liquid biopsy: monitoring cancer-genetics in the blood. *Nature reviews. Clinical oncology* 10, 472-484 (2013).

[32] A. M. Newman et al., An ultrasensitive method for quantitating circulating tumor DNA with broad patient coverage. *Nature medicine* 20, 548-554 (2014).

[33] J. D. Cohen et al., Detection and localization of surgically resectable cancers with a multi-analyte blood test. *Science* 359, 926-930 (2018).

[34] J. Moss et al., Comprehensive human cell-type methylation atlas reveals origins of circulating cell-free DNA in health and disease. *Nat Commun* 9, 5068 (2018).

[35] R. H. Xu et al., Circulating tumour DNA methylation markers for diagnosis and prognosis of hepatocellular carcinoma. *Nat Mater* 16, 1155-1161 (2017).

[36] G. R. Oxnard et al., LBA77 - Simultaneous multi-cancer detection and tissue of origin (TOO) localization using targeted bisulfite sequencing of plasma cell-free DNA (cfDNA). *Annals of Oncology* 30, v912 (2019).

[37] M. C. Liu et al., 50O - Plasma cell-free DNA (cfDNA) assays for early multi-cancer detection: The circulating cell-free genome atlas (CCGA) study. *Annals of Oncology* 29, viii14 (2018).

[38] M. C. Liu et al., Sensitive and specific multi-cancer detection and localization using methylation signatures in cell-free DNA. *Annals of Oncology* 31, 745-759 (2020).

[39] X. Chen et al., Non-invasive early detection of cancer four years before conventional diagnosis using a blood test. *Nature Communications* 11, 3475 (2020).

[40] S. Y. Shen et al., Sensitive tumour detection and classification using plasma cell-free DNA methylomes. *Nature* 563, 579-583 (2018).

[41] P. V. Nuzzo et al., Detection of renal cell carcinoma using plasma and urine cell-free DNA methylomes. *Nature medicine* 26, 1041-1043 (2020).

[42] F. Nassiri et al., Detection and discrimination of intracranial tumors using plasma cell-free DNA methylomes. *Nature medicine* 26, 1044-1047 (2020).

[43] C.-X. Song et al., 5-Hydroxymethylcytosine signatures in cell-free DNA provide information about tumor types and stages. *Cell Research* 27, 1231-1242 (2017).

[44] Matthew W. Snyder, M. Kircher, Andrew J. Hill, Riza M. Daza, J. Shendure, Cell-free DNA Comprises an In Vivo Nucleosome Footprint that Informs Its Tissues-Of-Origin. *Cell* 164, 57-68 (2016).

[45] S. Cristiano et al., Genome-wide cell-free DNA fragmentation in patients with cancer. *Nature* 570, 385-389 (2019).

[46] J.-F. Rahier et al., Circulating nucleosomes as new blood-based biomarkers for detection of colorectal cancer. *Clinical Epigenetics* 9, 53 (2017).

[47] H. C. Fan, Y. J. Blumenfeld, U. Chitkara, L. Hudgins, S. R. Quake, Noninvasive diagnosis of fetal aneuploidy by shotgun sequencing DNA from maternal blood. *Proceedings of the National Academy of Sciences* 105, 16266 (2008).

[48] R. W. Chiu et al., Noninvasive prenatal diagnosis of fetal chromosomal aneuploidy by massively parallel genomic sequencing of DNA in maternal plasma. *Proceedings of the National Academy of Sciences of the United States of America* 105, 20458-20463 (2008).

[49] J. Camunas-Soler et al., Noninvasive Prenatal Diagnosis of Single-Gene Disorders by Use of Droplet Digital PCR. *Clinical Chemistry* 64, 336-345 (2018).

[50] P. S. Mitchell et al., Circulating microRNAs as stable blood-based markers for cancer detection. *Proceedings of the National Academy of Sciences* 105, 10513 (2008).

[51] O. D. Murillo et al., exRNA Atlas Analysis Reveals Distinct Extracellular RNA Cargo Types and Their Carriers Present across Human Biofluids. *Cell* 177, 463-477.e415 (2019).

[52] S. Srinivasan et al., Small RNA Sequencing across Diverse Biofluids Identifies Optimal Methods for exRNA Isolation. *Cell* 177, 446-462.e416 (2019).

[53] J. Rozowsky et al., exceRpt: A Comprehensive Analytic Platform for Extracellular RNA Profiling. *Cell Systems* 8, 352-357.e353 (2019).

[54] S. Das et al., The Extracellular RNA Communication Consortium: Establishing Foundational Knowledge and Technologies for Extracellular RNA Research. *Cell* 177, 231-242 (2019).

[55] S. A. Hinger et al., Diverse Long RNAs Are Differentially Sorted into Extracellular Vesicles Secreted by Colorectal Cancer Cells. *Cell reports* 25, 715-725.e714 (2018).

[56] H. H. Pua et al., Increased Hematopoietic Extracellular RNAs and Vesicles in the Lung during Allergic Airway Responses. *Cell reports* 26, 933-944.e934 (2019).

[57] G. P. d. Oliveira et al., Detection of Extracellular Vesicle RNA Using Molecular Beacons. *iScience* 23, 100782 (2020).

[58] W. Koh et al., Noninvasive in vivo monitoring of tissue-specific global gene expression in humans. *Proceedings of the National Academy of Sciences of the United States of America* 111, 7361-7366 (2014).

[59] T. T. M. Ngo et al., Noninvasive blood tests for fetal development predict gestational age and preterm delivery. *Science* 360, 1133-1136 (2018).

[60] W. Pan et al., Simultaneously Monitoring Immune Response and Microbial Infections during Pregnancy through Plasma cfRNA Sequencing. *Clinical Chemistry* 63, 1695-1704 (2017).

[61] A. Ibarra et al., Non-invasive characterization of human bone marrow stimulation and reconstitution by cell-free messenger RNA sequencing. *Nature Communications* 11, 400 (2020).

[62] S. Yu et al., Plasma extracellular vesicle long RNA profiling identifies a diagnostic signature for the detection of pancreatic ductal adenocarcinoma. *Gut* 69, 540-550 (2020).

[63] J. D. Arroyo et al., Argonaute2 complexes carry a population of circulating microRNAs independent of vesicles in human plasma. *Proceedings of the National Academy of Sciences* 108, 5003 (2011).

[64] A. Thierry, S. El Messaoudi, P. Gahan, P. Anker, M. Stroun, Origins, structures, and functions of circulating DNA in oncology. *Cancer and Metastasis Reviews* 35, 347-376 (2016).

[65] Y. D. Lo et al., Presence of fetal DNA in maternal plasma and serum. *The lancet* 350, 485-487 (1997).

[66] F. Mauger, C. Dulary, C. Daviaud, J.-F. Deleuze, J. Tost, Comprehensive evaluation of methods to isolate, quantify, and characterize circulating cell-free DNA from small volumes of plasma. *Analytical and bioanalytical chemistry* 407, 6873-6878 (2015).

[67] L. Sorber et al., A comparison of cell-free DNA isolation kits: isolation and quantification of cell-free DNA in plasma. *The Journal of Molecular Diagnostics* 19, 162-168 (2017).

[68] J. H. van Ginkel et al., Preanalytical blood sample workup for cell-free DNA analysis using Droplet Digital PCR for future molecular cancer diagnostics. *Cancer medicine* 6, 2297-2307 (2017).

[69] R. J. Diefenbach, J. H. Lee, R. F. Kefford, H. Rizos, Evaluation of commercial kits for purification of circulating free DNA. *Cancer genetics* 228, 21-27 (2018).

[70] M. B. Giacona et al., Cell-free DNA in human blood plasma: length measurements in patients with pancreatic cancer and healthy controls. *Pancreas* 17, 89-97 (1998).

[71] P. Jiang et al., Lengthening and shortening of plasma DNA in hepato-cellular carcinoma patients. *Proceedings of the National Academy of Sciences* 112, E1317-E1325 (2015).

[72] F. Mouliere et al., Enhanced detection of circulating tumor DNA by fragment size analysis. *Science translational medicine* 10, (2018).

[73] X. Liu et al., Fragment Enrichment of Circulating Tumor DNA With Low-Frequency Mutations. *Frontiers in Genetics* 11, 147 (2020).

[74] S. Cristiano et al., Genome-wide cell-free DNA fragmentation in patients with cancer. *Nature* 570, 385-389 (2019).

[75] V. Kloten et al., Liquid biopsy in colon cancer: comparison of different circulating DNA extraction systems following absolute quantification of KRAS mutations using Intplex allele-specific PCR. *Oncotarget* 8, 86253 (2017).

[76] J. Solassol et al., Comparison of five cell-free DNA isolation meth-ods to detect the EGFR T790M mutation in plasma samples of patients with lung cancer. *Clinical Chemistry and Laboratory Medicine (CCLM)* 56, e243-e246 (2018).

[77] J. Park, K. H. Jo, H. Y. Park, J. H. Hahn, Spatially controlled silica coating in poly (dimethylsiloxane) microchannels with the sol-gel process. *Sensors and actuators B: chemical* 232, 428-433 (2016).

[78] S. M. Azimi, G. Nixon, J. Ahern, W. Balachandran, A magnetic bead-based DNA extraction and purification microfluidic device. *Microflu-idics and nanofluidics* 11, 157-165 (2011).

[79] H. Lee, C. Park, W. Na, K. H. Park, S. Shin, Precision cell-free DNA extraction for liquid biopsy by integrated microfluidics. *NPJ precision oncology* 4, 1-10 (2020).

[80] R. Zhang, H.-Q. Gong, X. Zeng, C. Lou, C. Sze, A microfluidic liquid phase nucleic acid purification chip to selectively isolate DNA or RNA from low copy/single bacterial cells in minute sample volume followed

by direct on-chip quantitative PCR assay. *Analytical chemistry* 85, 1484-1491 (2013).

[81] L. F. van Dessel et al., High-throughput isolation of circulating tumor DNA: a comparison of automated platforms. *Molecular Oncology* 13, 392-402 (2019).

[82] L. El Bali, A. Diman, A. Bernard, N. H. Roosens, S. C. De Keersmaecker, Comparative study of seven commercial kits for human DNA extraction from urine samples suitable for DNA biomarker-based public health studies. *Journal of biomolecular techniques: JBT* 25, 96 (2014).

[83] G. W. Tan, A. S. B. Khoo, L. P. Tan, Evaluation of extraction kits and RT-qPCR systems adapted to high-throughput platform for circulating miRNAs. *Scientific reports* 5, 9430 (2015).

[84] A. Brunet-Vega et al., Variability in microRNA recovery from plasma: Comparison of five commercial kits. *Analytical biochemistry* 488, 28-35 (2015).

[85] A. Al-Qatati et al., Plasma micro RNA signature is associated with risk stratification in prostate cancer patients. *International Journal of Cancer* 141, 1231-1239 (2017).

[86] Y. Guo et al., Comprehensive evaluation of extracellular small RNA isolation methods from serum in high throughput sequencing. *BMC genomics* 18, 1-9 (2017).

[87] J.-L. Park et al., Quantitative analysis of cell-free DNA in the plasma of gastric cancer patients. *Oncology letters* 3, 921-926 (2012).

[88] M. W. Snyder, M. Kircher, A. J. Hill, R. M. Daza, J. Shendure, Cell-free DNA comprises an in vivo nucleosome footprint that informs its tissues-of-origin. *Cell* 164, 57-68 (2016).

[89] A. S. Devonshire et al., Towards standardisation of cell-free DNA measurement in plasma: controls for extraction efficiency, fragment size bias and quantification. *Analytical and bioanalytical chemistry* 406, 6499-6512 (2014).

[90] Y. Zhou et al., Clinical factors associated with circulating tumor DNA (ct DNA) in primary breast cancer. *Molecular oncology* 13, 1033-1046 (2019).

[91] M. Elazezy, S. A. Joosse, Techniques of using circulating tumor DNA as a liquid biopsy component in cancer management. *Computational and structural biotechnology journal* 16, 370-378 (2018).

[92] E. M. Dauber et al., Quantitative PCR of INDEL s to measure donor-derived cell-free DNA—a potential method to detect acute rejection

in kidney transplantation: a pilot study. *Transplant International* 33, 298-309 (2020).

[93] A. Soave et al., Copy number variations of circulating, cell-free DNA in urothelial carcinoma of the bladder patients treated with radical cystectomy: a prospective study. *Oncotarget* 8, 56398 (2017).

[94] C. Newton et al., Analysis of any point mutation in DNA. The amplification refractory mutation system (ARMS). *Nucleic acids research* 17, 2503-2516 (1989).

[95] A. R. Thierry et al., Clinical validation of the detection of KRAS and BRAF mutations from circulating tumor DNA. *Nature medicine* 20, 430-435 (2014).

[96] H. Wang et al., Allele-specific, non-extendable primer blocker PCR (AS-NEPB-PCR) for DNA mutation detection in cancer. *The Journal of Molecular Diagnostics* 15, 62-69 (2013).

[97] T. Ehlert et al., Establishing PNB-qPCR for quantifying minimal ctDNA concentrations during tumour resection. *Scientific reports* 7, 1-8 (2017).

[98] Y. Nagai et al., Genetic heterogeneity of the epidermal growth factor receptor in non–small cell lung cancer cell lines revealed by a rapid and sensitive detection system, the peptide nucleic acid-locked nucleic acid PCR clamp. *Cancer research* 65, 7276-7282 (2005).

[99] K. Watanabe et al., EGFR mutation analysis of circulating tumor DNA using an improved PNA-LNA PCR clamp method. *Canadian respiratory journal* 2016, (2016).

[100] H.-R. Kim et al., Detection of EGFR mutations in circulating free DNA by PNA-mediated PCR clamping. *Journal of Experimental & Clinical Cancer Research* 32, 50 (2013).

[101] B. Boisselier et al., COLD PCR HRM: a highly sensitive detection method for IDH1 mutations. *Human mutation* 31, 1360-1365 (2010).

[102] M. B. Freidin et al., Circulating tumor DNA outperforms circulating tumor cells for KRAS mutation detection in thoracic malignancies. *Clinical chemistry* 61, 1299-1304 (2015).

[103] J. P. Schouten et al., Relative quantification of 40 nucleic acid sequences by multiplex ligation-dependent probe amplification. *Nucleic acids research* 30, e57-e57 (2002).

[104] B. J. Hindson et al., High-throughput droplet digital PCR system for absolute quantitation of DNA copy number. *Analytical chemistry* 83, 8604-8610 (2011).

[105] C. Demuth et al., Measuring KRAS mutations in circulating tumor DNA by droplet digital PCR and next-generation sequencing. *Translational oncology* 11, 1220-1224 (2018).

[106] H. Zhang et al., Advantage of next-generation sequencing in dynamic monitoring of circulating tumor DNA over droplet digital PCR in cetuximab treated colorectal cancer patients. *Translational oncology* 12, 426-431 (2019).

[107] H. Furuki et al., Evaluation of liquid biopsies for detection of emerging mutated genes in metastatic colorectal cancer. *European Journal of Surgical Oncology* 44, 975-982 (2018).

[108] Q.-m. Guo et al., Detection of plasma EGFR mutations in NSCLC Patients with a validated ddPCR lung cfDNA assay. *Journal of Cancer* 10, 4341 (2019).

[109] H. Ishii et al., Digital PCR analysis of plasma cell-free DNA for non-invasive detection of drug resistance mechanisms in EGFR mutant NSCLC: Correlation with paired tumor samples. *Oncotarget* 6, 30850 (2015).

[110] M. Lodrini et al., Using droplet digital PCR to analyze MYCN and ALK copy number in plasma from patients with neuroblastoma. *Oncotarget* 8, 85234 (2017).

[111] G. J. Weiss et al., Tumor cell–free DNA copy number instability predicts therapeutic response to immunotherapy. *Clinical Cancer Research* 23, 5074-5081 (2017).

[112] F. Diehl et al., BEAMing: single-molecule PCR on microparticles in water-in-oil emulsions. *Nature methods* 3, 551-559 (2006).

[113] C. Franczak et al., Evaluation of KRAS, NRAS and BRAF mutations detection in plasma using an automated system for patients with metastatic colorectal cancer. *PloS one* 15, e0227294 (2020).

[114] M. J. Higgins et al., Detection of tumor PIK3CA status in metastatic breast cancer using peripheral blood. *Clinical cancer research* 18, 3462-3469 (2012).

[115] B. O'Leary et al., Comparison of BEAMing and droplet digital PCR for circulating tumor DNA analysis. *Clinical chemistry* 65, 1405-1413 (2019).

[116] D. Vessies et al., Performance of four platforms for KRAS mutation detection in plasma cell-free DNA: ddPCR, Idylla, COBAS z480 and BEAMing. *Scientific reports* 10, 1-9 (2020).

[117] X. Li et al., Surface enhanced raman spectroscopy (SERS) for the multiplex detection of Braf, Kras, and Pik3ca mutations in plasma of colorectal cancer patients. *Theranostics* 8, 1678 (2018).

[118] E. J. Wee, Y. Wang, S. C.-H. Tsao, M. Trau, Simple, sensitive and accurate multiplex detection of clinically important melanoma DNA mutations in circulating tumour DNA with SERS nanotags. *Theranostics* 6, 1506 (2016).

[119] E. S. Gray et al., Genomic analysis of circulating tumor DNA using a melanoma-specific UltraSEEK Oncogene Panel. *The Journal of Molecular Diagnostics* 21, 418-426 (2019).

[120] B. M. Venkatesan, R. Bashir, Nanopore sensors for nucleic acid analysis. *Nature nanotechnology* 6, 615-624 (2011).

[121] J. Eid et al., Real-time DNA sequencing from single polymerase molecules. *Science* 323, 133-138 (2009).

[122] D. R. Bentley et al., Accurate whole human genome sequencing using reversible terminator chemistry. *nature* 456, 53-59 (2008).

[123] J. M. Rothberg et al., An integrated semiconductor device enabling non-optical genome sequencing. *Nature* 475, 348-352 (2011).

[124] E. J. Fox, K. S. Reid-Bayliss, M. J. Emond, L. A. Loeb, Accuracy of next generation sequencing platforms. *Next generation, sequencing & applications* 1, (2014).

[125] X. Ma et al., Analysis of error profiles in deep next-generation sequencing data. *Genome biology* 20, 1-15 (2019).

[126] S. R. Kennedy et al., Detecting ultralow-frequency mutations by Duplex Sequencing. *Nature protocols* 9, 2586 (2014).

[127] T. M. Butler et al., Circulating tumor DNA dynamics using patient-customized assays are associated with outcome in neoadjuvantly treated breast cancer. *Molecular Case Studies* 5, a003772 (2019).

[128] A. Ståhlberg et al., Simple, multiplexed, PCR-based barcoding of DNA enables sensitive mutation detection in liquid biopsies using sequencing. *Nucleic acids research* 44, e105-e105 (2016).

[129] S. Deng et al., TNER: a novel background error suppression method for mutation detection in circulating tumor DNA. *BMC bioinformatics* 19, 1-7 (2018).

[130] C. S. Kim et al., In silico error correction improves cfDNA mutation calling. *Bioinformatics* 35, 2380-2385 (2019).

[131] S. Manier et al., Whole-exome sequencing of cell-free DNA and circulating tumor cells in multiple myeloma. *Nature communications* 9, 1-11 (2018).

[132] T. D. Tailor et al., Whole exome sequencing of cell-free DNA for early lung cancer: a pilot study to differentiate benign from malignant CT-detected pulmonary lesions. *Frontiers in oncology* 9, 317 (2019).

[133] X. Chen et al., Low-pass whole-genome sequencing of circulating cell-free DNA demonstrates dynamic changes in genomic copy number in a squamous lung cancer clinical cohort. *Clinical Cancer Research* 25, 2254-2263 (2019).

[134] F. Mauger et al., comparison of commercially available whole-genome sequencing kits for variant detection in circulating cell-free DnA. *Scientific reports* 10, 1-11 (2020).

[135] R. Gyanchandani et al., Whole genome amplification of cell-free DNA enables detection of circulating tumor DNA mutations from fingerstick capillary blood. *Scientific reports* 8, 1-12 (2018).

[136] A. M. Bolger, M. Lohse, B. Usadel, Trimmomatic: a flexible trimmer for Illumina sequence data. *Bioinformatics*, btu170 (2014).

[137] S. Chen, Y. Zhou, Y. Chen, J. Gu, fastp: an ultra-fast all-in-one FASTQ preprocessor. *Bioinformatics* 34, i884-i890 (2018).

[138] B. Langmead, S. L. Salzberg, Fast gapped-read alignment with Bowtie 2. *Nature methods* 9, 357 (2012).

[139] H. Li, R. Durbin, Fast and accurate short read alignment with Burrows–Wheeler transform. *bioinformatics* 25, 1754-1760 (2009).

[140] H. Li, R. Durbin, Fast and accurate long-read alignment with Burrows–Wheeler transform. *Bioinformatics* 26, 589-595 (2010).

[141] N. Casiraghi et al., ABEMUS: platform-specific and data-informed detection of somatic SNVs in cfDNA. *Bioinformatics* 36, 2665-2674 (2020).

[142] K. Cibulskis et al., Sensitive detection of somatic point mutations in impure and heterogeneous cancer samples. *Nature biotechnology* 31, 213-219 (2013).

[143] C. H. Mermel et al., GISTIC2. 0 facilitates sensitive and confident localization of the targets of focal somatic copy-number alteration in human cancers. *Genome biology* 12, R41 (2011).

[144] R. Shen, V. E. Seshan, FACETS: allele-specific copy number and clonal heterogeneity analysis tool for high-throughput DNA sequencing. *Nucleic acids research* 44, e131-e131 (2016).

[145] R. J. Leary et al., Development of personalized tumor biomarkers using massively parallel sequencing. *Science translational medicine* 2, 20ra14-20ra14 (2010).

[146] J. Vermeulen et al., Measurable impact of RNA quality on gene expression results from quantitative PCR. *Nucleic acids research* 39, e63-e63 (2011).

[147] T. T. Ngo et al., Noninvasive blood tests for fetal development predict gestational age and preterm delivery. *Science* 360, 1133-1136 (2018).

[148] S. Shanker et al., Evaluation of commercially available RNA amplification kits for RNA sequencing using very low input amounts of total RNA. *Journal of biomolecular techniques: JBT* 26, 4 (2015).

[149] A. Dobin et al., STAR: ultrafast universal RNA-seq aligner. *Bioinformatics* 29, 15-21 (2013).

[150] B. Langmead, C. Trapnell, M. Pop, S. L. Salzberg, Ultrafast and memory-efficient alignment of short DNA sequences to the human genome. *Genome biology* 10, R25 (2009).

[151] M. Ziemann, A. Kaspi, A. El-Osta, Evaluation of microRNA alignment techniques. *RNA*, (2016).

[152] A. Kozomara, M. Birgaoanu, S. Griffiths-Jones, miRBase: from microRNA sequences to function. *Nucleic acids research* 47, D155-D162 (2019).

[153] M. I. Love, W. Huber, S. Anders, Moderated estimation of fold change and dispersion for RNA-seq data with DESeq2. *Genome biology* 15, 550 (2014).

[154] M. D. Robinson, D. J. McCarthy, G. K. Smyth, edgeR: a Bioconductor package for differential expression analysis of digital gene expression data. *Bioinformatics* 26, 139-140 (2010).

[155] C. D. Allis, T. Jenuwein, The molecular hallmarks of epigenetic control. *Nature Reviews Genetics* 17, 487-500 (2016).

[156] G. Cavalli, E. Heard, Advances in epigenetics link genetics to the environment and disease. *Nature* 571, 489-499 (2019).

[157] Mark A. Dawson, T. Kouzarides, Cancer Epigenetics: From Mechanism to Therapy. *Cell* 150, 12-27 (2012).

[158] K. D. Robertson, DNA methylation and human disease. *Nature Reviews Genetics* 6, 597-610 (2005).

[159] T. T. M. Ngo et al., Effects of cytosine modifications on DNA flexibility and nucleosome mechanical stability. *Nature Communications* 7, 10813 (2016).

[160] M. Frommer et al., A genomic sequencing protocol that yields a positive display of 5-methylcytosine residues in individual DNA strands. *PNAS* 89, 1827-1831 (1992).

[161] R. P. Darst, C. E. Pardo, L. Ai, K. D. Brown, M. P. Kladde, Bisulfite Sequencing of DNA. *Curr Protoc Mol Biol* CHAPTER, Unit-7.917 (2010).

[162] D. P. Genereux, W. C. Johnson, A. F. Burden, R. Stoger, C. D. Laird, Errors in the bisulfite conversion of DNA: modulating inappropriate- and failed-conversion frequencies. *Nucl Acids Res* 36, e150-e150 (2008).

[163] Y. Liu et al., Bisulfite-free direct detection of 5-methylcytosine and 5-hydroxymethylcytosine at base resolution. *Nature Biotechnology* 37, 424-429 (2019).

[164] C. Legendre et al., Whole-genome bisulfite sequencing of cell-free DNA identifies signature associated with metastatic breast cancer. *Clin Epigenet* 7, 100 (2015).

[165] L. Liu et al., Targeted methylation sequencing of plasma cell-free DNA for cancer detection and classification. *Ann Oncol* 29, 1445-1453 (2018).

[166] S. W. Walsh et al., Increased expression of toll-like receptors 2 and 9 is associated with reduced DNA methylation in spontaneous preterm labor. *J Reprod Immunol* 121, 35-41 (2017).

[167] J. G. Herman, J. R. Graff, S. Myöhänen, B. D. Nelkin, S. B. Baylin, Methylation-specific PCR: a novel PCR assay for methylation status of CpG islands. *Proc Natl Acad Sci U S A* 93, 9821-9826 (1996).

[168] L. Sigalotti, A. Covre, F. Colizzi, E. Fratta, in *Cell-free DNA as Diagnostic Markers: Methods and Protocols,* V. Casadio, S. Salvi, Eds. (Springer, New York, NY, 2019), pp. 137-162.

[169] S. Y. Shen, J. M. Burgener, S. V. Bratman, D. D. De Carvalho, Preparation of cfMeDIP-seq libraries for methylome profiling of plasma cell-free DNA. *Nature Protocols* 14, 2749-2780 (2019).

[170] J. P. Ross, J. M. Shaw, P. L. Molloy, Identification of differentially methylated regions using streptavidin bisulfite ligand methylation enrichment (SuBLiME), a new method to enrich for methylated DNA prior to deep bisulfite genomic sequencing. *Epigenetics* 8, 113-127 (2013).

[171] E. Kriukienė et al., DNA unmethylome profiling by covalent capture of CpG sites. *Nature Communications* 4, 2190 (2013).

[172] K. Hashimoto, S. Kokubun, E. Itoi, H. I. Roach, Improved quantification of DNA methylation using methylation-sensitive restriction enzymes and real-time PCR. *Epigenetics* 2, 86-91 (2007).

[173] M. P. Ball et al., Targeted and genome-scale strategies reveal gene-body methylation signatures in human cells. *Nature biotechnology* 27, 361-368 (2009).

[174] A. K. Maunakea et al., Conserved role of intragenic DNA methylation in regulating alternative promoters. *Nature* 466, 253-257 (2010).

[175] A. L. Brunner et al., Distinct DNA methylation patterns characterize differentiated human embryonic stem cells and developing human fetal liver. *Genome research* 19, 1044-1056 (2009).

[176] G. Beikircher, W. Pulverer, M. Hofner, C. Noehammer, A. Wein-haeusel, in *DNA Methylation Protocols,* J. Tost, Ed. (Springer New York, New York, NY, 2018), vol. 1708, pp. 407-424.

[177] J. A. Hyett et al., Reduction in diagnostic and therapeutic interventions by non-invasive determination of fetal sex in early pregnancy. *Prenatal Diagnosis* 25, 1111-1116 (2005).

[178] M. G. Forest, Y. Morel, M. David, Prenatal Treatment of Congenital Adrenal Hyperplasia. *Trends in Endocrinology & Metabolism* 9, 284-289 (1998).

[179] G. Breveglieri, E. D'Aversa, A. Finotti, M. Borgatti, Non-invasive Prenatal Testing Using Fetal DNA. *Mol Diagn Ther* 23, 291-299 (2019).

[180] S. Illanes, M. Denbow, C. Kailasam, K. Finning, P. W. Soothill, Early detection of cell-free fetal DNA in maternal plasma. *Early Human Development* 83, 563-566 (2007).

[181] Y. M. Lo, Fetal DNA in maternal plasma: biology and diagnostic applications. *Clin Chem* 46, 1903-1906 (2000).

[182] A. Sekizawa et al., Apoptosis in fetal nucleated erythrocytes circulating in maternal blood. *Prenatal Diagnosis* 20, 886-889 (2000).

[183] A. Sekizawa et al., Evaluation of bidirectional transfer of plasma DNA through placenta. *Hum Genet* 113, 307-310 (2003).

[184] L. Jackson, Fetal cells and DNA in maternal blood. *Prenatal Diagnosis* 23, 837-846 (2003).

[185] Y. Zhou et al., Effects of Maternal and Fetal Characteristics on Cell-Free Fetal DNA Fraction in Maternal Plasma. *Reprod. Sci.* 22, 1429-1435 (2015).

[186] S. S. C. Chim et al., Detection of the placental epigenetic signature of the <em>maspin</em> gene in maternal plasma. *Proceedings of the National Academy of Sciences of the United States of America* 102, 14753 (2005).

[187] J. Harraway, Non-invasive prenatal testing. *Aust Fam Physician* 46, 735-739 (2017).

[188] X. Y. Zhong, W. Holzgreve, S. Hahn, The Levels of Circulatory Fetal Dna in Maternal Plasma Are Elevated Prior to the Onset of Preeclampsia. *Hypertension in Pregnancy* 21, 77-83 (2002).

[189] S. S. C. Chim et al., Detection of the placental epigenetic signature of the maspin gene in maternal plasma. *PNAS* 102, 14753-14758 (2005).

[190] G. Zhu et al., Highly Sensitive Droplet Digital PCR Method for Detection of EGFR-Activating Mutations in Plasma Cell–Free DNA from Patients with Advanced Non–Small Cell Lung Cancer. *The Journal of Molecular Diagnostics* 17, 265-272 (2015).

[191] Y. Tang et al., Maximum allele frequency observed in plasma: A potential indicator of liquid biopsy sensitivity. *Oncol Lett* 18, 2118-2124 (2019).

[192] I. S. Haque, O. Elemento, Challenges in Using ctDNA to Achieve Early Detection of Cancer. *bioRxiv*, 237578 (2017).

[193] E. Heitzer, I. S. Haque, C. E. S. Roberts, M. R. Speicher, Current and future perspectives of liquid biopsies in genomics-driven oncology. *Nature Reviews Genetics* 20, 71-88 (2019).

[194] R. B. Lanman et al., Analytical and Clinical Validation of a Digital Sequencing Panel for Quantitative, Highly Accurate Evaluation of Cell-Free Circulating Tumor DNA. *PloS one* 10, e0140712 (2015).

[195] I. Kinde, J. Wu, N. Papadopoulos, K. W. Kinzler, B. Vogelstein, Detection and quantification of rare mutations with massively parallel sequencing. *Proceedings of the National Academy of Sciences* 108, 9530 (2011).

[196] M. Gao et al., Next Generation-Targeted Amplicon Sequencing (NG-TAS): an optimised protocol and computational pipeline for cost-effective profiling of circulating tumour DNA. *Genome medicine* 11, 1 (2019).

[197] T. Forshew et al., Noninvasive Identification and Monitoring of Cancer Mutations by Targeted Deep Sequencing of Plasma DNA. *Science translational medicine* 4, 136ra168 (2012).

[198] D. E. Piccioni et al., Analysis of cell-free circulating tumor DNA in 419 patients with glioblastoma and other primary brain tumors. *CNS oncology* 8, Cns34 (2019).

[199] J. A. Bellairs, R. Hasina, N. Agrawal, Tumor DNA: an emerging biomarker in head and neck cancer. *Cancer Metastasis Rev* 36, 515-523 (2017).

[200] P. Meng et al., Targeted sequencing of circulating cell-free DNA in stage II-III resectable oesophageal squamous cell carcinoma patients. *BMC Cancer* 19, 818 (2019).

[201] H. Pasternack et al., Somatic alterations in circulating cell-free DNA of oesophageal carcinoma patients during primary staging are indicative for post-surgical tumour recurrence. *Scientific Reports* 8, 14941 (2018).

[202] J. M. Fussey et al., The Clinical Utility of Cell-Free DNA Measurement in Differentiated Thyroid Cancer: A Systematic Review. *Frontiers in oncology* 8, 132 (2018).

[203] J. A. Shaw, J. Stebbing, Circulating free DNA in the management of breast cancer. *Annals of translational medicine* 2, 3 (2014).

[204] K. Lasseter et al., Plasma cell-free DNA variant analysis compared with methylated DNA analysis in renal cell carcinoma. *Genetics in Medicine* 22, 1366-1373 (2020).

[205] M. Iqbal, A. Roberts, J. Starr, K. Mody, P. M. Kasi, Feasibility and clinical value of circulating tumor DNA testing in patients with gastric adenocarcinomas. *Journal of gastrointestinal oncology* 10, 400-406 (2019).

[206] F. Bi et al., Circulating tumor DNA in colorectal cancer: opportunities and challenges. *American journal of translational research* 12, 1044-1055 (2020).

[207] D.-B. Asante, L. Calapre, M. Ziman, T. M. Meniawy, E. S. Gray, Liquid biopsy in ovarian cancer using circulating tumor DNA and cells: Ready for prime time? *Cancer Letters* 468, 59-71 (2020).

[208] D. Gasi Tandefelt, J. de Bono, Circulating cell-free DNA: Translating prostate cancer genomics into clinical care. *Molecular Aspects of Medicine* 72, 100837 (2020).

[209] M. L. Hemming et al., Detection of Circulating Tumor DNA in Patients With Leiomyosarcoma With Progressive Disease. *JCO precision oncology* 2019, (2019).

[210] L. Buedts, P. Vandenberghe, Circulating cell-free DNA in hematological malignancies. *Haematologica* 101, 997-999 (2016).

[211] A. Zviran et al., Genome-wide cell-free DNA mutational integration enables ultra-sensitive cancer monitoring. *Nature medicine* 26, 1114-1124 (2020).

[212] J. D. Cohen et al., Detection and localization of surgically resectable cancers with a multi-analyte blood test. *Science (New York, N.Y.)* 359, 926-930 (2018).

[213] A. M. Lennon et al., Feasibility of blood testing combined with PET-CT to screen for cancer and guide intervention. *Science* 369, (2020).

[214] C. Abbosh, C. Swanton, N. J. Birkbak, Clonal haematopoiesis: a source of biological noise in cell-free DNA analyses. *Annals of oncology : official journal of the European Society for Medical Oncology* 30, 358-359 (2019).

[215] N. T. Potter et al., Validation of a real-time PCR-based qualitative assay for the detection of methylated SEPT9 DNA in human plasma. *Clin Chem* 60, 1183-1191 (2014).

[216] J. Moss et al., Comprehensive human cell-type methylation atlas reveals origins of circulating cell-free DNA in health and disease. *Nature Communications* 9, 5068 (2018).

[217] D. Demircioğlu et al., A Pan-cancer Transcriptome Analysis Reveals Pervasive Regulation through Alternative Promoters. *Cell* 178, 1465-1477.e1417 (2019).

[218] O. A. Kent, J. T. Mendell, A small piece in the cancer puzzle: microRNAs as tumor suppressors and oncogenes. *Oncogene* 25, 6188-6196 (2006).

[219] A. Ventura, T. Jacks, MicroRNAs and cancer: short RNAs go a long way. *Cell* 136, 586-591 (2009).

[220] J. T. Wagner et al., Diurnal stability of cell-free DNA and cell-free RNA in human plasma samples. *Scientific reports* 10, 1-8 (2020).

[221] K. Cuk et al., Circulating microRNAs in plasma as early detection markers for breast cancer. *International journal of cancer* 132, 1602-1612 (2013).

[222] S. Suryawanshi et al., Plasma microRNAs as novel biomarkers for endometriosis and endometriosis-associated ovarian cancer. *Clinical cancer research : an official journal of the American Association for Cancer Research* 19, 1213-1224 (2013).

[223] K. Asakura et al., A miRNA-based diagnostic model predicts resectable lung cancer in humans with high accuracy. *Communications Biology* 3, 134 (2020).

[224] F. Liu et al., Circulating miRNAs as novel potential biomarkers for esophageal squamous cell carcinoma diagnosis: a meta-analysis update. *Diseases of the Esophagus* 30, 1-9 (2017).

[225] D. Bidarra et al., Circulating MicroRNAs as Biomarkers for Prostate Cancer Detection and Metastasis Development Prediction. *Frontiers in oncology* 9, 900 (2019).

[226] A. H. Alhasan et al., Circulating microRNA signature for the diagnosis of very high-risk prostate cancer. *Proceedings of the National Academy of Sciences* 113, 10655 (2016).

[227] Y. Ding, J. L. Yan, A. N. Fang, W. F. Zhou, L. Huang, Circulating miRNAs as novel diagnostic biomarkers in hepatocellular carcinoma detection: a meta-analysis based on 24 articles. *Oncotarget* 8, 66402-66413 (2017).

[228] C. Clancy, M. R. Joyce, M. J. Kerin, The use of circulating microRNAs as diagnostic biomarkers in colorectal cancer. *Cancer biomarkers : section A of Disease markers* 15, 103-113 (2015).

[229] G. E. D. Petrescu, A. A. Sabo, L. I. Torsin, G. A. Calin, M. P. Dragomir, MicroRNA based theranostics for brain cancer: basic principles. *Journal of experimental & clinical cancer research : CR* 38, 231 (2019).

[230] Y. Lin, Q. Leng, M. Zhan, F. Jiang, A Plasma Long Noncoding RNA Signature for Early Detection of Lung Cancer. *Translational Oncology* 11, 1225-1231 (2018).

[231] S. Anfossi, A. Babayan, K. Pantel, G. A. Calin, Clinical utility of circulating non-coding RNAs — an update. *Nature Reviews Clinical Oncology* 15, 541-563 (2018).

[232] J. R. Brown, A. M. Chinnaiyan, The Potential of Circular RNAs as Cancer Biomarkers. *Cancer epidemiology, biomarkers & prevention : a publication of the American Association for Cancer Research, cosponsored by the American Society of Preventive Oncology* 29, 2541-2555 (2020).

[233] J. A. Moscow, T. Fojo, R. L. Schilsky, The evidence framework for precision cancer medicine. *Nature Reviews Clinical Oncology* 15, 183-192 (2018).

[234] J. Rawluk, C. F. Waller, in *Small Molecules in Oncology,* U. M. Martens, Ed. (Springer International Publishing, Cham, 2018), pp. 235-246.

[235] F. A. Shepherd et al., Erlotinib in previously treated non-small-cell lung cancer. *The New England journal of medicine* 353, 123-132 (2005).

[236] T. Krzyszkowski, R. Czepko, D. Adamek, Stereotactic biopsy in surgically inaccessible tumors with the use of "P.N." type frame. *Annales Academiae Medicae Stetinensis* 53, 27-32 (2007).

[237] J. D. Cunningham et al., Accessible or Inaccessible? Diagnostic Efficacy of CT-Guided Core Biopsies of Head and Neck Masses. *CardioVascular and Interventional Radiology* 38, 422-429 (2015).

[238] J. Sun et al., Examination of Plasma Cell-Free DNA of Glioma Patients by Whole Exome Sequencing. *World Neurosurgery* 125, e424-e428 (2019).

[239] K. Takeda et al., Analysis of colorectal cancer-related mutations by liquid biopsy: Utility of circulating cell-free DNA and circulating tumor cells. *Cancer Science* 110, 3497-3509 (2019).

[240] G. Vandekerkhove et al., Circulating Tumor DNA Abundance and Potential Utility in De Novo Metastatic Prostate Cancer. *European Urology* 75, 667-675 (2019).

[241] F. Janku et al., BRAF Mutation Testing in Cell-Free DNA from the Plasma of Patients with Advanced Cancers Using a Rapid, Automated Molecular Diagnostics System. *Mol Cancer Ther* 15, 1397-1404 (2016).

[242] Z.-Y. Hu et al., Identifying Circulating Tumor DNA Mutation Profiles in Metastatic Breast Cancer Patients with Multiline Resistance. *EBioMedicine* 32, 111-118 (2018).

[243] A. D. Choudhury et al., Tumor fraction in cell-free DNA as a biomarker in prostate cancer. *JCI Insight* 3.

[244] T. Powrózek et al., Methylation of the DCLK1 promoter region in circulating free DNA and its prognostic value in lung cancer patients. *Clin Transl Oncol* 18, 398-404 (2016).

[245] W. Tang et al., CircRNA microarray profiling identifies a novel circulating biomarker for detection of gastric cancer. *Mol. Cancer* 17, 137 (2018).

[246] Z. Kang, S. Stevanović, C. S. Hinrichs, L. Cao, Circulating Cell-free DNA for Metastatic Cervical Cancer Detection, Genotyping, and Monitoring. *Clin Cancer Res* 23, 6856-6862 (2017).

[247] A. S. C. Machado et al., Circulating cell-free and Epstein–Barr virus DNA in pediatric B-non-Hodgkin lymphomas. *Leukemia & Lymphoma* 51, 1020-1027 (2010).

[248] J. Sato, T. Koyama, T. Shimizu, N. Yamamoto, 489P - High performance of serial tumour biopsies in first in human (FIH) phase I trials. *Annals of Oncology* 30, v185 (2019).

[249] I. Boeddinghaus, S. R. Johnson, Serial biopsies/fine-needle aspirates and their assessment. *Methods in molecular medicine* 120, 29-41 (2006).

[250] C. Abbosh, N. J. Birkbak, C. Swanton, Early stage NSCLC - challenges to implementing ctDNA-based screening and MRD detection. *Nature Reviews. Clinical Oncology* 15, 577-586 (2018).

[251] K. Page et al., Next generation sequencing of circulating cell-free DNA evaluating mutations and gene amplification in metastatic breast cancer. *Clin Chem* 63, 532-541 (2017).

[252] L. V. Schøler et al., Clinical Implications of Monitoring Circulating Tumor DNA in Patients with Colorectal Cancer. *Clin Cancer Res* 23, 5437-5445 (2017).

[253] P. M. Torbati, F. Asadi, P. Fard-Esfahani, Circulating miR-20a and miR-26a as Biomarkers in Prostate Cancer. *Asian Pac J Cancer Prev* 20, 1453-1456 (2019).

[254] S. B. Goldberg et al., Early Assessment of Lung Cancer Immunotherapy Response via Circulating Tumor DNA. *Clin Cancer Res* 24, 1872 (2018).

[255] E. Tormo et al., MicroRNA Profile in Response to Doxorubicin Treatment in Breast Cancer. *Journal of Cellular Biochemistry* 116, 2061-2073 (2015).

[256] D. M. Allin et al., Circulating tumour DNA is a potential biomarker for disease progression and response to targeted therapy in advanced thyroid cancer. *European Journal of Cancer* 103, 165-175 (2018).

[257] J. S. Bhangu et al., Circulating Free Methylated Tumor DNA Markers for Sensitive Assessment of Tumor Burden and Early Response Monitoring in Patients Receiving Systemic Chemotherapy for Colorectal Cancer Liver Metastasis. *Annals of Surgery* 268, (2018).

[258] Y. M. Lo et al., Presence of donor-specific DNA in plasma of kidney and liver-transplant recipients. *Lancet* 351, 1329-1330 (1998).

[259] F. Dengu, Next-generation sequencing methods to detect donor-derived cell-free DNA after transplantation. *Transplantation Reviews* 34, 100542 (2020).

[260] T. M. Snyder, K. K. Khush, H. A. Valantine, S. R. Quake, Universal noninvasive detection of solid organ transplant rejection. *PNAS* 108, 6229-6234 (2011).

[261] R. D. Bloom et al., Cell-Free DNA and Active Rejection in Kidney Allografts. *JASN* 28, 2221-2232 (2017).

[262] J. Beck et al., Digital Droplet PCR for Rapid Quantification of Donor DNA in the Circulation of Transplant Recipients as a Potential Universal Biomarker of Graft Injury. *Clin Chem* 59, 1732-1741 (2013).

[263] I. De Vlaminck et al., Noninvasive monitoring of infection and rejection after lung transplantation. *Proc Natl Acad Sci U S A* 112, 13336-13341 (2015).

[264] I. De Vlaminck et al., Circulating cell-free DNA enables noninvasive diagnosis of heart transplant rejection. *Science Translational Medicine* 6, 241ra277 (2014).

[265] I. De Vlaminck et al., Temporal response of the human virome to immunosuppression and antiviral therapy. *Cell* 155, 1178-1187 (2013).

[266] C. Vollmers et al., Monitoring pharmacologically induced immunosuppression by immune repertoire sequencing to detect acute allograft rejection in heart transplant patients: a proof-of-concept diagnostic accuracy study. *PLoS Med.* 12, e1001890 (2015).

[267] D. J. Castillo, R. F. Rifkin, D. A. Cowan, M. Potgieter, The healthy human blood microbiome: fact or fiction? *Frontiers in Cellular and Infection Microbiology* 9, 148 (2019).

[268] K. Califf, A. Gonzalez, R. Knight, J. G. Caporaso, The human microbiome: getting personal. *Microbe* 9, 410-415 (2014).

[269] P. Burnham et al., Separating the signal from the noise in metagenomic cell-free DNA sequencing. *Microbiome* 8, 1-9 (2020).

[270] M. Kowarsky et al., Numerous uncharacterized and highly divergent microbes which colonize humans are revealed by circulating cell-free DNA. *Proceedings of the National Academy of Sciences* 114, 9623-9628 (2017).

[271] Y.-F. Huang et al., Analysis of microbial sequences in plasma cell-free DNA for early-onset breast cancer patients and healthy females. *BMC medical genomics* 11, 16 (2018).

[272] G. D. Poore et al., Microbiome analyses of blood and tissues suggest cancer diagnostic approach. *Nature* 579, 567-574 (2020).

[273] T. Ou et al., Increased Preoperative Plasma Level of Microbial 16S rDNA Translocation Is Associated With Relapse After Prostatectomy in Prostate Cancer Patients. *Frontiers in Oncology* 9, 1532 (2020).

[274] K. P. Goggin et al., Evaluation of plasma microbial cell-free DNA sequencing to predict bloodstream infection in pediatric patients with relapsed or refractory cancer. *JAMA oncology* 6, 552-556 (2020).

[275] S. L. R. Barrett et al., Cell free DNA from respiratory pathogens is detectable in the blood plasma of Cystic Fibrosis patients. *Scientific Reports* 10, 1-6 (2020).

[276] W. Pan et al., Simultaneously monitoring immune response and microbial infections during pregnancy through plasma cfRNA sequencing. *Clinical Chemistry* 63, 1695-1704 (2017).

[277] M. Stumvoll, B. J. Goldstein, T. W. van Haeften, Type 2 diabetes: principles of pathogenesis and therapy. *The Lancet* 365, 1333-1346 (2005).

[278] T. L. Van Belle, K. T. Coppieters, M. G. Von Herrath, Type 1 Diabetes: Etiology, Immunology, and Therapeutic Strategies. *Physiological Reviews* 91, 79-118 (2011).

[279] F. J. Ortega et al., Profiling of Circulating MicroRNAs Reveals Common MicroRNAs Linked to Type 2 Diabetes That Change With Insulin Sensitization. *Dia Care* 37, 1375-1383 (2014).

[280] C. Zhao et al., Early Second-Trimester Serum MiRNA Profiling Predicts Gestational Diabetes Mellitus. *PLOS ONE* 6, e23925 (2011).

[281] N. Pescador et al., Serum Circulating microRNA Profiling for Identification of Potential Type 2 Diabetes and Obesity Biomarkers. *PLOS ONE* 8, e77251 (2013).

[282] Y. Zhu et al., Profiling maternal plasma microRNA expression in early pregnancy to predict gestational diabetes mellitus. *International Journal of Gynecology & Obstetrics* 130, 49-53 (2015).

[283] X. Wang et al., Determination of 14 Circulating microRNAs in Swedes and Iraqis with and without Diabetes Mellitus Type 2. *PLOS ONE* 9, e86792 (2014).

[284] L. B. Nielsen et al., Erratum to "Circulating Levels of MicroRNA from Children with Newly Diagnosed Type 1 Diabetes and Healthy Controls: Evidence That miR-25 Associates to Residual Beta-Cell Function and Glycaemic Control during Disease Progression". *Experimental Diabetes Research* 2012, 672865 (2012).

[285] M. Párrizas et al., Circulating miR-192 and miR-193b Are Markers of Prediabetes and Are Modulated by an Exercise Intervention. *The Journal of Clinical Endocrinology & Metabolism* 100, E407-E415 (2015).

[286] M. Cross et al., The global burden of rheumatoid arthritis: estimates from the Global Burden of Disease 2010 study. *Ann Rheum Dis* 73, 1316-1322 (2014).

[287] J. Widdifield, J. M. Paterson, A. Huang, S. Bernatsky, Causes of Death in Rheumatoid Arthritis: How Do They Compare to the General Population? *Arthritis Care Res* 70, 1748-1755 (2018).

[288] L. Martinez-Prat et al., Comparison of Serological Biomarkers in Rheumatoid Arthritis and Their Combination to Improve Diagnostic Performance. *Front Immunol* 9, 1113 (2018).

[289] M. Dunaeva et al., Decreased serum cell-free DNA levels in rheumatoid arthritis. *Autoimmun Highlights* 6, 23-30 (2015).

[290] E. Rykova et al., Circulating DNA in rheumatoid arthritis: pathological changes and association with clinically used serological markers. *Arthritis Res Ther* 19, 85 (2017).

[291] T. Hashimoto et al., Circulating cell free DNA: a marker to predict the therapeutic response for biological DMARDs in rheumatoid arthritis. *Int J Rheum Dis* 20, 722-730 (2017).

[292] M. Filková et al., Association of circulating miR-223 and miR-16 with disease activity in patients with early rheumatoid arthritis. *Ann Rheum Dis* 73, 1898-1904 (2014).

[293] C. Castro-Villegas et al., Circulating miRNAs as potential biomarkers of therapy effectiveness in rheumatoid arthritis patients treated with anti-TNFα. *Arthritis Res Ther* 17, 49 (2015).

[294] M. Komori et al., Cerebrospinal fluid markers reveal intrathecal inflammation in progressive multiple sclerosis. *Ann. Neurol.* 78, 3-20 (2015).

[295] D. Ferraro et al., Cerebrospinal fluid oligoclonal IgM bands predict early conversion to clinically definite multiple sclerosis in patients with clinically isolated syndrome. *J. Neuroimmunol.* 257, 76-81 (2013).

[296] A. Petzold, The prognostic value of CSF neurofilaments in multiple sclerosis at 15-year follow-up. *J. Neurol. Neurosurg. Psychiatry* 86, 1388-1390 (2015).

[297] D. Luo et al., Identification and functional analysis of specific MS risk miRNAs and their target genes. *Mult Scler Relat Disord* 41, 102044 (2020).

[298] I. Manna et al., Exosome-associated miRNA profile as a prognostic tool for therapy response monitoring in multiple sclerosis patients. *FASEB J.* 32, 4241-4246 (2018).

[299] G. Lugli et al., Plasma Exosomal miRNAs in Persons with and without Alzheimer Disease: Altered Expression and Prospects for Biomarkers. *PLOS ONE* 10, e0139233 (2015).

[300] H. Wei et al., Serum Exosomal miR-223 Serves as a Potential Diagnostic and Prognostic Biomarker for Dementia. *Neuroscience* 379, 167-176 (2018).

[301] H. Dong et al., Serum MicroRNA Profiles Serve as Novel Biomarkers for the Diagnosis of Alzheimer's Disease. *Dis. Markers* 2015, 1-11 (2015).

[302] S. Absalon, D. M. Kochanek, V. Raghavan, A. M. Krichevsky, MiR-26b, Upregulated in Alzheimer's Disease, Activates Cell Cycle Entry, Tau-Phosphorylation, and Apoptosis in Postmitotic Neurons. *Journal of Neuroscience* 33, 14645-14659 (2013).

[303] C.-G. Liu, J. Song, Y.-Q. Zhang, P.-C. Wang, MicroRNA-193b is a regulator of amyloid precursor protein in the blood and cerebrospinal fluid derived exosomal microRNA-193b is a biomarker of Alzheimer's disease. *Molecular Medicine Reports* 10, 2395-2400 (2014).

[304] M. Mustapic et al., Plasma Extracellular Vesicles Enriched for Neuronal Origin: A Potential Window into Brain Pathologic Processes. *Front Neurosci* 11, 278 (2017).

[305] T. T. Yang, C. G. Liu, S. C. Gao, Y. Zhang, P. C. Wang, The Serum Exosome Derived MicroRNA-135a, -193b, and -384 Were Potential Alzheimer's Disease Biomarkers. *Biomed. Environ. Sci.* 31, 87-96 (2018).

3

Animal Models and Techniques Used in Cardiovascular Research

Jimmy Zhang[1], Samantha Lauren Laboy-Segarra[2], Anh H. Nguyen[3], Juhyun Lee[2], and Hung Cao[1,3,4]

[1]Department of Biomedical Engineering, UC Irvine, USA
[2]Department of Bioengineering, University of Texas at Arlington, USA
[3]Department of Electrical Engineering and Computer Science, UCI Irvine, USA
[4]Department of Computer Science, UC Irvine, USA

Abstract

Despite extensive efforts in cardiovascular disease (CVD) research, no cure has been developed to induce human cardiac regeneration after an injury caused by CVDs. In an effort to accelerate the development of a cure, this chapter outlines various animal models utilized in current CVD research as well as the pertinent methodologies devised to injure the animal models based on specific CVDs. Mammalian models are emphasized in this chapter due to numerous similarities to humans in anatomy, physiology, and genetics. Zebrafish models are also described due to their intrinsic capability to regenerate cardiac tissue, providing an intriguing model to understand the fundamentals of inherent regeneration. Lastly, this chapter provides an overview on the use of these animal models in research on specific CVDs, highlighting important findings as well as limitations to address in future research.

Keywords: Cardiovascular diseases, Animal injury models, Mammalian models, Zebrafish Ischemia, Cardiomyopathy, Metabolism, Arrhythmia, Atherosclerosis, Genetic disorders

3.1 Introduction

Despite numerous efforts to understand and treat cardiovascular diseases (CVDs), they are still the leading cause of death in the US and worldwide. In the most recent report from the American Heart Association (AHA), cardiovascular diseases were responsible for almost 860,000 deaths in the United States. Overall, CVDs accounted for approximately $351.2 billion in costs [8]. In the European Nation, CVDs were responsible for roughly 40% of all deaths [11]. While the efforts of treating CVDs were well documented in developed nations, statistics indicate that 80% of worldwide CVD-related deaths were located in developing nations [12]. Studies have extensively outlined the risk factors associated with the etiology of CVDs. These include uncontrollable factors such as age, sex, and race, as well as controllable factors such as blood pressure, cholesterol levels, and smoking [13]. Indeed, preventative treatments have been devised to mitigate the controllable risk factors, which include dietary modifications, exercise implementation, surgical intervention, and medication to restore physiological blood pressure and/or cholesterol levels [14].

However, current treatments are limited in their efficacy. While issues caused by social disparities and deficiencies in the prevention or modification of cardiovascular risk factors were mentioned, a major concern that is not readily addressed is the long-term prognosis of CVDs, such as the prevalence of recurring cardiovascular events [15, 16]. Generally, medications such as statins and beta-blockers are prescribed after palliative intervention to improve patient outcomes, but these procedures do not encompass the multifaceted nature of CVDs and might even create additional problems of their own [17, 18]. For example, research has determined that beta-blockers could accelerate the progression of chronic kidney disease [19]. Therefore, more attention should be focused on resolving cardiovascular issues on a fundamental level. One such endeavor is the analysis of intrinsic heart regeneration capabilities.

To investigate cardiovascular regeneration, studies have employed various animal injury models to determine their regenerative capabilities. Each animal injury model has its specific advantages and disadvantages based on availability, maintenance, reliability, and applicability to human treatments. These will be discussed in the following section below. Additionally, the biological discoveries and implications will be explored afterward.

3.2 Mouse Models

First scientifically documented in the seventeenth century, the mouse (*Mus musculus*) model has since grown into one of the most popular animal models for biological and medical research [20]. They are chosen because of their similar features in anatomy, physiology, and disease pathology, especially in regard to the cardiovascular system [21]. Mice also share a remarkable degree of genetic homology with humans compared with other models [22]. With the advent of genetic research, numerous genetic strains are produced to determine of the role of genetic pathways in the research of various pathologies. Indeed, the development of robust genetic manipulation in mice has led to their usage in a great variety of studies. In addition to their ease of maintenance and fecundity, it should come as no surprise that researchers have published a great number of studies on the application of mice in cardiac regeneration research.

In order to properly conduct experiments on mice cardiac regeneration, a robust and reproducible method of cardiac injury must first be established. These methods could be categorized into the following: physical surgery, chemical treatment, and genetic models.

3.2.1 Physical surgery

Currently, the scientific community has yielded several main methods of physical surgery to injure the mouse heart. The first established method is the ligation of the left anterior descending (LAD) artery. Branching off the left coronary artery, the LAD artery is responsible for the perfusion of blood throughout the left ventricle. Thus, the ligation of the blood vessel walls within the LAD artery would sever blood flow to the left ventricle, resulting in the development of an infarct area and corresponding injury analogous to that seen in myocardial infarction. The general procedure of LAD ligation would be as follows [23]. After the setup of the operating area, which generally includes the anesthetizing agent, ventilator, heating pad, and surgical instruments. The anesthetizing agent, which can include isoflurane, ketamine, and xylazine, is injected into the mouse. After checking for consciousness by pinching the tail, the throat, and chest areas of the mouse are then shaved to expose the incision areas. The mouse is affixed to the heating pad and its mouth is held open to facilitate ventilation. An incision is made along the thorax for the insertion of a ventilation tube into the throat. After turning on the ventilator and repositioning the mice, an incision is made on the chest in the 3rd intercoastal space between the 3rd and 4th ribs to

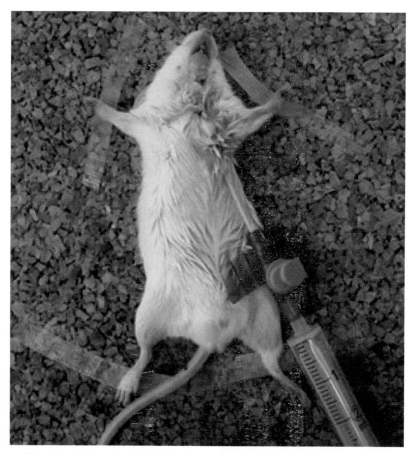

Figure 3.1   General surgical setup for mice LAD ligation prior to incision, showing mice affixed onto the surface by the limbs with tape and the ventilation tube inserted into the thorax [9].

expose the heart. The pericardium is then pulled off. A suture is placed around the LAD artery approximately 3 mm below the left atrium and subsequently tied. The left ventricle should be noticeably paler due to the occlusion of blood flow. The incision areas are then closed, and the mouse is injected with analgesic agents, such as buprenorphine, as well as other agents to assist in recovery. The mouse is then observed for at least one hour for any abnormal symptoms such as difficulty breathing or bleeding.

A variation of the LAD ligation procedure is the ischemia-reperfusion (I/R) procedure to induce temporary ischemia in the left ventricle of the heart. The I/R procedure follows the same methodology as the LAD ligation procedure until the suturing of the LAD artery. In LAD ligation, the suture will remain permanently around the LAD. However, in the I/R procedure, the suture will eventually be untied from the LAD after 20–60 min of occlusion, allowing for the reperfusion of blood in the ventricle [24]. The suture will also be accompanied by a tubing, which will be tied alongside the LAD to temporarily close blood flow into the LAD. Unlike LAD ligation, which provides a long-term prognosis of the effects in myocardial ischemia, the I/R procedure is utilized for short-term analysis of myocardial infarction. It also provides insight into the consequences of reperfusion injury, as the sudden influx of oxygen could result in oxidative damage in cardiomyocytes [25]. As treatment for myocardial infarction involves the urgent need to reperfuse the heart, the I/R procedure is representative of the general protocol for myocardial infarction diagnosis and immediate treatment.

As the LAD and I/R surgeries focus on left ventricular injury, others have proposed ideas on the analysis of injuries in other areas of the heart. Sicard et al. detailed a procedure to ligate the right coronary artery to analyze how right ventricular ischemia would affect overall cardiac function [26]. After ligating the region of the right coronary artery about 3–5 mm from its origin, the study found noticeable right ventricular damage, which led to left ventricular diastolic dysfunction. This injury model demonstrates that new research directions could uncover a more comprehensive picture of cardiac diseases.

While the LAD ligation and I/R procedures have shown significant infarct areas and associated cardiac dysfunction, researchers have also documented shortcomings of these procedures, including the variable sizes of infarct areas, potential aneurysm formations due to modifications in the vasculature, as well as the degree of training and difficulty in performing the procedure

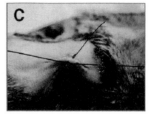

Figure 3.2 I/R injury model in ,mice. This panel shows the progression of ischemic portion of procedure, where the LAD is tied in panel A, slipped back into thoracic cavity in B, and having needle sutures remaining outside shown in C [10].

to limit mortality. Modifications of the procedures have been implemented to overcome these shortcomings. Gao et al. described a method to decrease the procedure time for both LAD ligation and I/R procedures by forgoing ventilation and a large chest incision [10]. Mice were anesthetized by isoflurane inhalation and a small chest opening was made to allow the heart to "pop out" from the chest for easier and faster access. They stated that this procedure required the experience to perform the LAD suturing in under 3 min, but the survival rates are significantly increased with their developed method while achieving similar cardiac dysfunction symptoms. Reichert et al. introduced their methodology of LAD ligation, which included modifications such as ventilation intubation through the mouth into the trachea, retractors to open the chest cavity for better visualization, application of gauze to prevent lung damage, and saline injection immediately after surgery to improve mouse recovery [27]. de Andrade et al. utilized surgical clips to quickly close the LAD, expediting the surgery procedure [28]. Evidently, researchers have attempted and reported various strategies to decrease difficulty and improve surgical outcomes for LAD ligation and I/R surgeries.

Another common surgical model for mice cardiac injury is the transverse aortic constriction (TAC) model. Whereas the LAD and I/R surgeries are more representative of the pathology for myocardial infarctions, TAC surgery is associated with pressure overload of the heart as well as aortic stenosis and subsequent hypertrophy. This surgery is performed in a similar manner to the LAD and I/R surgeries in regard to preparation, anesthesia, and ventilation.

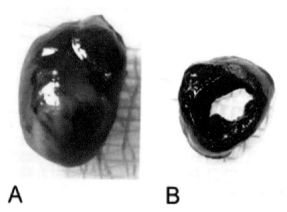

A　　　　**B**

Figure 3.3　Representative explanted mice heart after LAD ligation. Panel A shows the whole heart, with infarcted area indicated by a pale whitish color. Panel B is a cross-section of the heart, with the infarct area shown in the pale section on the top-right [9].

During chest incision, the second intercoastal space near the left upper sternal border is opened instead, exposing the aortic region of the heart [29]. The transverse aorta is then narrowed by approximately 0.4 mm in diameter with a suture. Rockman et al. demonstrated that this model led to significant cardiac hypertrophy without leading to excessive overload [30]. The TAC model does have limitations as well, such as the higher difficulty in performing TAC than LAD surgery as well as potential variability in left ventricle remodeling [30].

Besides the physical suturing of arteries, another surgical method to inducing cardiac injury is cryosurgery. After preparation of the mouse in a similar manner to that in the LAD ligation (anesthesia, ventilation, and incision), a liquid-nitrogen chilled ~3 mm cryoprobe composed of copper, aluminum, or vanadium is applied to the left ventricle of the heart for 5–10 s [32, 32]. Cryoprobes could be purchased commercially, such as the Brymill Cryo–Gun apparatus, or fabricated in the lab [31]. When fabricating, a thin metal filament is held in place by a custom probe, with adequate insulating layers to maintain a consistent temperature as well as to protect its user. Vanadium seems to be the optimal choice for the metal filament, as its low thermal conductivity allows it to retain a colder temperature for a longer time. The application of the cryoprobe induces cryoinjury to the ventricle wall, causing the death of cardiomyocytes, endothelial cells, and nodal cells. These contribute to eventual scarring and cardiac dysfunction. van den Bos et al. conducted a comparison study between cryosurgery and LAD ligation procedures and found that cryosurgery yielded a more uniform infarct size as well as a higher survival rate while maintaining similar cardiac dysfunction [33]. The histology of the infarct area also resembles that seen in myocardial infarction, as significant adverse remodeling was discovered after the cryosurgery procedure. Cryosurgery also bypasses the need to locate the LAD and thread the suture around it, decreasing the difficulty of the surgery operation. However, this study also noted that cryosurgery generally yielded a smaller infarct size than LAD ligation, indicating that cryosurgery should be utilized as a more modest cardiac remodeling procedure, not as a heart failure model.

An injury model more commonly seen in neonatal mice is apical resection. In the study by Porrello et al., P1 (postnatal day 1) mice are first placed in a sheet-covered ice bed for approximately 4 min to induce anesthesia [34]. (Anesthesia agents are not effective on neonatal mice.) After fixing the mice under a stereomicroscope, an incision is made on the 4th or 5th intercoastal space to expose the heart. Using forceps to hold the heart steady, the apex of the heart is then amputated with iridectomy scissors. Approximately 15%

of the heart is amputated with this method, which would expose the left ventricular chamber. The study by Porrello et al. was conducted to develop a regenerative mammalian cardiac model, serving as the mammalian equivalent to that in zebrafish and other species that could regenerate cardiac tissue. Neonatal mice were shown to fully regenerate their apex within 21 days after surgery. While the model was effective in demonstrating that neonatal mice can generate a similar regenerative response to that in zebrafish, such as epicardial activation, the apical resection procedure has numerous limitations. Porrello et al. stated that myocyte hypertrophy and fibrotic scarring, two hallmarks in myocardial infarction pathology, were not apparent after apical resection. Therefore, the histological and functional injury seen in apical resection would not be representative as an adequate myocardial infarction or cardiac injury model. However, most interestingly, subsequent studies have yielded varied results. Andersen et al. discovered that their neonatal mice had very limited regeneration potential, highlighted by extensive fibrotic scarring [35]. Bryant et al. indicated that scarring did occur mainly after large resections, but they did not provide an explanation why a regenerative response would differ based on resection size [36]. As a result, the apical resection model in mice should not be utilized unless it serves as a verification of certain cardiac regenerative processes first seen in zebrafish and other regenerative species.

3.2.2 Chemical treatment

Chemical intraperitoneal injection into mice is also a reliable method of inducing cardiac injury. Generally, chemical models are used to investigate potential cardiac side effects of certain drugs or to determine multifactorial effects in relation to other diseases commonly encountered in human medical practice. A common intraperitoneal injection procedure is as follows [37]. The mouse is held steady by pinching the neck with the thumb and index finger while the tail is held along the ring and pinky fingers. Using a 26 gauge needle, the chemical is injected at a 40°–45° into the lower left or right quadrant of the abdomen between the hips and genitals. Careful attention must be directed toward preventing accidental punctures into surrounding organs to avoid bleeding and excessive tissue damage.

The most common chemical utilized in these studies is doxorubicin (DOX). Used mainly as a chemotherapeutic agent for the treatment of cancer, DOX was discovered to generate detrimental side effects to the cardiovascular system, including arrhythmia, ventricular dysfunction, and congestive heart failure [38]. Currently, the pathology of DOX-induced cardiomyopathy is not fully understood, although it is suggested that DOX leads to mitochondrial dysfunction and oxidative damage in cardiomyocytes [39]. Two methods of administering DOX treatment in mice are utilized: an acute model and a chronic model. The acute model dictates that mice will be treated once via a single intraperitoneal injection of a high dose of DOX. The mice would then be monitored for up to 12 weeks. Conversely, the chronic model outlines an 8- to 12-week regimen where the mouse will receive a weekly injection of a smaller dose of DOX. The acute model is generally utilized to identify relevant genetic pathways and protein markers that are significant in DOX-induced cardiac injury [40]. The chronic model is more representative of the chemotherapeutic treatment for cancer patients and is more reliable as a disease model for observing histological and functional effects of DOX on the cardiovascular system, which can include tissue staining and echocardiography. The chronic model could also be utilized to test potential treatment methods for DOX-induced cardiac injury [41].

Other chemicals have also been utilized to investigate cardiovascular-associated injury. For example, liposaccharide (LPS), derived from bacteria and responsible for the development of cytokine storms in sepsis, has been employed to induce sepsis-induced cardiac injury, caused by the accumulation of reactive oxygen species and the initiation of apoptosis in cardiomyocytes. Xianchu et al. investigated sepsis-induced cardiac injury by injecting a one-time dose of 10 mg/kg LPS into wild-type mice [42]. They observed that LPS upregulated the expression of inflammatory cytokines such as TNF-α and IL-6 as well as increased the concentration of reactive oxygen species, which induced damage to cardiac tissue. Abdel-Daim et al. conducted a similar study with methotrexate by injecting a single dose of 20 mg/kg into mice to observe cardiac necrotic damage due to oxidative stress and inflammatory cytokine upregulation [43].

Overall, cardiac injury in mice due to chemical treatment is mainly used to investigate the cardiotoxic effects of certain chemicals and to explore potential drugs to mitigate those cardiotoxic effects. While they should not be used as the sole cardiac injury model due to the inability to accurately gauge the extent of injury, they could be used as a supplement to other injury models as a source of comparison and further exploration into future research directions.

3.2.3 Genetic models

As genetic engineering technology become increasingly developed, several researchers have employed various genetic models to simulate cardiac injury and to investigate associated genetic pathways. The number of genetic models for cardiac injury is constantly evolving as more genetic pathways associated with cardiac development and regeneration are discovered. To develop mice genetic models, the most common procedure is as follows. The gene of interest is cloned into a genetic vector containing cassettes that assist in the integration of the gene of interest into the mouse's genome. After determining that the gene is successfully incorporated in the vector through a chemical selection process specific to the vector, the complete vector is electroporated or chemically inserted into mouse embryonic stem cells and transferred into a surrogate mouse. These cells are then raised and subsequently breed to produce a complete transgenic model [44]. Successful transgenic mutation can be verified by genetic screening such as southern blotting [45]. Through the development of mouse genetic models, researchers are able to analyze the roles that certain genetic pathways play in cardiac pathology. For example, Arber et al. delineated a mouse genetic model with deficient MLP [46]. MLP is an integral protein in actin cytoskeleton development, and MLP deficiency leads to disruption in cardiomyocyte structure and organization, causing dilated cardiomyopathy and heart failure.

Two other common mouse transgenic models utilized in cardiac injury studies are Apolipoprotein E (*apoe*) and low-density lipoprotein receptor (*ldlr*) double knockout models [47, 48]. Both models are responsible for the development of atherosclerosis, as both associated genes are crucial in the uptake of low-density lipoproteins into the liver. Removal of these genes by knockout causes an excessive accumulation of fat in blood vessels, leading to atherosclerosis. High fat and/or cholesterol diets could also be supplemented to create an enhanced effect of atherosclerotic-induced injury. While these injury models are relatively indirect in the induction of cardiac injury and are not representative of conventional cardiac pathologies, studies have shown that they are pivotal in understanding the multifactorial nature of cardiovascular diseases. A deeper analysis of various genetic models employed will be discussed in a later section.

3.3 Other Mammalian Models

3.3.1 Rats

Similar to mice, rats are also small mammals with similar cardiac anatomy and physiology to humans. Additionally, husbandry and fecundity are also comparable between rats and mice. Rats also possess similar anatomical

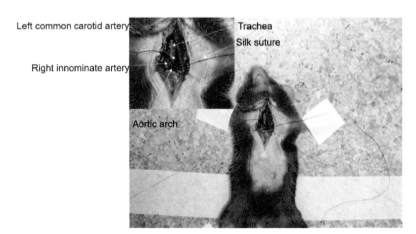

Figure 3.4 Surgical layout of rat undergoing transverse aortic constriction. A black silk suture is thread under the aortic arch [5].

features to humans [49]. Therefore, the rat is seen as a robust alternative model to the mouse. The major advantage that rats have over mice is their larger size, making rats more conducive for physical surgery. Indeed, numerous cardiac injury models in rats were developed via physical surgery, some of which are similar to those in mice. Pfeffer et al. described a rat LAD ligation model where the left coronary artery ligated between the pulmonary artery outflow tract and the left atrium [50]. The overall procedure, including anesthesia, ventilation, incision, and suturing, is similar to that in mice.

Aside from the LAD ligation models, researchers have also developed several cardiac pressure overload models to induce cardiac hypertrophy in rats. The most common pressure overload model is the ascending aortic constriction model. The ascending aortic constriction model was developed due to the easier access to the ascending aorta than the transverse aorta in rats. Ajith Kumar et al. detailed a procedure where constricting the ascending aorta leads to hypertrophy and subsequent heart failure [51]. After anesthesia and intubation of the rat, the second intercoastal space is opened to reveal the ascending aorta. The aorta is then constricted by approximately 50% of its original diameter by the application of a titanium surgical clip. After postoperative care, resulting hypertension caused by constriction can be measured by Doppler echocardiography. Thus, this model is useful in measuring the degree of constriction in relation to the amount of resultant cardiac injury. Another pressure overload model utilized is the abdominal aorta constriction model, where the abdominal aorta near the midsection of the abdomen is constricted [52]. The abdominal aorta constriction model is comparable to the ascending aorta constriction model in that the abdominal aorta is also fairly easy to access compared with the transverse aorta. Additionally, due to the surgical proximity to the renal system, the abdominal aorta constriction model could also be used as a kidney hypoperfusion disease model, which is linked to the pathologic development of hypertension [53].

Aside from physical cardiac injury models, other injury models induced through diet and breeding have been introduced. The Dahl salt-sensitive rat is a model that is derived via a high-salt diet. Lewis Dahl developed this model by selectively breeding rats after being fed a high salt diet (8% NaCl) [54]. After several generations of breeding, Dahl discovered a colony of rats that consistently exhibits high blood pressure after salt intake. Currently, Dahl salt-sensitive rats are used to understand the role of high sodium content on renal and cardiac injury such as left ventricle hypertrophy via hypertension and oxidative damage [55, 56]. Similarly, the spontaneous hypertensive rat, first established by Okamoto et al., was produced by selectively inbreeding rats based on high blood pressure, which could be used as an alternative for the hypertensive pathology model [57]. While cardiac hypertrophy has been documented in spontaneously hypertensive rats [58], the genetic basis for their development is not fully understood. Therefore, future studies should elaborate on the underlying genetic pathways that significantly differ between the wild-type rat models and the selectively inbred models.

Arrhythmogenic models were also introduced to provide a resource for understanding the etiology and pathology of arrhythmias. Lee et al.

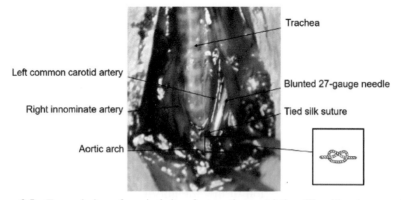

Figure 3.5   Zoomed view of surgical site of rat aortic constriction. The silk suture was tied around the aortic arch to induce constriction [5].

documented a method of inducing AV block in rats by injecting 5–10 μL of 70% ethanol into the atrioventricular junction, which is located just below a fat pad between the aortic root and medial wall of the right atrium [59]. AV block is verified by electrocardiogram (ECG) recording. Hemodynamic and histological data also indicated ventricular dysfunction and necrotic damage along the atrioventricular node.

While rats are generally larger than mice, making them more accessible for surgeries and subsequent observation and imaging, researchers did not have many options in terms of genetic engineering technologies for rats [60]. However, these technologies are steadily developing, such as the advent of zinc-finger nucleases and CRISPR-Cas9 editing in rats [61, 62]. Hopefully with the further advancement of genetic approaches for rats, more comprehensive studies could be formulated to analyze the pathology of various cardiac diseases in rats and to discern the underlying genetic pathways involved with the progression of these diseases.

3.3.2 Rabbits

Although rabbits do not match mice or rats in the ease of husbandry and high fecundity, rabbits can also be valuable as a comparable mammalian model to human CVD research. Rabbits serve as an intermediate mammalian model in terms of size, as rabbits are larger than mice and rats but smaller than dogs, pigs, and other mammalian models. Rabbits are also the largest mammalian model available that does not require central animal committee approval seen with dogs, pigs, and other large mammalian models [63]. Therefore, rabbits offer similar ease of maintenance as mice and rats while providing larger anatomical structures for experimentation.

A major application of rabbits in CVD research is the hyperlipidemic model. First established by Watanabe in 1973, the Watanabe heritable hyperlipidemic (WHHL) rabbit was selectively bred to exhibit hypercholesterolemia and abnormal accumulation of β-lipoprotein [64]. Shiomi et al. eventually continued to selectively breed WHHL rabbits to produce coronary atherosclerosis–prone WHHL strain, also known as the WHHLMI strain [65]. These are produced due to the high propensity in developing atherosclerotic lesions in coronary arteries, causing myocardial infarction symptoms seen both histologically and electrophysiologically [66]. Rabbits became a popular model in studying the prognosis of atherosclerosis due to the similarity in lipid composition and metabolism to humans, more so than in mice and rats [64].

Aside from hyperlipidemic studies, rabbits are also utilized in several cardiac injury models. Similar to mice and rats, rabbits have undergone analogous procedures to induce cardiovascular diseases such as LAD ligation [67], I/R surgery [68], aortic constriction [69], and doxorubicin treatment [70]. Additional procedures for rabbits also include electrical ablation of cardiac tissue. Arnolda et al. established an electrical ablation model where catheters were fed into the left ventricle through the right carotid artery to induce up to four 5 J shocks to the myocardial tissue [71]. Subsequent hemodynamic measurements revealed significant ventricular dysfunction in the cardiac system. Lorell et al. introduced a method of wrapping the left kidney with a cellophane dialysis membrane while surgically removing the right kidney to induce hypertension and ventricular hypertrophy in rabbits [72]. Tsuji et al. performed an AV block on rabbits by injecting 37% formaldehyde into the atrioventricular junction and then embedding a pacing electrode in the right ventricle and a pacemaker in the back of the rabbit [73]. This model allows for artificial ventricular pacing as well as observation of arrhythmic symptoms.

Overall, the rabbit is an intriguing model for analyzing the progression of CVDs. Rabbits are not as popular as mice and rats due to their higher cost of maintenance, but researchers have found that rabbits exhibit more similar physiology to humans than mice and rats do, particularly in lipid metabolism and calcium gradient behaviors throughout the myocardium [74]. Additionally, rabbits are larger in size than mice and rats, allowing for better visualization in surgery, imaging, and measurements. Indeed, more surgical techniques have been employed in rabbits that are currently not feasible in smaller mammals. Future studies could further integrate rabbits into genetic studies to improve genotype-phenotype juxtapositions.

3.3.3 Pigs

The pig is one of the largest mammalian models utilized in cardiovascular research. In general, large models require a great amount of husbandry, but the methodology in inducing, monitoring, and treating CVDs in these models is also more representative and possibly more translatable to human research. Indeed, numerous treatment methods have been devised and tested in pig models, including pharmacological administrations and stem cell injections [75]. Conducting experiments in pig models have also discerned possible limitations of other models that are less similar to humans. For example, Kaiser et al. documented a comparative study between mice and pigs in the

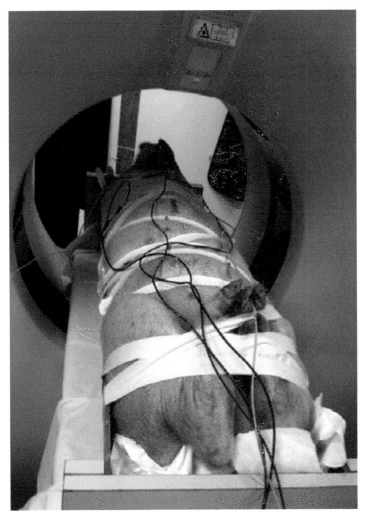

Figure 3.6 Pig undergoing computerized tomography scan after anesthesia to locate AV node [6].

role of p38 inhibition in reducing myocardial infarction injury, ultimately concluding that p38 inhibition was effective in mice but not in pigs [76]. LAD ligation has been previously performed in pigs, but the more common strategy implemented currently involves the use of balloon catheters that are fed into the coronary artery to administer a temporary I/R occlusion of the artery [77, 78]. Catheters are inserted through the femoral artery and guided to the coronary artery by ultrasound, where the balloon catheter is then inflated for at least 90 min [79]. Semenas et al. implemented a cardiac arrest model where controlled hemorrhage is induced by a catheter in the femoral artery, followed by five minutes of ventricular fibrillation by a 50 Hz transthoracic alternating current application [80]. Other physical injury models were also introduced due to the feasibility of implementing such techniques in a larger model. Couret et al. described a blunt chest trauma model where pneumatic chest injury is induced by applying piston drivers onto aluminum disks covering the second to fifth intercoastal spaces on the chest [81]. Approximately 70 J of energy is applied for every instance, and five instances of pneumatic force are usually applied on the chest to cause sufficient injury, indicated by reduced respiratory function and developing symptoms of heart failure. Dietary experiments have also been implemented in pig models, such as the study described by Nargesi et al. where pigs were fed a high cholesterol/high carbohydrate diet for 6 weeks before undergoing histological and echocardiographic measurement [82].

3.3.4 Dogs

Dogs provide an intriguing model for studying CVD pathology in large mammalian species. Dogs were chosen as a model of CVD research because of some similar qualities also seen in human cardiovascular systems. For example, the canine heart tends to possess collateralized coronary vessels, a characteristic that is commonly seen in adult human hearts [83]. This quality is not apparent in other mammalian models, such as pigs, displaying an advantage that the canine model has over other models in modeling myocardial infarction and other CVDs. Perhaps not surprisingly, a LAD ligation model has been developed in dogs, and it was specifically mentioned that all surrounding collateral vessels along the epicardial surface should be ligated as well [84]. LAD occlusion with polystyrene latex microspheres has also been performed in dogs as a closed chest alternative. The microspheres, approximately 100 μm in diameter, are injected into the coronary artery via catheters once per week for up to 9 weeks [85]. Cao et al. introduced a method of inducing cardiac arrest in dogs by ligating the LAD and ablating along the AV node [86]. An implantable cardioverter-defibrillator was implanted to monitor the electrophysiological signals following surgical injury. Similar to other models, ascending aortic constriction has been performed in dogs as a pressure overload model for cardiac hypertrophy. Koide et al. introduced a graded aortic constriction model by using a balloon catheter

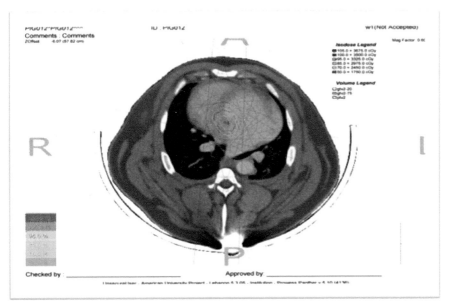

Figure 3.7   Representative CT scan locating AV node (shown in red circle), as well as estimated radiation exposure in peripheral areas (shown as rings of varied colors) [6].

[87]. After inserting the catheter through the carotid or femoral arteries, the balloon was gradually inflated over the course of 6 weeks through successive catheterization surgeries. Implantation of pacemakers was also utilized in dogs for arrhythmogenic studies. Nishijima et al. described a study where pacemakers were implanted with leads attached at the right ventricle apex to induce tachycardia over a period of 12 months [88]. Functional tests and left ventricular dysfunctions were analyzed after the experimental period. Canine models in research have emphasized the use of catheters and implantable devices to induce long-term CVD injuries, studies more conducive in large animal models.

3.3.5 Sheep

The sheep (*Ovis aries*) has been utilized as an alternative large mammalian model. The sheep tends to be regarded as a docile animal capable of long-term, repeated experimentation of cardiovascular structures [89]. Aside from possessing advantages found in other large models, some studies have also indicated that it might be easier to reproduce myocardial infarction and observe angiogenic properties in sheep [90]. Several studies have documented the use of LAD ligation in both fetal and adult sheep. Lock et al. conducted a comparative study of inducing LAD ligation injury between fetal and adult sheep to assess the regenerative properties after injury, including cardiomyocyte growth and scar formation [91]. Emmert et al. discussed a proof-of-concept study where fetal sheep heart LAD was ligated for approximately 20 min before receiving an intra-myocardial and intra-peritoneal injection of human mesenchymal stem cells [92]. This procedure was conducted due to the ability of the sheep immune system to adapt to human stem cell injections for cardiac recovery. Rienzo et al. introduced a coronary alcoholization model by injecting ethanol into the LAD to induce myocardial injury, citing that cardiogenic shock via ethanol could provide a closed-chest, reproducible method, although the overall ethanol injection took approximately 85 min per animal [90]. Killingsworth et al. developed a cardiac arrest model with a 150-minute occlusion of the LAD by a balloon catheter and atrioventricular node ablation, documenting arrhythmogenic symptoms after injury [93].

3.4 Zebrafish

Zebrafish (*Danio rerio*) is a freshwater fish with the intrinsic ability to regenerate certain parts of its body such as its heart, brain, retina, pancreas, spine, fins, and kidneys. This unique ability allows zebrafish to become a valuable vertebrate model to elucidate cardiovascular development. In addition to having this valuable intrinsic ability, zebrafish can epitomize human cardiac pathophysiology due to their similar pathways. Zebrafish as a cardiovascular model for human cardiac pathologies has given us knowledge surrounding the architecture of the mechanisms of regeneration and development of the heart up to the molecular level [94]. Notably, the zebrafish electrocardiogram signals or the cardiac action potential is comparable to the human action potential of the heart [95]. They both have a long plateau phase; however, there are still significant differences to consider. Throughout the last few decades, zebrafish have proved to be one of the main animal models that can provide significant advantages surrounding cardiovascular research and disease causality [94, 96].

To study the complexities of cardiovascular development and disease, the use of zebrafish as a complete in vivo model is logical [97]. Some of the main benefits surrounding zebrafish as a cardiac model are their small size, rapid embryonic development, the optical transparency of embryos, high genetic homology with humans, and the ability to have transgenic models, amongst many others [98]. They also provide many additional advantages over other mammalian models for the elucidation of vertebrate development and organogenesis [99]. In direct comparison to the mouse model, zebrafish can regenerate their heart indefinitely if the injury does not exceed >25% for area of injury, as opposed to intrinsic mouse cardiac regeneration only lasting around 7 days after birth [100].

Due to these advantages, there have been numerous zebrafish injury models that mimic human cardiac pathologies [101]. In this section, we will review the methodologies of these injury models including the mechanical resection model, targeted laser injury, cryoinjury, ventricular apical transection model, and genetic ablation model.

3.4.1 Mechanical resection model

One of the main and most common injury models is the mechanical resection model or the ventricular apical transection [102, 103]. The first studies of zebrafish heart regeneration utilized the mechanical resection model that was initially used on amphibians and later developed into the model we know today [103]. The main procedure of this cardiac injury model is to create an incision in the chest/ventral surface of the zebrafish to be able to expose the

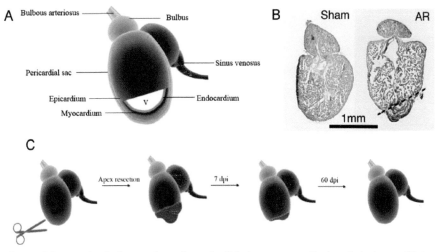

Figure 3.8 Mechanical resection of a zebrafish heart, steps displayed during a 60-day period. (a) Anatomical description of the zebrafish heart consisting of the bulbous arteriosus, bulbus, pericardial sac, sinus venosus, epicardium, endocardium, myocardium, atrium (A) and ventricle (V). (b) Histological analysis of resection. (c) Schematic displaying the levels of regeneration from 1 dpi to 60 dpi [3].

heart. Once the heart is exposed, the apex of the ventricle is transected with a pair of iridectomy scissors. With this injury method, many research groups were able to reproduce a transection of between 15 and 20% of the apex that was able to be fully regenerated within several weeks after the initial procedure [103–107].

After the mechanical injury, cardiomyocyte proliferation is elevated in the adult zebrafish, in addition, this model can show how zebrafish regenerate their hearts without having lasting scarring [96]. After 2 months post-transection or amputation a study done by Dickover et al. displayed that the ventricular apex was almost fully regenerated [96, 108]. This injury model has been able to be replicated in other animal models including mice due to its high reliability and reproducibility [34].

3.4.2 Targeted laser injury

Another type of cardiac injury model is the targeted laser injury. Targeted laser injury is a relatively novel cardiac model compared to the mechanical resection model. The main procedure of this injury model is to use a multipurpose Infrared Laser Ablator that was traditionally used for the injection of cells into the mouse blastocyst. Zebrafish embryos are used for this method and are anesthetized with 20 μM of Tricaine. After embryos are

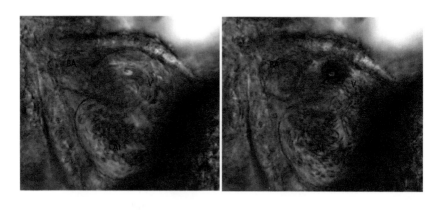

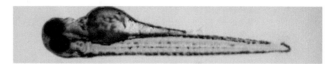

Figure 3.9 Zebrafish at 72 hpf, before and after targeted laser ablation [4].

successfully anesthetized, they are placed on a glass slide, which is placed under a microscope that has a built-in collimated red-light beam that allows the user to see the position of the laser beam. A target of injury is chosen and then an optimized pulse is delivered to this area. One of the main target areas used is the mid-cavity of the ventricle due to its facile and consistent injury on the embryos without damaging a larger portion of the cardiac tissues [109].

3.4.3 Cryoinjury

Cryoinjury or cryocauterization is another method that has been developed to mainly simulate myocardial infarction-like cardiac injuries. This method is mainly used in zebrafish but has been also used in fetal mice to study cardiac regeneration [110, 111]. The overall method of this technique is making slight contact with the exposed heart with a copper filament that has been dipped in liquid nitrogen.

This procedure first starts with anesthetizing approximately 6-month-old zebrafish with a 0.032% (wt/vol) of Tricaine solution. If the zebrafish is anesthetized for too long, this has been shown to affect the recovery and regeneration process of the injured tissue [112]. The cryoprobe is made from copper filament, which will be precooled by immersing it in liquid nitrogen for around 1 min to allow the filament to stabilize. Using iridectomy scissors, the ventral scales at the level of the heart are removed. This incision should be made superficially to prevent unnecessary injury to the internal organs. After creation of this incision, the cardiac sac should be exposed, and this sac should be torn to expose the heart. Application of the precooled cryoprobe is done over the ventricular surface. The probe should be kept in its initial contact location until the probe is visually thawed. This imposes necrotic death of the surrounding cardiomyocytes, resulting in fibrotic scar tissue that is commonly seen in myocardial infarctions. This scar tissue eventually gets reabsorbed by the body to allow new cardiomyocytes into the damaged myocardium. The cryoprobe is then removed, and zebrafish are placed in a freshwater tank [96]. When the zebrafish is in the tank, water is sprayed onto the gills with a pipette for several minutes until the fish has reanimated [110]. After cryoinjury, zebrafish hearts are dissected and analyzed for gene expression, electrocardiographs, and fluorescent staining.

Even though the aforementioned injury techniques are invasive, they can create a robust and consistent injury model. However, these injury models are extensive and incommodious and required mastery of the technique to achieve desirable results. Without mastery, the structural integrity of the

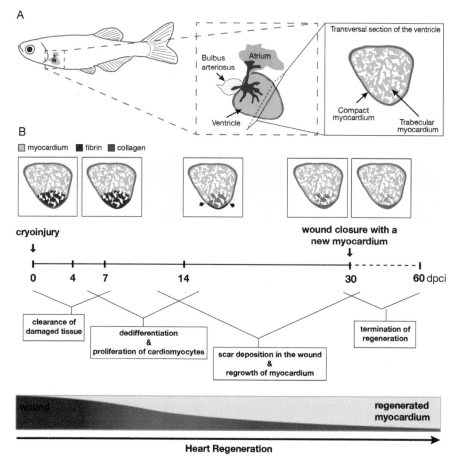

Figure 3.10   Schematic summary of the steps taken during the cryoinjury model and the regeneration that takes place after. (a) Schematic of anatomical target area of the heart of the zebrafish (b) Regeneration steps taken in the zebrafish heart from 1 dpci to 60 dpci [2].

heart could be destroyed along with potential damage to other organs and tissues.

3.4.4 Genetic ablation model

More recently, genetic models were developed to resolve some of the issues that were present with the aforementioned mechanical techniques. One of these genetic injury models is genetic ablation which can non-invasively injure the targeted area of interest. Due to the ease of genetic manipulation in

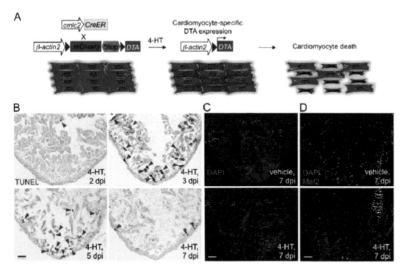

Figure 3.11 Genetic ablation of adult zebrafish cardiomyocytes. (a) Illustration of the specific transgenes used for zebrafish cardiomyocyte ablation. Histological examination of the zebrafish heart after vehicle or 4-HT injection from 2 dpi to 7 dpi with TUNEL staining (b), Myosin heavy chain (MHC) staining (c), and Mef2 staining to indicate cardiomyocyte nuclei (d) [7].

zebrafish, there have been several techniques developed and with time established genetic ablation models have been able to reach an injury of 60% of the cardiomyocytes in the zebrafish heart-inducing heart failure. After this severe injury using the genetic ablation model scientist were able to observe the efficient regeneration of the lost cardiomyocytes which in turn reversed the heart failure [113]. With genetic ablation, there have been numerous studies that have been done that include cardiomyocyte-specific ablations [113]. The main methodology used in this model is to obtain two or three transgenic zebrafish of interest to mate them and acquire the positive zebrafish. After the target gene is identified the use of a drug to suppress the expression of that target gene is applied. The regeneration is then slowly monitored, an assay of cardiomyocyte proliferation is analyzed, and the histological analysis is also performed.

3.4.5 Tissue clearing technique

In addition to these cardiac injury models in zebrafish there are tissue-clearing techniques that can be used for better visualization of the heart or other target areas. This is done through a series of procedures to convert the opaque and dense tissues transparent. There are numerous techniques that researchers have developed to clear tissue, including X-CLARITY, uDISCO, CUBIC, and PEGASOS. According to current research regarding tissue-clearing techniques, the newly developed PEGASOS method has been proven to combat some of the faults in current methods. It provides greater effectiveness in clearing, and nearly all types of tissues are able to be made transparent, including bone, teeth, and the more highly colorized organs [114].

The most recently published methodology for the PEGASOS method involves the preparation of decalcification, decolorization, delipidation, dehydration, and clearing solutions. A 20% (w/v) solution of EDTA is used for decalcification. The decolorization solution includes a 25% (*v/v*) solution of Quadrol and a 5% (*v/v*) solution of ammonium. For the delipidation solutions, they are composed in a gradient of 30% (*v/v*), 50% (*v/v*), and 70% (*v/v*) of tert-butanol (tB). The dehydrating solution is a mixture of 70% (*v/v*) tert-butanol, 27% (*v/v*) PEG methacrylate, and 3% Quadrol. The clearing solution is a recipe composed of benzyl benzoate, PEGMMA, and Quadrol (BB-PEG) [114].

Before the tissue is fully submerged in these solutions, it is preferred to perfuse and prepare the tissue accordingly. Transcardiac perfusion can be used on the animal and anesthetized prior. After perfusion, organs can be harvested, collected, and fixed with 4% PFA [102]. Another method is to fix the animal model directly, keeping organs intact using a transcardial infusion of the 4% PFA. In this fixed animal model skin is then removed to better improve the penetration of the solutions into the tissue. If needed, organs, such as the intestines and stomach, can be flushed out with a neural buffer solution such as PBS. For whole-mount clearing, the animal is transferred into a perfusion chamber containing all reagents within an incubator set at 37°C. Each reagent will have a different recirculation time which will be optimized for the tissue. On average the whole recirculation procedure takes about 2 weeks. Once clearing is performed with the BB-PEG (about 24 h), samples can be stored in the BB-PEG at room temperature [114].

If a perfusion system is not available, then passive immersion of the reagents can be used. This involves immersing the tissues in the reagents and shaking them for an optimized amount of time depending on the weight and thickness of the tissue samples. For any soft tissue samples, the decalcification step can be skipped to keep the integrity of the tissue [114].

This clearing process is pertinent to injury models because it expands the methods of imaging and observing the injury. Clearing as described above with the PEGASOS method is a great tool that can look at different animal models including zebrafish and be able to combine it with staining to observe different aspects such as genetic, or cellular processes that occur to be able to elucidate the cardiac regeneration pathways. Overall, the models mentioned in this section currently do not garner as much attention as mice due to the relative lack of versatility, especially in terms of genetic engineering and associated studies. While a great variety of procedures involving mice have been widely established and could be easily accessed by many labs, more effort should be placed on establishing other mammalian models to provide comparable specimens for histological, functional, and genetic research in CVDs.

3.5 Current Outlook of CVD Research with Injury Models

As discussed in the previous sections, a great spectrum of injury models has been developed to analyze the large number of pathologies associated with the cardiovascular system. Injury models tend to be specific in their applications in terms of the nature of the injury-induced, the duration of the injury (short-term or long-term), as well as the degree of similarity between humans and the animal model of choice. In this section, the current progress in researching prominent CVDs will be discussed regarding the available injury models in the literature.

3.5.1 Myocardial infarction

Fundamentally, myocardial infarction refers to the ischemic myocardial damage that occurs due to obstruction of blood flow in one or multiple coronary blood vessels. However, this simple definition of myocardial infarction belies the multifactorial nature of its pathology. The American Heart Association has published extensive guidelines for diagnosing myocardial infarction, which include formulating myocardial infarctions into multiple categories based on etiologies and outlining various symptoms from biomarker detection to ECG changes [115]. Correspondingly, researchers on myocardial infarction have employed multiple animal injury models to provide comprehensive knowledge and to develop potential treatment methods.

A great emphasis placed on current studies is to understand the genetic and molecular basis behind the pathology of myocardial infarction. An episode of myocardial infarction can induce several physiological responses, including myocardial cell death, immune response, and cardiac remodeling.

With the advent of more advanced genetic analysis techniques, especially in mice, more studies have shifted toward elucidating the genetic pathways that are potentially involved in the prognosis and treatment of myocardial infarction. For example, Pan et al. investigated the role of *miR-1* in cardiac injury by overexpressing and knocking down *miR-1* [116]. After inducing I/R injury in adult mice, they discovered that *miR-1* overexpression exacerbates injury by inhibiting cardioprotective genes *PKC-ε* and *HSP60* and upregulating apoptosis in cardiomyocytes. These results were reversed when knocking down *miR-1* expression. Kan et al. demonstrated that knocking out *IL-12p35* attenuates inflammation while increasing angiogenesis in mice after LAD ligation, resulting in improved cardiac function [117]. Rios-Navarro et al. conducted LAD occlusion and subsequent reperfusion in pigs to analyze the genes and proteins most pertinent to myocardial remodeling, such as metalloproteinases and connective tissue fibers [118]. They noted that coronary reperfusion can help reverse pathological remodeling.

In hopes of creating treatments or being able to predict myocardial infarctions, many animal injury models have been developed with this cardiac pathology in mind. One of the currently uprising animal models includes the zebrafish. As mentioned in this chapter, zebrafish elicit cardiac regeneration capabilities that are of current interest in the field of bioengineering and medicine. As explained in detail above, there are many zebrafish injury models that can be used to invoke similar human cardiac mechanisms during a myocardial infarction. Non-mammalian vertebrates capable of heart regeneration, such as urodele amphibians and teleost fish, reconstitute the myocardium with only moderate scarring in newts, or with little or no scarring in zebrafish [119, 120]. In contrast to mammals, adult newt and zebrafish cardiomyocytes can dedifferentiate and re-enter the cell cycle. However, the molecular and cellular mechanisms underlying heart regeneration in these model organisms are still very poorly understood. Comparative analysis between animals with different capacities to regenerate their heart will advance our understanding and allow the development of strategies to limit scarring and enhance myocyte restoration after heart injuries in humans. Inducing a myocardial infarction in zebrafish is still a very novel technique and studies still need to understand the mechanism in a developing heart.

While the underlying mechanisms of myocardial infarction are still being elucidated, some studies have devised potential treatment options for myocardial infarctions through demonstration in an animal injury model. One major option is the transplantation of cells (i.e. stem cells and progenitor cells) into the infarct regions. The goal is to allow the cells to integrate into the cardiac environment and replace infarcted tissue to restore cardiac function. For example, Rojas et al. demonstrated that implanted mice iPSC-derived cardiomyocytes attained mature cardiomyocyte features and improved cardiac morphology and function after mice LAD ligation [121]. Another major option is the use of currently available medications or chemical agents in the treatment of infarct symptoms in injury models, such as the use of rivaroxaban and Xin-Ke-Shu in the reduction of cardiac dysfunction and overload [122, 123]. In addition to the two aforementioned options, studies have also ventured into less-explored treatment methods. Holfeld et al. documented the use of shock-wave therapy on LAD-ligated pigs, stating that inducing shocks on the epicardial surface promoted angiogenesis and improved ejection fraction after injury [124]. Aghajanian et al. discussed that the induction of hypoxia reduces infarct size and promotes angiogenesis after mice LAD ligation [125].

While studies that elucidate the pathological mechanisms of myocardial infarction or introduce potential treatment methods provide legitimate evidence for their claims, future studies should conduct comparative studies between animal models in order to account for possible discrepancies. A particular issue regarding myocardial infarction models is the difference in coronary vasculature between animal models, which could result in disparate infarct locations and sizes [126]. Careful attention must be placed to understand the limitations of specific animal and injury models in order to properly translate research discoveries into future methods for diagnosing and treating myocardial infarctions.

3.5.2 Arrhythmias

In simplest terms, arrhythmias are defined by abnormalities in the cardiac conduction system. Arrhythmias are generally diagnosed by electrocardiograms (ECGs), which display the propagation of the conduction signals as they travel through the heart. While research has provided numerous insights into the cardiac conduction system, it is still not completely understood, and processes such as ion channel dysfunction, ion deregulation, conduction system damage, and cardiac remodeling have yet to be fully clarified. Similar to myocardial infarctions, arrhythmias are multifaceted, which necessitates the use of various injury models to discover arrhythmogenic mechanisms and treatment methods.

Arrhythmogenic models tend to be most common in larger animal models, such as dogs and pigs, due to the ability to pinpoint the nodal cells that comprise the cardiac conduction system. Studies have utilized ablation catheters to induce arrhythmogenic models by inserting catheter-based electrodes into a specific location on the conduction system, usually along the atrioventricular node near the septum of the heart. Larger animal models also provide easier means of monitoring arrhythmic symptoms through the implantation of pacemakers and implantable cardioverter-defibrillators [80, 86]. These models provide insight into the role of arrhythmic dysfunction on cardiac physiology, such as the presence of volume overload, as well as the etiology of other cardiovascular diseases including heart failure and cardiomyopathy [86, 127]. Dog models are generally outstanding in arrhythmic studies due to the discovery of certain dog breeds that are more susceptible to arrhythmias, such as Boxer dogs [128]. However, more recent studies have focused more on the genetic and molecular basis of arrhythmic diseases. As the mice model is most robust in genetic protocols, mice have been the model of choice in studying the effect of genetic mutations in the pathology of arrhythmia. Alvarado et al. generated RyR2-P1124L knock-in mice to investigate the role of ryanodine receptor 2 in modulating the susceptibility to ventricular arrhythmia [129]. Originally chosen due to its first discovery in cardiomyopathic patients, the RyR2-P1124L mutation was found to induce ventricular tachycardia due to abnormal increases in Ca^{2+} release, disrupting the physiological cardiac rhythm. Similar studies have identified specific ion channel mutations associated with arrhythmogenic diseases, such as the development of sodium channel mutations in the pathology of long QT syndrome [130]. While the methodology of identifying the prognosis of specific mutations has been established, future studies should continue to delineate the effects of mutations on phenotypic outcomes. For example, over 600 variants of the gene *KCNQ1* corresponding to a potassium channel have been identified, but the exact arrhythmic change has not been elucidated thus far [131]. As shown in other ion channel mutation studies, specific mutations on the same gene can produce distinct arrhythmic symptoms [132, 133].

Various arrhythmic treatment options have also been tested, including the use of drugs, cell transplantation, and radiotherapy. While pacemakers remain the most common method of treating arrhythmia, they also have significant limitations, such as electrode lead disconnections, pneumothorax, and potential immune reactions [134]. Thus, researchers have sought to uncover new potential treatment options. Perhaps the most pervasively tested option currently is the use of medication to treat arrhythmia, such as the use of sodium channel-blocking drugs in the treatment of long QT syndrome in mice [135]. While certain drugs such as beta-blockers and calcium channel blockers have seen successful clinical use, research has discovered that drugs can produce unintended side effects, including drug-induced arrhythmia [136]. Cell transplantation presents another area of research for arrhythmia treatment, as a biological replacement of dysfunctional tissue in the conduction system would provide the most natural means of recovery. Gorabi et al. described the use of Tbx18-inserted induced pluripotent stem cells in treating rats with complete heart block [137]. *Tbx18* is a gene specifically found in sinoatrial node pacemaker cells, the primary group of cells responsible for initiating cardiac conduction. After the cultivation of the stem cells expressing *Tbx18*, the stem cells are injected into the left ventricular wall of rat hearts. While results indicate electrophysiological and histological recovery near the injection site, more studies need to be completed to analyze the overall recovery of the heart. Radiotherapy is the last major option explored for arrhythmia treatment. While ablation using a transcatheter procedure has been developed previously, current testing of radiotherapy involves the use of stereotactic techniques to transmit radiofrequency ablation without surgical incisions. For example, Bode et al. documented a study of conducting radiosurgery on pulmonary veins in pigs as a potential treatment for atrial fibrillation, the most common arrhythmic disease [138]. While the radiosurgery had achieved the intended result of isolating the pulmonary veins for the attenuation of arrhythmias, the study also noted the high dosage required to induce significant results as well as the possible ablation of areas outside the intended target, such as the atrioventricular node. Thus, the positioning of the dose administration is most vital in improving radiotherapy, as it is critical to note that radiofrequency ablation, especially along the atrioventricular node, has also been utilized to induce arrhythmic symptoms [6]. Large animal models, such as pigs, sheep, and dogs, would be ideal for future radiotherapeutic studies as they would be more representative of the spatial accuracy needed to conduct radiosurgery in humans.

3.5.3 Cardiomyopathy

Similar to other CVDs, cardiomyopathy can be categorized into multiple forms based on the symptoms present for each specific case, but it is generally defined by a progressive enlargement and/or stiffening of the ventricles and subsequent reduction of cardiac function [139]. During cardiomyopathy, the myocardium of the heart become thickened and weakened in different areas leading to increased susceptibility to other CVDs, including heart failure. Due to the wide spectrum of cardiomyopathic pathologies, models that exhibit cardiomyopathy are just as diverse. The most common physically-induced cardiomyopathic models used in research are pressure overload models, such as those caused by aortic constrictions in various mammalian species. The increased pressure on the ventricles eventually leads to the presence of numerous symptoms associated with cardiomyopathy. Studies utilizing constriction models documented signs of hypertrophy, fibrosis, remodeling, and cardiac dysfunction [140–142].

Cardiomyopathies can also be derived from a hereditary nature, as dysfunction of a myriad of genes, from myocardial structure to ion channels, can contribute to the etiology [143]. Consequently, research on treating cardiomyopathies is focused on analyzing the genetic pathways associated with cardiomyopathy. For example, Le et al. used the aortic constriction on both wild-type and *Dec1*KO mice to determine the role of *Dec1* in the fibrotic development in cardiomyopathy [144]. *Dec1* was previously documented to upregulate inflammation and fibrosis, and their study showed that *Dec1*KO did lead to lower fibrosis and downregulation of downstream inflammatory genes such as *TGFβ* and *TNFα*. Song et al. used the method of aortic constriction on mice to demonstrate that downregulating *BRD4* can attenuate *TGFβ* induced fibrosis and remodeling after cardiomyopathy [142]. On a similar front, experiments have also been conducted to identify drugs and agents that could target relevant genes and pathways to reduce the detrimental effects of cardiomyopathy. Silva et al. demonstrated that alamandine administration to mice that underwent aortic constriction led to decreased hypertrophy, collagen deposition, and remodeling [145]. Ruan et al. tested the efficacy of QiShenYiQi dripping pills (QSYQ) on improving symptoms, ultimately reporting that QSYQ reduced fibrosis and remodeling, as well as upregulated angiogenesis for an improved prognosis after aortic constriction [146].

Even though there are significant anatomical differences between mammals and zebrafish, zebrafish hearts and their innate regeneration ability allow for many research groups to be able to assess the overall mechanisms behind many human cardiomyopathies. One of the main approaches to study these cardiac diseases in zebrafish is through either forward or reverse genetics [147]. There have been many studies that detail the discovery of similar mutations that cause cardiomyopathy in both zebrafish and humans [148, 149]. Therefore, zebrafish remains a popular model for studying underlying pathways for cardiomyopathy due to their established genetic relevance and their ease of maintenance.

While the understanding of molecular and genetic pathways associated with cardiomyopathy are constantly updated, and the potential treatments are subsequently devised, it is evident that more research and testing needs to be conducted to provide efficient treatments in the future. Current knowledge of the pathways encompassing all forms of cardiomyopathy is incomplete, and potential drug treatments are still being tested in small animal models. Future studies should expand the testing of these treatments to other models for a more comprehensive understanding.

3.5.4 Atherosclerosis

Atherosclerosis is described as a chronic disease that causes inflammation, myocardial infarction, and neurovascular episodes. Atherosclerosis has many causes and onsets including high blood pressure, hypertriglyceridemia, and genetic disorders. One of the main developments that are caused by this cardiac disease is plaques that build up in the sub-endothelial arterial wall. These plaques cause inflammatory responses in the heart, the stability of the plaques decreases over time and leads to ruptures which can trigger thromboses. Many animal models have been used to elucidate this disease and its pathomechanics in the human body. Certain genetic models, such as the *apoe(-/-)* and *ldlr(-/-)* mice models, and dietary models, such as the WHHL rabbit model, were developed to induce atherosclerotic-associated cardiac dysfunction due to the accumulation of fats and cholesterol pertinent to the etiology of atherosclerosis. Additionally, zebrafish have also been utilized as a model for their similar metabolic and immune responses [150].

The zebrafish models used to recreate a similar impact on the heart include genetic ablation models and certain dietary regiments that initiate the development of atherosclerotic lesions. This development will show the impact it has on the zebrafish cardiomyocytes and associated histological and

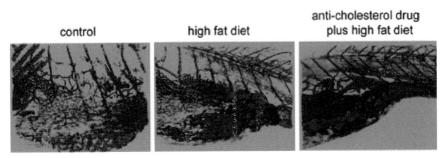

Figure 3.12 Zebrafish model image of cholesterol accumulation in the digestive system after high fat diet. Lipid accumulation often leads to atherosclerosis [1].

Table 3.1   List of common cardiac injury models described in literature

Animal model (species)	Injury/disease model name	Applications (i.e., for what disease)	Advantages	Disadvantages
Mouse	-LAD ligation and I/R -Cryosurgery -Apical resection -Aortic constriction -Chemical (e.g., DOX) -Genetic Models (*apoe, ldlr*)	Myocardial infarction, atherosclerosis, chemotoxicity	-Low maintenance -High fecundity -Well established among literature -High flexibility with genetics	-Small size -Inability to regenerate heart after neonatal period -Some discrepancies in mouse and human disease phenotypes
Rat	-LAD and I/R -Aortic constriction -Dietary (e.g. - Dahl sensitive, spontaneously hypersensitive) -AV block by ethanol	Myocardial infarction, hypertension, arrhythmia, atherosclerosis	-Low maintenance -High fecundity -Larger size (than mouse) -Fairly well established -Comparable anatomy structure to humans	-Genetic models not widely established -Small size compared with other models, limits
Rabbit	-LAD and I/R -Aortic constriction -WHHL (hyperlipidemic) -Electrical ablation (on LV) -Kidney removal -AV block by 37% formaldehyde	Myocardial infarction, renal disease, arrhythmia (AV block), cardiomyopathy	-Larger than popular models (mouse and rat) -Relatively low maintenance -Does not require central animal committee approval	-Not readily established in literature -Requires specialized training -Phylogenetically more distant to humans than mice and rats
Pig	-LAD and I/R -Hemorrhage and ventricular fibrillation -Blunt chest trauma -High cholesterol/carbohydrate	Myocardial infarction, cardiac arrest, physical injury, atherosclerotic hypertension	-Large animal model, comparable size to humans -Widely established large animal model -Longitudinal studies possible -Less ethical restrictions than companion animals, such as dogs	-High costs, maintenance -Studies at a long time scale -Requires special facilities, training

Table 3.1   (*Continued*).

Animal model (species)	Injury/disease model name	Applications (i.e., for what disease)	Advantages	Disadvantages
Dog	-LAD occlusion -AV ablation -Aortic constriction -Pacemaker implantation -Mutation models	Myocardial infarction, cardiac arrest, hypertension, arrhythmia	-Large animal model, providing more physical procedures -Presence of natural breeds with predisposed cardiac diseases -Similarities in cardiac structure to humans	-High maintenance -Many ethical restrictions -Long studies -Lack of direct genetic manipulation (despite being extensively used in genetic studies) -Requires special facilities, training
Sheep	-LAD occlusion -Cardiogenic shock	Myocardial infarction, cardiomyopathy	-Similar anatomical size to humans -Applicable to physical surgery similar to those in humans -Relatively low cost (for large animal model) -Less ethical obstacles than other large animal models	-High maintenance -Experiments requires long duration of time -Not widely established genetically
Zebrafish	-Mechanical resection (ventricular apical transection) -Targeted laser injury -Cryoinjury -Genetic ablation	Myocardial infarction, cardiomyopathy, atherosclerosis, congenital heart disease	-Innate regeneration abilities -High fecundity -Small size relative to other models -Transparent until 30 dpf -Transgenic options -Rapid development	-Small size make models very tedious -High maintenance -Requires special facilities, training

metabolic shifts. Which further allows study into the etiologies of cardio-vascular disease. One of the main benefits of using zebrafish is their inherit translucency until about 30 days post-fertilization (dpf). This allows for the observation of their cardiovascular system in real-time allowing for studies of development [151–153]. Similar to mammals, zebrafish process lipids in

Table 3.2   Current progress of CVD research via animal models

Disease category	Applicable injury models	Major discoveries	Limitations/future outlook
Myocardial Infarction	-LAD Ligation/Occlusion -I/R surgery -Cryosurgery -Electrical ablation (on ventricle) -Chemical ablation	- Elucidation of Potential Genetic Pathways pertinent to associated pathologic processes (i.e. cardioprotection, cell death, and cardiac remodeling -Development of novel potential treatment methods with models (i.e., stem cell therapy, drug targets, and shock therapy)	- Validation with multiple animal models needed - Full elucidation of cellular and molecular mechanisms of the heart needed - Continued testing of potential treatment methods
Arrhythmogenic Diseases	-AV Block -Pacemaker Intervention -Radioablation -Genetic models	- Determination of ion channels crucial to maintaining physiological heart rhythm - Use of gene therapy, radiosurgery	- Identifying variants of ion channels associated with arrhythmogenic symptoms - Continued testing of limitations of potential treatments
Cardiomyopathy	-Genetic ablation -Pressure Overload (i.e. aortic constriction)	- Determination of relevant genes and pathways crucial in cardiomyopathic pathology - Testing of potential drugs and agents to alleviate symptoms	- Elucidating pathways for different forms of cardiomyopathy -Developing relevant genetic models besides those in small animal models - Continued testing of drug treatments

the same way through their intestinal tract and can be used to study vascular lipid accumulations that lead to atherosclerosis [154].

3.6 Concluding Remarks

CVDs are a complex set of diseases, as these diseases can possess multiple etiologies and result in variable prognoses. Understandably, elucidation of the various pathologies is still ongoing, and the development of treatments has proven to be an arduous task, despite the amount of effort dedicated to cardiovascular research. Because of numerous ethical restrictions and practical limitations in studying CVDs in humans, animal models have been employed as an avenue to devise experiments in understanding and treating CVDs. While no animal model can serve as a perfect substitute for a human, each unique model possesses advantages and disadvantages for the purposes of research. Mammalian models tend to be most utilized due to their high degree of anatomical, physiological, and genetic similarity to those in humans. Indeed, the majority of animal cardiac injury models are derived from mammalian species. These mammalian models vary greatly in size, from small models such as mice to large models such as sheep. However, most mammalian models lack the ability to regenerate cardiac function after simulated cardiac injury, which necessitates the use of non-mammalian models. Zebrafish serves as a popular non-mammalian model for cardiovascular research due to their intrinsic ability in regenerating cardiac anatomy and physiology. With that said, the success of discovering new pathologies and treatments hinges on the use of multiple animal models in order to provide a comprehensive knowledge base on CVDs. Indeed, different animal models can be utilized for similar injury techniques, such as LAD ligation/occlusion, AV block, and genetic manipulation of homologous genes. Devising new injury models based on newfound knowledge would always be crucial in progressing CVD research, but future studies should also focus on validating results with multiple animal models to improve the success of implementing future treatments.

References

[1] Y. Miller, "A Zebrafish Model Of Atherosclerosis," United States Patent 9220243, 2015.

[2] T. Bise, P. Sallin, C. Pfefferli, and A. Jaźwińska, "Multiple cryoinjuries modulate the efficiency of zebrafish heart regeneration," *Scientific Reports,* vol. 10, p. 11551, 2020/07/14 2020.

[3] H. Juul Belling, W. Hofmeister, and D. C. Andersen, "A Systematic Exposition of Methods used for Quantification of Heart Regeneration after Apex Resection in Zebrafish," *Cells,* vol. 9, p. 548, 2020.

[4] G. Matrone, J. M. Taylor, K. S. Wilson, J. Baily, G. D. Love, J. M. Girkin, et al., "Laser-targeted ablation of the zebrafish embryonic ventricle: A novel model of cardiac injury and repair," *International Journal of Cardiology,* vol. 168, pp. 3913-3919, 2013.

[5] R. Au - Tavakoli, S. Au - Nemska, P. Au - Jamshidi, M. Au - Gassmann, and N. Au - Frossard, "Technique of Minimally Invasive Transverse Aortic Constriction in Mice for Induction of Left Ventricular Hypertrophy," *JoVE,* p. e56231, 2017/09/25/ 2017.

[6] M. M. Refaat, J. A. Ballout, P. Zakka, M. Hotait, K. A. A. Feghali, I. A. Gheida, et al., "Swine Atrioventricular Node Ablation Using Stereotactic Radiosurgery: Methods and In Vivo Feasibility Investigation for Catheter‐Free Ablation of Cardiac Arrhythmias," *Journal of the American Heart Association,* vol. 6, p. e007193, 2017.

[7] J. Wang, D. Panáková, K. Kikuchi, J. E. Holdway, M. Gemberling, J. S. Burris, et al., "The regenerative capacity of zebrafish reverses cardiac failure caused by genetic cardiomyocyte depletion," *Development,* vol. 138, pp. 3421-3430, 2011.

[8] S. S. Virani, A. Alonso, E. J. Benjamin, M. S. Bittencourt, C. W. Callaway, A. P. Carson, et al., "Heart Disease and Stroke Statistics—2020 Update: A Report From the American Heart Association," *Circulation,* vol. 141, pp. e139-e596, 2020.

[9] M. V. V. Kolk, D. Meyberg, T. Deuse, K. R. Tang-Quan, R. C. Robbins, H. Reichenspurner, et al., "LAD-ligation: a murine model of myocardial infarction," *Journal of visualized experiments : JoVE,* p. 1438, 2009.

[10] E. Gao, Y. H. Lei, X. Shang, Z. M. Huang, L. Zuo, M. Boucher, et al., "A novel and efficient model of coronary artery ligation and myocardial infarction in the mouse," *Circulation research,* vol. 107, pp. 1445-1453, 2010.

[11] N. Townsend, M. Nichols, P. Scarborough, and M. Rayner, "Cardiovascular disease in Europe 2015: epidemiological update," *Eur Heart J,* vol. 36, pp. 2673-4, Oct 21 2015.

[12] J. Perk, G. De Backer, H. Gohlke, I. Graham, Z. Reiner, M. Verschuren, et al., "European Guidelines on cardiovascular disease prevention in clinical practice (version 2012). The Fifth Joint Task Force of the European Society of Cardiology and Other Societies on Cardiovascular Disease Prevention in Clinical Practice (constituted by representatives of nine societies and by invited experts)," *Eur Heart J,* vol. 33, pp. 1635-701, Jul 2012.

[13] M. J. Pencina, A. M. Navar, D. Wojdyla, R. J. Sanchez, I. Khan, J. Elassal, et al., "Quantifying Importance of Major Risk Factors for Coronary Heart Disease," *Circulation,* vol. 139, pp. 1603-1611, 2019.

[14] J. Stewart, G. Manmathan, and P. Wilkinson, "Primary prevention of cardiovascular disease: A review of contemporary guidance and literature," *JRSM cardiovascular disease,* vol. 6, pp. 2048004016687211-2048004016687211, 2017.

[15] M. McClellan, N. Brown, R. M. Califf, and J. J. Warner, "Call to Action: Urgent Challenges in Cardiovascular Disease: A Presidential Advisory From the American Heart Association," *Circulation,* vol. 139, pp. e44-e54, 2019.

[16] F.-J. Lin, W.-K. Tseng, W.-H. Yin, H.-I. Yeh, J.-W. Chen, and C.-C. Wu, "Residual Risk Factors to Predict Major Adverse Cardiovascular Events in Atherosclerotic Cardiovascular Disease Patients with and without Diabetes Mellitus," *Scientific Reports,* vol. 7, p. 9179, 2017/08/23 2017.

[17] J. Rockberg, L. Jørgensen, B. Taylor, P. Sobocki, and G. Johansson, "Risk of mortality and recurrent cardiovascular events in patients with acute coronary syndromes on high intensity statin treatment," *Preventive Medicine Reports,* vol. 6, pp. 203-209, 2017/06/01/ 2017.

[18] Y. Hammer, Z. Iakobishvili, D. Hasdai, I. Goldenberg, N. Shlomo, M. Einhorn, et al., "Guideline‐Recommended Therapies and Clinical Outcomes According to the Risk for Recurrent Cardiovascular Events After an Acute Coronary Syndrome," *Journal of the American Heart Association,* vol. 7, p. e009885, 2018.

[19] H. Tomiyama and A. Yamashina, "Beta-Blockers in the Management of Hypertension and/or Chronic Kidney Disease," *International journal of hypertension,* vol. 2014, pp. 919256-919256, 2014.

[20] E. Pennisi, "A Mouse Chronology," *Science,* vol. 288, p. 248, 2000.

[21] A. Krishnan, R. Samtani, P. Dhanantwari, E. Lee, S. Yamada, K. Shiota, et al., "A detailed comparison of mouse and human cardiac development," *Pediatr Res,* vol. 76, pp. 500-7, Dec 2014.

[22] A. T. Chinwalla, L. L. Cook, K. D. Delehaunty, G. A. Fewell, L. A. Fulton, R. S. Fulton, et al., "Initial sequencing and comparative analysis of the mouse genome," *Nature,* vol. 420, pp. 520-562, 2002/12/01 2002.

[23] J. Au - Lugrin, R. Au - Parapanov, T. Au - Krueger, and L. Au - Liaudet, "Murine Myocardial Infarction Model using Permanent Ligation of Left Anterior Descending Coronary Artery," *JoVE,* p. e59591, 2019/08/16/ 2019.

[24] Z. Au - Xu, J. Au - Alloush, E. Au - Beck, and N. Au - Weisleder, "A Murine Model of Myocardial Ischemia-reperfusion Injury through Ligation of the Left Anterior Descending Artery," *JoVE,* p. e51329, 2014/04/10/ 2014.

[25] P. Cowled and R. Fitridge, "Pathophysiology of Reperfusion Injury," in *Mechanisms of Vascular Disease: A Reference Book for Vascular Specialists*, R. Fitridge and M. Thompson, Eds., ed Adelaide: University of Adelaide Press, 2011.

[26] P. Sicard, T. Jouitteau, T. Andrade-Martins, A. Massad, G. R. d. Araujo, H. David, et al., "Right coronary artery ligation in mice: a novel method to investigate right ventricular dysfunction and biventricular interaction," *American Journal of Physiology-Heart and Circulatory Physiology,* vol. 316, pp. H684-H692, 2019.

[27] K. Reichert, B. Colantuono, I. McCormack, F. Rodrigues, V. Pavlov, and M. R. Abid, "Murine Left Anterior Descending (LAD) Coronary Artery Ligation: An Improved and Simplified Model for Myocardial Infarction," *Journal of visualized experiments : JoVE,* p. 55353, 2017.

[28] J. N. B. M. d. Andrade, J. Tang, M. T. Hensley, A. Vandergriff, J. Cores, E. Henry, et al., "Rapid and Efficient Production of Coronary Artery Ligation and Myocardial Infarction in Mice Using Surgical Clips," *PLOS ONE,* vol. 10, p. e0143221, 2015.

[29] H. A. Rockman, R. S. Ross, A. N. Harris, K. U. Knowlton, M. E. Steinhelper, L. J. Field, et al., "Segregation of atrial-specific and inducible expression of an atrial natriuretic factor transgene in an in vivo murine model of cardiac hypertrophy," *Proceedings of the National Academy of Sciences,* vol. 88, p. 8277, 1991.

[30] R. D. Patten and M. R. Hall-Porter, "Small Animal Models of Heart Failure," *Circulation: Heart Failure,* vol. 2, pp. 138-144, 2009.

[31] B. D. Polizzotti, B. Ganapathy, B. J. Haubner, J. M. Penninger, and B. Kühn, "A cryoinjury model in neonatal mice for cardiac translational and regeneration research," *Nat Protoc,* vol. 11, pp. 542-52, Mar 2016.

[32] W. Roell, T. Lewalter, P. Sasse, Y. N. Tallini, B.-R. Choi, M. Breit-bach, et al., "Engraftment of connexin 43-expressing cells prevents post-infarct arrhythmia," *Nature,* vol. 450, pp. 819-824, 2007/12/01 2007.

[33] E. J. v. d. Bos, B. M. E. Mees, M. C. d. Waard, R. d. Crom, and D. J. Duncker, "A novel model of cryoinjury-induced myocardial infarction in the mouse: a comparison with coronary artery ligation," *American Journal of Physiology-Heart and Circulatory Physiology,* vol. 289, pp. H1291-H1300, 2005.

[34] E. R. Porrello, A. I. Mahmoud, E. Simpson, J. A. Hill, J. A. Richardson, E. N. Olson, et al., "Transient regenerative potential of the neonatal mouse heart," *Science (New York, N.Y.),* vol. 331, pp. 1078-1080, 2011.

[35] Ditte C. Andersen, S. Ganesalingam, Charlotte H. Jensen, and Søren P. Sheikh, "Do Neonatal Mouse Hearts Regenerate following Heart Apex Resection?," *Stem Cell Reports,* vol. 2, pp. 406-413, 2014/04/08/ 2014.

[36] D. M. Bryant, C. C. O'Meara, N. N. Ho, J. Gannon, L. Cai, and R. T. Lee, "A systematic analysis of neonatal mouse heart regeneration after apical resection," *Journal of molecular and cellular cardiology,* vol. 79, pp. 315-318, 2015.

[37] J. M. Baek, S. C. Kwak, J.-Y. Kim, S.-J. Ahn, H. Y. Jun, K.-H. Yoon, et al., "Evaluation of a novel technique for intraperitoneal injections in mice," *Lab Animal,* vol. 44, pp. 440-444, 2015/11/01 2015.

[38] X. Yi, R. Bekeredjian, N. J. DeFilippis, Z. Siddiquee, E. Fernandez, and R. V. Shohet, "Transcriptional analysis of doxorubicin-induced cardiotoxicity," *American Journal of Physiology-Heart and Circulatory Physiology,* vol. 290, pp. H1098-H1102, 2006.

[39] C. S. Abdullah, S. Alam, R. Aishwarya, S. Miriyala, M. A. N. Bhuiyan, M. Panchatcharam, et al., "Doxorubicin-induced cardiomyopathy associated with inhibition of autophagic degradation process and defects in mitochondrial respiration," *Scientific Reports,* vol. 9, p. 2002, 2019/02/14 2019.

[40] C. Yarana, D. Carroll, J. Chen, L. Chaiswing, Y. Zhao, T. Noel, et al., "Extracellular Vesicles Released by Cardiomyocytes in a Doxorubicin-Induced Cardiac Injury Mouse Model Contain Protein Biomarkers of Early Cardiac Injury," *Clinical Cancer Research,* vol. 24, p. 1644, 2018.

[41] V. W. Dolinsky, K. J. Rogan, M. M. Sung, B. N. Zordoky, M. J. Haykowsky, M. E. Young, et al., "Both aerobic exercise and resveratrol supplementation attenuate doxorubicin-induced cardiac

injury in mice," *American Journal of Physiology-Endocrinology and Metabolism,* vol. 305, pp. E243-E253, 2013.

[42] L. Xianchu, P. Z. Lan, L. Qiufang, L. Yi, R. Xiangcheng, H. Wenqi, et al., "Naringin protects against lipopolysaccharide-induced cardiac injury in mice," *Environmental Toxicology and Pharmacology,* vol. 48, pp. 1-6, 2016/12/01/ 2016.

[43] M. M. Abdel-Daim, H. A. Khalifa, A. I. Abushouk, M. A. Dkhil, and S. A. Al-Quraishy, "Diosmin Attenuates Methotrexate-Induced Hepatic, Renal, and Cardiac Injury: A Biochemical and Histopathological Study in Mice," *Oxidative Medicine and Cellular Longevity,* vol. 2017, p. 3281670, 2017/07/27 2017.

[44] C. B. Gurumurthy and K. C. K. Lloyd, "Generating mouse models for biomedical research: technological advances," *Disease models & mechanisms,* vol. 12, p. dmm029462, 2019.

[45] J. E. Sligh, Jr., C. M. Ballantyne, S. S. Rich, H. K. Hawkins, C. W. Smith, A. Bradley, et al., "Inflammatory and immune responses are impaired in mice deficient in intercellular adhesion molecule 1," *Proc Natl Acad Sci U S A,* vol. 90, pp. 8529-33, Sep 15 1993.

[46] S. Arber, J. J. Hunter, J. Ross, M. Hongo, G. Sansig, J. Borg, et al., "MLP-Deficient Mice Exhibit a Disruption of Cardiac Cytoarchitec-tural Organization, Dilated Cardiomyopathy, and Heart Failure," *Cell,* vol. 88, pp. 393-403, 1997/02/07/ 1997.

[47] S. H. Zhang, R. L. Reddick, J. A. Piedrahita, and N. Maeda, "Spon-taneous hypercholesterolemia and arterial lesions in mice lacking apolipoprotein E," *Science,* vol. 258, p. 468, 1992.

[48] S. Ishibashi, M. S. Brown, J. L. Goldstein, R. D. Gerard, R. E. Hammer, and J. Herz, "Hypercholesterolemia in low density lipoprotein receptor knockout mice and its reversal by adenovirus-mediated gene delivery," *The Journal of clinical investigation,* vol. 92, pp. 883-893, 1993.

[49] P. M. Iannaccone and H. J. Jacob, "Rats!," *Disease models & mecha-nisms,* vol. 2, pp. 206-210, May-Jun 2009.

[50] M. A. Pfeffer, J. M. Pfeffer, M. C. Fishbein, P. J. Fletcher, J. Spadaro, R. A. Kloner, et al., "Myocardial infarct size and ventricular function in rats," *Circ Res,* vol. 44, pp. 503-12, Apr 1979.

[51] A. K. Gs, B. Raj, K. S. Santhosh, G. Sanjay, and C. C. Kartha, "Ascending aortic constriction in rats for creation of pressure overload cardiac hypertrophy model," *Journal of visualized experiments : JoVE,* pp. e50983-e50983, 2014.

[52] H.-C. Ku, S.-Y. Lee, Y.-K. A. Wu, K.-C. Yang, and M.-J. Su, "A Model of Cardiac Remodeling Through Constriction of the Abdominal Aorta in Rats," *Journal of visualized experiments : JoVE,* p. 54818, 2016.

[53] B. Rodriguez-Iturbe, Y. Quiroz, C. H. Kim, and N. D. Vaziri, "Hypertension induced by aortic coarctation above the renal arteries is associated with immune cell infiltration of the kidneys," *Am J Hypertens,* vol. 18, pp. 1449-56, Nov 2005.

[54] J. P. Rapp and M. R. Garrett, "Will the real Dahl S rat please stand up?," *American Journal of Physiology-Renal Physiology,* vol. 317, pp. F1231-F1240, 2019.

[55] F. Salehpour, Z. Ghanian, C. Yang, N. N. Zheleznova, T. Kurth, R. K. Dash, et al., "Effects of p67phox on the mitochondrial oxidative state in the kidney of Dahl salt-sensitive rats: optical fluorescence 3-D cryoimaging," *American Journal of Physiology-Renal Physiology,* vol. 309, pp. F377-F382, 2015.

[56] D. Feng, C. Yang, A. M. Geurts, T. Kurth, M. Liang, J. Lazar, et al., "Increased expression of NAD(P)H oxidase subunit p67(phox) in the renal medulla contributes to excess oxidative stress and salt-sensitive hypertension," *Cell Metab,* vol. 15, pp. 201-8, Feb 8 2012.

[57] K. Okamoto and K. Aoki, "Development of a Strain of Spontaneously Hypertensive Rats," *Japanese Circulation Journal,* vol. 27, pp. 282-293, 1963.

[58] O. H. Bing, W. W. Brooks, K. G. Robinson, M. T. Slawsky, J. A. Hayes, S. E. Litwin, et al., "The spontaneously hypertensive rat as a model of the transition from compensated left ventricular hypertrophy to failure," *J Mol Cell Cardiol,* vol. 27, pp. 383-96, Jan 1995.

[59] R. J. Lee, R. E. Sievers, G. J. Gallinghouse, and P. C. Ursell, "Development of a model of complete heart block in rats," *Journal of Applied Physiology,* vol. 85, pp. 758-763, 1998.

[60] E. C. Bryda, "The Mighty Mouse: the impact of rodents on advances in biomedical research," *Mo Med,* vol. 110, pp. 207-11, May-Jun 2013.

[61] A. M. Geurts, G. J. Cost, S. Rémy, X. Cui, L. Tesson, C. Usal, et al., "Generation of gene-specific mutated rats using zinc-finger nucleases," *Methods Mol Biol,* vol. 597, pp. 211-25, 2010.

[62] Y. Shao, Y. Guan, L. Wang, Z. Qiu, M. Liu, Y. Chen, et al., "CRISPR/Cas-mediated genome editing in the rat via direct injection of one-cell embryos," *Nat Protoc,* vol. 9, pp. 2493-512, Oct 2014.

[63] M. Mapara, B. S. Thomas, and K. M. Bhat, "Rabbit as an animal model for experimental research," *Dental research journal,* vol. 9, pp. 111-118, 2012.

[64] M. Shiomi and T. Ito, "The Watanabe heritable hyperlipidemic (WHHL) rabbit, its characteristics and history of development: A tribute to the late Dr. Yoshio Watanabe," *Atherosclerosis,* vol. 207, pp. 1-7, 2009/11/01/ 2009.

[65] M. Shiomi, T. Ito, S. Yamada, S. Kawashima, and J. Fan, "Development of an Animal Model for Spontaneous Myocardial Infarction (WHHLMI Rabbit)," *Arteriosclerosis, Thrombosis, and Vascular Biology,* vol. 23, pp. 1239-1244, 2003.

[66] K. Hatanaka, T. Ito, M. Shiomi, A. Yamamoto, and Y. Watanabe, "Ischemic heart disease in the WHHL rabbit: A model for myocardial injury in genetically hyperlipidemic animals," *American Heart Journal,* vol. 113, pp. 280-288, 1987/02/01/ 1987.

[67] G. A. Ng, S. M. Cobbe, and G. L. Smith, "Non-uniform prolongation of intracellular Ca2+ transients recorded from the epicardial surface of isolated hearts from rabbits with heart failure," *Cardiovasc Res,* vol. 37, pp. 489-502, Feb 1998.

[68] M. Zeng, H. Yan, Y. Chen, H.-j. Zhao, Y. Lv, C. Liu, et al., "Suppression of NF-κB Reduces Myocardial No-Reflow," *PLOS ONE,* vol. 7, p. e47306, 2012.

[69] K. Mohammadi, P. Rouet-Benzineb, M. Laplace, and B. Crozatier, "Protein kinase C activity and expression in rabbit left ventricular hypertrophy," *J Mol Cell Cardiol,* vol. 29, pp. 1687-94, Jun 1997.

[70] R. B. Wanless, I. S. Anand, P. A. Poole-Wilson, and P. Harris, "An experimental model of chronic cardiac failure using adriamycin in the rabbit: central haemodynamics and regional blood flow," *Cardiovasc Res,* vol. 21, pp. 7-13, Jan 1987.

[71] L. Arnolda, B. P. McGrath, and C. I. Johnston, "Systemic and regional effects of vasopressin and angiotensin in acute left ventricular failure," *Am J Physiol,* vol. 260, pp. H499-506, Feb 1991.

[72] B. H. Lorell, W. N. Grice, and C. S. Apstein, "Influence of hypertension with minimal hypertrophy on diastolic function during demand ischemia," *Hypertension,* vol. 13, pp. 361-370, 1989.

[73] Y. Tsuji, T. Opthof, K. Yasui, Y. Inden, H. Takemura, N. Niwa, et al., "Ionic Mechanisms of Acquired QT Prolongation and Torsades de Pointes in Rabbits With Chronic Complete Atrioventricular Block," *Circulation,* vol. 106, pp. 2012-2018, 2002.

[74] S. M. Pogwizd and D. M. Bers, "Rabbit models of heart disease," *Drug discovery today. Disease models,* vol. 5, pp. 185-193, Fall 2008.

[75] A. Baehr, N. Klymiuk, and C. Kupatt, "Evaluating Novel Targets of Ischemia Reperfusion Injury in Pig Models," *International journal of molecular sciences,* vol. 20, p. 4749, 2019.

[76] R. A. Kaiser, J. M. Lyons, J. Y. Duffy, C. J. Wagner, K. M. McLean, T. P. O'Neill, et al., "Inhibition of p38 reduces myocardial infarction injury in the mouse but not pig after ischemia-reperfusion," *American Journal of Physiology-Heart and Circulatory Physiology,* vol. 289, pp. H2747-H2751, 2005.

[77] R. B. Arora and D. S. Sivappa, "Ectopic ventricular rhythms and myocardial infarction in the domestic pig and their response to nialamide, a monoamine oxidase inhibitor," *Br J Pharmacol Chemother,* vol. 19, pp. 394-404, Dec 1962.

[78] F. C. McCall, K. S. Telukuntla, V. Karantalis, V. Y. Suncion, A. W. Heldman, M. Mushtaq, et al., "Myocardial infarction and intramyocardial injection models in swine," *Nature protocols,* vol. 7, pp. 1479-1496, 2012.

[79] O. Bikou, S. Watanabe, R. J. Hajjar, and K. Ishikawa, "A Pig Model of Myocardial Infarction: Catheter-Based Approaches," in *Experimental Models of Cardiovascular Diseases*, K. Ishikawa, Ed., ed New York: Humana Press, 2018, pp. 281-294.

[80] E. Semenas, A. Nozari, and L. Wiklund, "Sex differences in cardiac injury after severe haemorrhage and ventricular fibrillation in pigs," *Resuscitation,* vol. 81, pp. 1718-1722, 2010/12/01/ 2010.

[81] D. Couret, S. de Bourmont, N. Prat, P. Y. Cordier, J. B. Soureau, D. Lambert, et al., "A pig model for blunt chest trauma: no pulmonary edema in the early phase," *Am J Emerg Med,* vol. 31, pp. 1220-5, Aug 2013.

[82] A. A. Nargesi, M. C. Farah, X.-Y. Zhu, L. Zhang, H. Tang, K. L. Jordan, et al., "Renovascular Hypertension Induces Myocardial Mitochondrial Damage, Contributing to Cardiac Injury and Dysfunction in Pigs With Metabolic Syndrome," *American Journal of Hypertension,* vol. 34, pp. 172-182, 2021.

[83] W. Flameng, F. Schwarz, and W. Schaper, "Coronary collaterals in the canine heart: development and functional significance," *American Heart Journal,* vol. 97, pp. 70-77, 1979/01/01/ 1979.

[84] J. Bartunek, J. D. Croissant, W. Wijns, S. Gofflot, A. d. Lavareille, M. Vanderheyden, et al., "Pretreatment of adult bone marrow mesenchymal stem cells with cardiomyogenic growth factors and repair of the chronically infarcted myocardium," *American Journal of Physiology-Heart and Circulatory Physiology,* vol. 292, pp. H1095-H1104, 2007.

[85] H. N. Sabbah, P. D. Stein, T. Kono, M. Gheorghiade, T. B. Levine, S. Jafri, et al., "A canine model of chronic heart failure produced by multiple sequential coronary microembolizations," *Am J Physiol,* vol. 260, pp. H1379-84, Apr 1991.

[86] J.-M. Cao, L. S. Chen, B. H. KenKnight, T. Ohara, M.-H. Lee, J. Tsai, et al., "Nerve Sprouting and Sudden Cardiac Death," *Circulation Research,* vol. 86, pp. 816-821, 2000.

[87] M. Koide, M. Nagatsu, M. R. Zile, M. Hamawaki, M. M. Swindle, G. Keech, et al., "Premorbid Determinants of Left Ventricular Dysfunction in a Novel Model of Gradually Induced Pressure Overload in the Adult Canine," *Circulation,* vol. 95, pp. 1601-1610, 1997.

[88] Y. Nishijima, D. S. Feldman, J. D. Bonagura, Y. Ozkanlar, P. J. Jenkins, V. A. Lacombe, et al., "Canine Nonischemic Left Ventricular Dysfunction: A Model of Chronic Human Cardiomyopathy," *Journal of Cardiac Failure,* vol. 11, pp. 638-644, 2005/10/01/ 2005.

[89] L. DiVincenti, Jr., R. Westcott, and C. Lee, "Sheep (Ovis aries) as a model for cardiovascular surgery and management before, during, and after cardiopulmonary bypass," *J Am Assoc Lab Anim Sci,* vol. 53, pp. 439-48, Sep 2014.

[90] M. Rienzo, J. Imbault, Y. El Boustani, A. Beurton, C. Carlos Sampedrano, P. Pasdois, et al., "A total closed chest sheep model of cardiogenic shock by percutaneous intracoronary ethanol injection," *Scientific Reports,* vol. 10, p. 12417, 2020/07/24 2020.

[91] M. C. Lock, J. R. T. Darby, J. Y. Soo, D. A. Brooks, S. R. Perumal, J. B. Selvanayagam, et al., "Differential Response to Injury in Fetal and Adolescent Sheep Hearts in the Immediate Post-myocardial Infarction Period," *Frontiers in physiology,* vol. 10, pp. 208-208, 2019.

[92] M. Y. Emmert, B. Weber, P. Wolint, T. Frauenfelder, S. M. Zeisberger, L. Behr, et al., "Intramyocardial Transplantation and Tracking of Human Mesenchymal Stem Cells in a Novel Intra-Uterine Pre-Immune Fetal Sheep Myocardial Infarction Model: A Proof of Concept Study," *PLOS ONE,* vol. 8, p. e57759, 2013.

[93] C. R. Killingsworth, G. P. Walcott, T. L. Gamblin, L. V. T. Girouard, D. Steven, W. M. Smith, and R. E. Ideker, "Chronic

Myocardial Infarction is a Substrate for Bradycardia-Induced Spontaneous Tachyarrhythmias and Sudden Death in Conscious Animals," *Journal of Cardiovascular Electrophysiology,* vol. 17, pp. 189-197, 2006.

[94] P. Giardoglou and D. Beis, "On Zebrafish Disease Models and Matters of the Heart," *Biomedicines,* vol. 7, p. 15, 2019.

[95] M. Vornanen and M. Hassinen, "Zebrafish heart as a model for human cardiac electrophysiology," *Channels (Austin, Tex.),* vol. 10, pp. 101-110, 2016.

[96] M. S. Dickover, R. Zhang, P. Han, and N. C. Chi, "Zebrafish cardiac injury and regeneration models: a noninvasive and invasive in vivo model of cardiac regeneration," *Methods Mol Biol,* vol. 1037, pp. 463-73, 2013.

[97] R. Ryan, B. R. Moyse, and R. J. Richardson, "Zebrafish cardiac regeneration-looking beyond cardiomyocytes to a complex microenvironment," *Histochem Cell Biol,* vol. 154, pp. 533-548, Nov 2020.

[98] T.-Y. Choi, T.-I. Choi, Y.-R. Lee, S.-K. Choe, and C.-H. Kim, "Zebrafish as an animal model for biomedical research," *Experimental & Molecular Medicine,* vol. 53, pp. 310-317, 2021/03/01 2021.

[99] D. Bournele and D. Beis, "Zebrafish models of cardiovascular disease," *Heart Fail Rev,* vol. 21, pp. 803-813, Nov 2016.

[100] B. J. Haubner, M. Adamowicz-Brice, S. Khadayate, V. Tiefenthaler, B. Metzler, T. Aitman, et al., "Complete cardiac regeneration in a mouse model of myocardial infarction," *Aging (Albany NY),* vol. 4, pp. 966-77, Dec 2012.

[101] X. Shi, R. Chen, Y. Zhang, J. Yun, K. Brand-Arzamendi, X. Liu, et al., "Zebrafish heart failure models: opportunities and challenges," *Amino Acids,* vol. 50, pp. 787-798, 2018/07/01 2018.

[102] J. O. Oberpriller and J. C. Oberpriller, "Response of the adult newt ventricle to injury," *J Exp Zool,* vol. 187, pp. 249-53, Feb 1974.

[103] K. D. Poss, L. G. Wilson, and M. T. Keating, "Heart Regeneration in Zebrafish," *Science,* vol. 298, pp. 2188-2190, 2002.

[104] A. Lepilina, A. N. Coon, K. Kikuchi, J. E. Holdway, R. W. Roberts, C. G. Burns, et al., "A dynamic epicardial injury response supports progenitor cell activity during zebrafish heart regeneration," *Cell,* vol. 127, pp. 607-19, Nov 3 2006.

[105] C. Jopling, E. Sleep, M. Raya, M. Martí, A. Raya, and J. C. I. Belmonte, "Zebrafish heart regeneration occurs by cardiomyocyte

dedifferentiation and proliferation," *Nature,* vol. 464, pp. 606-609, 2010/03/01 2010.

[106] K. Kikuchi, J. E. Holdway, A. A. Werdich, R. M. Anderson, Y. Fang, G. F. Egnaczyk, et al., "Primary contribution to zebrafish heart regeneration by gata4(+) cardiomyocytes," *Nature,* vol. 464, pp. 601-5, Mar 25 2010.

[107] C. L. Lien, M. Schebesta, S. Makino, G. J. Weber, and M. T. Keating, "Gene expression analysis of zebrafish heart regeneration," *PLoS Biol,* vol. 4, p. e260, Aug 2006.

[108] J. Cao and K. D. Poss, "The epicardium as a hub for heart regeneration," *Nature reviews. Cardiology,* vol. 15, pp. 631-647, 2018.

[109] P. Jagadeeswaran, M. Carrillo, U. P. Radhakrishnan, S. K. Rajpurohit, and S. Kim, "Chapter 9 - Laser-Induced Thrombosis in Zebrafish," in *Methods in Cell Biology.* vol. 101, H. W. Detrich, M. Westerfield, and L. I. Zon, Eds., ed: Academic Press, 2011, pp. 197-203.

[110] J. M. González-Rosa, V. Martín, M. Peralta, M. Torres, and N. Mercader, "Extensive scar formation and regression during heart regeneration after cryoinjury in zebrafish," *Development,* vol. 138, pp. 1663-1674, 2011.

[111] T. E. Robey and C. E. Murry, "Absence of regeneration in the MRL/MpJ mouse heart following infarction or cryoinjury," *Cardiovasc Pathol,* vol. 17, pp. 6-13, Jan-Feb 2008.

[112] D. G. Ellman, I. M. Slaiman, S. B. Mathiesen, K. S. Andersen, W. Hofmeister, E. A. Ober, et al., "Apex Resection in Zebrafish (Danio rerio) as a Model of Heart Regeneration: A Video-Assisted Guide," *Int J Mol Sci,* vol. 22, May 30 2021.

[113] F. Sun, A. R. Shoffner, and K. D. Poss, "A Genetic Cardiomyocyte Ablation Model for the Study of Heart Regeneration in Zebrafish," *Methods Mol Biol,* vol. 2158, pp. 71-80, 2021.

[114] D. Jing, S. Zhang, W. Luo, X. Gao, Y. Men, C. Ma, et al., "Tissue clearing of both hard and soft tissue organs with the PEGASOS method," *Cell Research,* vol. 28, pp. 803-818, 2018/08/01 2018.

[115] K. Thygesen, J. S. Alpert, and H. D. White, "Universal Definition of Myocardial Infarction," *Circulation,* vol. 116, pp. 2634-2653, 2007.

[116] Z. Pan, X. Sun, J. Ren, X. Li, X. Gao, C. Lu, et al., "miR-1 Exacerbates Cardiac Ischemia-Reperfusion Injury in Mouse Models," *PLOS ONE,* vol. 7, p. e50515, 2012.

[117] X. Kan, Y. Wu, Y. Ma, C. Zhang, P. Li, L. Wu, et al., "Deficiency of IL-12p35 improves cardiac repair after myocardial infarction

by promoting angiogenesis," *Cardiovascular Research,* vol. 109, pp. 249-259, 2016.

[118] C. Rios-Navarro, M. Ortega, V. Marcos-Garces, J. Gavara, E. de Dios, N. Perez-Sole, et al., "Interstitial changes after reperfused myocardial infarction in swine: morphometric and genetic analysis," *BMC Veterinary Research,* vol. 16, p. 262, 2020/07/29 2020.

[119] K. D. Poss, "Getting to the heart of regeneration in zebrafish," *Seminars in Cell & Developmental Biology,* vol. 18, pp. 36-45, 2007/02/01/ 2007.

[120] B. N. Singh, N. Koyano-Nakagawa, J. P. Garry, and C. V. Weaver, "Heart of Newt: A Recipe for Regeneration," *Journal of Cardiovascular Translational Research,* vol. 3, pp. 397-409, 2010/08/01 2010.

[121] S. V. Rojas, G. Kensah, A. Rotaermel, H. Baraki, I. Kutschka, R. Zweigerdt, et al., "Transplantation of purified iPSC-derived cardiomyocytes in myocardial infarction," *PLOS ONE,* vol. 12, p. e0173222, 2017.

[122] M. F. Bode, A. C. Auriemma, S. P. Grover, Y. Hisada, A. Rennie, W. D. Bode, et al., "The factor Xa inhibitor rivaroxaban reduces cardiac dysfunction in a mouse model of myocardial infarction," *Thrombosis Research,* vol. 167, pp. 128-134, 2018/07/01/ 2018.

[123] Y. Yang, H. Jia, M. Yu, C. Zhou, L. Sun, Y. Zhao, et al., "Chinese patent medicine Xin-Ke-Shu inhibits Ca2+ overload and dysfunction of fatty acid β-oxidation in rats with myocardial infarction induced by LAD ligation," *Journal of Chromatography B,* vol. 1079, pp. 85-94, 2018/03/15/ 2018.

[124] J. Holfeld, D. Zimpfer, K. Albrecht-Schgoer, A. Stojadinovic, P. Paulus, J. Dumfarth, et al., "Epicardial shock-wave therapy improves ventricular function in a porcine model of ischaemic heart disease," *Journal of Tissue Engineering and Regenerative Medicine,* vol. 10, pp. 1057-1064, 2016.

[125] A. Aghajanian, H. Zhang, B. K. Buckley, E. S. Wittchen, W. Y. Ma, and J. E. Faber, "Decreased inspired oxygen stimulates de novo formation of coronary collaterals in adult heart," *Journal of Molecular and Cellular Cardiology,* vol. 150, pp. 1-11, 2021/01/01/ 2021.

[126] C. Yen and P. C. Hsieh, "Pathology of permanent, LAD-ligation induced myocardial infarction differs across small (mice, rat) and large (pig) animal models," *Frontiers Bioengineering Conference Abstract: 10th World Biomaterials Congress,* 2016.

[127] A. Smoczynska, H. D. M. Beekman, R. W. Chui, S. Rajamani, and M. A. Vos, "Atrial dilation as a substrate for atrial fibrillation in the canine complete chronic atrioventricular block model," *European Heart Journal,* vol. 41, 2020.

[128] A. S. Vischer, D. J. Connolly, C. J. Coats, V. L. Fuentes, W. J. McKenna, S. Castelletti, et al., "Arrhythmogenic right ventricular cardiomyopathy in Boxer dogs: the diagnosis as a link to the human disease," *Acta myologica : myopathies and cardiomyopathies : official journal of the Mediterranean Society of Myology,* vol. 36, pp. 135-150, 2017.

[129] F. J. Alvarado, J. M. Bos, Z. Yuchi, C. R. Valdivia, J. J. Hernández, Y. T. Zhao, et al., "Cardiac hypertrophy and arrhythmia in mice induced by a mutation in ryanodine receptor 2," *JCI Insight,* vol. 5, Mar 5 2019.

[130] J. S. Lowe, D. M. Stroud, T. Yang, L. Hall, T. C. Atack, and D. M. Roden, "Increased late sodium current contributes to long QT-related arrhythmia susceptibility in female mice," *Cardiovascular Research,* vol. 95, pp. 300-307, 2012.

[131] H. Huang, G. Kuenze, J. A. Smith, K. C. Taylor, A. M. Duran, A. Hadziselimovic, et al., "Mechanisms of KCNQ1 channel dysfunction in long QT syndrome involving voltage sensor domain mutations," *Science Advances,* vol. 4, p. eaar2631, 2018.

[132] S. G. Priori and C. Napolitano, "53 - Inheritable Phenotypes Associated With Altered Intracellular Calcium Regulation," in *Cardiac Electrophysiology: From Cell to Bedside (Sixth Edition),* D. P. Zipes and J. Jalife, Eds., ed Philadelphia: W.B. Saunders, 2014, pp. 521-528.

[133] Y. T. Zhao, C. R. Valdivia, G. B. Gurrola, P. P. Powers, B. C. Willis, R. L. Moss, et al., "Arrhythmogenesis in a catecholaminergic polymorphic ventricular tachycardia mutation that depresses ryanodine receptor function," *Proc Natl Acad Sci U S A,* vol. 112, pp. E1669-77, Mar 31 2015.

[134] M. R. Carrión-Camacho, I. Marín-León, J. M. Molina-Doñoro, and J. R. González-López, "Safety of Permanent Pacemaker Implantation: A Prospective Study," *Journal of clinical medicine,* vol. 8, p. 35, 2019.

[135] J. A. Herrera, C. S. Ward, M. R. Pitcher, A. K. Percy, S. Skinner, W. E. Kaufmann, et al., "Treatment of cardiac arrhythmias in a mouse model of Rett syndrome with Na+-channel-blocking antiepileptic drugs," *Dis Model Mech,* vol. 8, pp. 363-71, Apr 2015.

[136] J. E. Tisdale, M. K. Chung, K. B. Campbell, M. Hammadah, J. A. Joglar, J. Leclerc, et al., "Drug-Induced Arrhythmias: A Scientific

Statement From the American Heart Association," *Circulation,* vol. 142, pp. e214-e233, 2020.

[137] A. M. Gorabi, S. Hajighasemi, V. Khori, M. Soleimani, M. Rajaei, S. Rabbani, et al., "Functional biological pacemaker generation by T-Box18 protein expression via stem cell and viral delivery approaches in a murine model of complete heart block," *Pharmacological Research,* vol. 141, pp. 443-450, 2019/03/01/ 2019.

[138] F. Bode, O. Blanck, M. Gebhard, P. Hunold, M. Grossherr, S. Brandt, et al., "Pulmonary vein isolation by radiosurgery: implications for non-invasive treatment of atrial fibrillation," *Europace,* vol. 17, pp. 1868-74, Dec 2015.

[139] R. K. Wexler, T. Elton, A. Pleister, and D. Feldman, "Cardiomyopathy: an overview," *American family physician,* vol. 79, pp. 778-784, 2009.

[140] H.-I. Lu, F.-Y. Lee, C. G. Wallace, P.-H. Sung, K.-H. Chen, J.-J. Sheu, et al., "SS31 therapy effectively protects the heart against transverse aortic constriction-induced hypertrophic cardiomyopathy damage," *American journal of translational research,* vol. 9, pp. 5220-5237, 2017.

[141] A. M. Nicks, S. H. Kesteven, M. Li, J. Wu, A. Y. Chan, N. Naqvi, et al., "Pressure overload by suprarenal aortic constriction in mice leads to left ventricular hypertrophy without c-Kit expression in cardiomy-ocytes," *Scientific Reports,* vol. 10, p. 15318, 2020/09/18 2020.

[142] S. Song, L. Liu, Y. Yu, R. Zhang, Y. Li, W. Cao, et al., "Inhibition of BRD4 attenuates transverse aortic constriction- and TGF-β-induced endothelial-mesenchymal transition and cardiac fibrosis," *Journal of Molecular and Cellular Cardiology,* vol. 127, pp. 83-96, 2019/02/01/ 2019.

[143] D. Jacoby and W. J. McKenna, "Genetics of inherited cardiomyopa-thy," *European heart journal,* vol. 33, pp. 296-304, 2012.

[144] H. T. Le, F. Sato, A. Kohsaka, U. K. Bhawal, T. Nakao, Y. Mura-gaki, et al., "Dec1 Deficiency Suppresses Cardiac Perivascular Fibrosis Induced by Transverse Aortic Constriction," *International Journal of Molecular Sciences,* vol. 20, p. 4967, 2019.

[145] M. M. Silva, F. P. d. Souza-Neto, I. C. G. d. Jesus, G. K. Gonçalves, M. d. C. Santuchi, B. d. L. Sanches, et al., "Alamandine improves cardiac remodeling induced by transverse aortic constriction in mice," *American Journal of Physiology-Heart and Circulatory Physiology,* vol. 320, pp. H352-H363, 2021.

[146] G. Ruan, H. Ren, C. Zhang, X. Zhu, C. Xu, and L. Wang, "Cardio-protective Effects of QiShenYiQi Dripping Pills on Transverse Aortic Constriction-Induced Heart Failure in Mice," *Frontiers in Physiology,* vol. 9, 2018-April-03 2018.

[147] A. Asimaki, S. Kapoor, E. Plovie, A. Karin Arndt, E. Adams, Z. Liu, et al., "Identification of a new modulator of the intercalated disc in a zebrafish model of arrhythmogenic cardiomyopathy," *Sci Transl Med,* vol. 6, p. 240ra74, Jun 11 2014.

[148] B. Gerull, M. Gramlich, J. Atherton, M. McNabb, K. Trombitás, S. Sasse-Klaassen, et al., "Mutations of TTN, encoding the giant muscle filament titin, cause familial dilated cardiomyopathy," *Nat Genet,* vol. 30, pp. 201-4, Feb 2002.

[149] X. Xu, S. E. Meiler, T. P. Zhong, M. Mohideen, D. A. Crossley, W. W. Burggren, et al., "Cardiomyopathy in zebrafish due to mutation in an alternatively spliced exon of titin," *Nature Genetics,* vol. 30, pp. 205-209, 2002/02/01 2002.

[150] C. Zaragoza, C. Gomez-Guerrero, J. L. Martin-Ventura, L. Blanco-Colio, B. Lavin, B. Mallavia, et al., "Animal models of cardiovascular diseases," *Journal of biomedicine & biotechnology,* vol. 2011, pp. 497841-497841, 2011.

[151] L. Fang and Y. I. Miller, "Emerging applications for zebrafish as a model organism to study oxidative mechanisms and their roles in inflammation and vascular accumulation of oxidized lipids," *Free radical biology & medicine,* vol. 53, pp. 1411-1420, 2012.

[152] R. N. Wilkinson and F. J. van Eeden, "The zebrafish as a model of vascular development and disease," *Prog Mol Biol Transl Sci,* vol. 124, pp. 93-122, 2014.

[153] L. Fang, C. Liu, and Y. I. Miller, "Zebrafish models of dyslipidemia: relevance to atherosclerosis and angiogenesis," *Translational research: the journal of laboratory and clinical medicine,* vol. 163, pp. 99-108, 2014.

[154] D. R. Brown, L. A. Samsa, L. Qian, and J. Liu, "Advances in the Study of Heart Development and Disease Using Zebrafish," *J Cardiovasc Dev Dis,* vol. 3, Jun 2016.

4

Phage-based Biosensors

Aarcha Shanmugha Mary[1], Vinodhini Krishnakumar[2], and Kaushik Rajaram[1]

[1]Department of Microbiology, Central University of Tamil Nadu,
Thiruvarur, Tamil Nadu, India
[2]Department of Lifescience, Central University of Tamil Nadu, Thiruvarur,
Tamil Nadu, India

Abstract

Bacteriophages, popularly known as phages, are parasitic to bacteria and archaea. Phages were used in developing phage therapy and cloning vectors for a long time. Recently, phages are used in biosensors and are considered as a wonderful probe molecule for detecting biomarkers and specific bacterial cells. The high throughput molecular technique termed phage display is being used as a tool for target-specific display and screening of peptides, proteins, and antibodies. Currently, phage display has proven to be an alternative to conventional antibody-based assays, and it allows conjugation or manipulation at the genetic level. Phage-displayed ligands are widely used as an affinity reagent in diagnostic biosensors. Phage-based ligands are easy to produce, less expensive, highly specific, and highly stable to different extreme conditions. To date, there are many different phage display-based sensors and assays that have been developed including optical sensors such as surface plasmon resonance (SPR), piezoelectric devices such as quartz crystal microbalance (QCM-D), and electrochemical sensors based on impedimetric, etc. In this chapter, we are focusing on different biosensing platforms where phage entities are used as ligands or receptors for the biosensing purpose, with an emphasis on the phage display technique.

Keywords: Phage probe, Phage display, bio-sensors, diagnosis, SPR, QCM

4.1 Introduction

In the present scenario of the biomedical sector, the need for research and the emergence of novel techniques and equipment for the rapid detection of microorganisms and other antigens is of great importance. There are several technologies that have been developed to detect a particular analyte which further helps in diagnosis, prognosis, and also therapeutic drug or vaccine development. Any substance or chemical of interest can be considered as an analyte when it can bind to a specific receptor. These bindings can be interpreted to determine either the presence or the amount of analyte, that is, they provide qualitative and quantitative analysis. The development of new technologies with more affinity toward binding can increase the performance and efficiency of the equipment. One latest approach toward this is the use of biosensors; biosensors can be developed using various approaches and are classified either based on the type of bio-recognition element/bio-receptor or based on the type of signal transduction. The biological element used for the detection of an analyte in a biosensor is mostly biologically derived materials, whereas certain biomimetic components are also utilized [1]. Biological interactions like antigen–antibody, substrate–enzyme, nucleic acid interactions, etc., are employed as bio-recognition components. Antigen–antibody interactions are highly specific and have a high binding affinity, henceforth used in immune sensors but are limited by their stringent assay conditions like pH and temperature. The high molecular weight of the antibody is also a major drawback which was resolved by the development of recombinant antibody fragments like Fab, Fc/scFv. Enzyme–substrate interactions are also utilized since they are highly specific and have many advantages due to their catalytic activity. The high specificity of the interaction often decreases the limit of detection of the biosensor, which in turn ensures high and efficient performance. The affinity of the binding can be increased by genetically modifying the bio-recognition element that is used. A higher affinity probe can be selected by the high throughput technique called phage display, in which cost-effective, stable affinity reagents are produced followed by affinity maturation in order to increase the sensitivity and specificity. This technique ensures in obtaining high-affinity receptors for a particular analyte.

4.1.1 Phage display

Phage display (PD) was introduced by Prof. George Smith in the early 1980s, and he was awarded with Nobel Prize for his efforts in the year 2018 [2, 3]. In PD, the gene sequence that encodes for the foreign peptide of interest will be introduced into one of the coat proteins of the phage, and the foreign peptide is displayed on the surface of the phage as a fusion product connected to the same coat protein. Hence, it provides a physical link between genotype and phenotype. Based on this approach, up to 10^{10} variants of phage displaying peptide/antibody library can be prepared. PD gets its importance in displaying the small molecular ligands as well as protecting them from getting denatured in unfavorable conditions. Phages that specifically infect bacterial cells are defined as bacteriophages [4]. Compared to other viruses, phage has a very simple structural and genetic organization that enables them to grow easily in the host bacterium. Phage contains a protein coat that covers most of its structure and carries either single-stranded or double-stranded DNA or RNA as genetic material in the core. Phage infects the bacteria via the F pilus and injects their genetic material into the bacteria, which utilizes bacterial machinery to produce phage protein coats and other structural proteins. It may produce at least 100–300 progeny phage particles within the bacterial system and it may go for the lytic cycle (which kills bacterial cells and releases the phage particles) or lysogenic cycle (which does not kill bacterial cells and releases the phage particles upon physical distribution of cell). M13 filamentous phage is the lysogenic phage-type that is most commonly used for phage display.

In order to understand the phage display completely, we should understand the M13 phage in detail. M13 phage has almost 7400 nucleotides resulting in a length of about 800–900 nm while the diameter is 6−7 nm and the molecular weight is estimated to be 16,300 kDa. M13 contains 11 genes, in which five are called coat protein genes and the remaining are structural genes. Gene VIII (pVIII) is called major coat protein which displays approximately 2700 copies and it spreads throughout the phage body. Each end has two coat proteins (pVI, pIII and pIX, and pVII) and can display up to five copies each. Although all coat proteins could be used for phage display, the pIII and pVIII are majorly used for the display of different peptides and proteins due to the presence of naturally occurring signaling peptides [5]. Other coat proteins could also be modified to display proteins and peptides by inserting a signaling peptide with the protein of interest. The combination of the signaling peptide and the protein of interest can be called as a fusion protein.

4.1.2 Vectors used in phage display

Mainly two different vectors are used in the phage display technique namely phage and phagemid. When a phage vector is genetically modified to display a protein of interest, it produces genetically and phenotypically homogeneous phage but this type of vector system could go for deletion mutation quite often and larger proteins may not be feasible to display [6, 7]. In the phagemid system, which is specially engineered for effective and stable phage display, phagemid could allow for larger protein to be displayed, and it would not go for deletion mutation; due to these features, phagemid is very commonly used as a phage display vector. Phagemid contains the origin of replication (Ori) of a phage that is a single-stranded DNA and Ori of double-stranded DNA as of plasmids. And it is a constrained one having only one or few coat protein genes essential for PD and always needs helper phage's superinfection in order to produce complete phage particles. Helper phage has a weaker packaging signal, which preferably packs the phagemid's fusion peptide but not the helper phage's peptide. Though the phagemid system can only produce a monovalent display system, the efficiency of the PD will be better than the phage vector-based display system.

The display level of a fusion protein relies upon the size of the protein. A larger protein could be displayed only as a single copy while the smaller ones can be displayed in multiple numbers [4]. Phagemids can be modified to display multiple proteins at the same time on two (dual display) or three (triple display) coat proteins [8]. This multifunctional/bifunctional phage can be a very good probe molecule in the development of diagnostic biosensors. The phage library of peptides/antibodies will have at least 10^{9-10} variants from which the highest affinity ones are selected. Customized phage libraries could be generated using the isolated cDNA from the targeted antigen present in tissues, tumors, and organs. Phage libraries are the key to selecting high-specific probes against target molecules. In order to select probes, a high throughput technique called bio-panning is used. The screened highly sensitive probes can be used as biomarkers or receptor molecules in biosensor platforms.

4.1.3 Bio-panning

Bio-panning is the key screening step for target-specific high-affinity phage or phage-derived probe molecules, the following are the steps involved in the process of bio-panning. First, the phage library containing at least 10^{9-10} phage variants, each carrying a unique peptide/protein/antibody on their protein coat is treated against immobilized target molecules. After 1–2 hours of incubation, unbound phages were washed out while the bound were eluted and amplified by infecting bacterial host (*Escherichia coli* for M13 phage) to

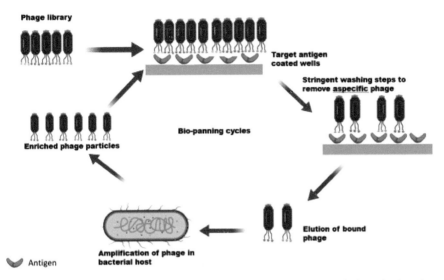

Figure 4.1 Biopanning: the selection of a high-affinity probe displayed phage is done by repeated cycles of stringent washing, followed by elution and amplification of the bound phages also called as biopanning.

prepare the enriched library for the second round of bio-panning. Likewise, several rounds of bio-panning will be carried out with different conditions and stringency [9]. The selected enriched clones were assessed using ELISA and sequenced to identify the encoded sequences. In vivo bio-panning is also performed in order to find highly specific targets (Figure 4.1). Cung and his colleagues in 2012 have performed the microfluidic-based controlled bio-panning, which was considered as the basis of automated and rapid screening methods for detecting affinity molecules [10]. The major setback in the phage display system is in the non-uniform amplification of phage libraries, in which the propagation of fast growers dominates the slow growers in the enrichment process while performing bio-panning [11]. It leads to a reduction in the number of clones which can eventually reduce the chances of getting high-affinity clones. In order to overcome the problem, Matochko et al. 2012 have developed phage libraries in the monodisperse emulsion which provides uniform amplification of each clone [12]. Affinity enhancement or maturation is an important step followed in phage display in order to provide high affinity or specificity to the selected clones especially when the antibody library is used [13]. From a wide range of library, a high-affinity peptide against the target can be identified and can be used in developing biosensors or with other similar techniques. For instance, Chen et al., from a library of random triplets in the CDR regions of camel nanobody consensus sequences, screened and obtained several candidate nanobodies that were further characterized to develop specific nanobodies for the detection of peanut allergen [14]. Various molecular techniques such as error-prone PCR, site-directed mutagenesis, randomization of amino acids, CDR walking, in silico guided mutagenesis, etc., are employed in order to provide affinity maturation [15].

4.1.4 Biosensors

Biosensors are analytical devices composed of a biological receptor that binds to an analyte, a transducer, and a system to process, amplify, and display the signals. The bio-receptor should be of high selectivity and the binding should produce a signal that is strong enough to be detected by the signal detector. The bio-receptor interactions are of several types including antigen/antibody, enzyme/ligand, nucleic acid/DNA, bio-mimetic materials, etc. For an immune sensor, the antigen–antibody specificity is utilized for the binding; the binding event generates signals like fluorescence and conductance. Another approach is using artificial binding proteins as bio-receptors in order to overcome the limitations of high molecular weights of antibodies and these proteins are selected by in vitro display techniques such as phage display, ribosome display, yeast display, or mRNA display [16]. Enzyme substrate interactions and nucleic acid interactions are also utilized for the development of bio-recognition elements in biosensors. Gunay et al. 2015 described the display of peptides that can specifically bind to natural, synthetic, and organic polymers also referred to as soft matter [17].

Biosensors can also be classified on the basis of the type of transducer used that includes electrochemical [18], optical [19], electronic [20], piezo-electric [21], gravimetric [22], and pyroelectric [23]. The field of biosensors relies on the discovery and development of highly specific and efficient molecular probes such as small molecules, peptides, proteins, antibodies, aptamers, etc. The bioreceptor molecules mostly determine the transducing platforms. One of the important, efficient, rapid, economical, and powerful approaches to developing these probes is phage display. The efficiency of a biosensor is evaluated by the minimum amount of sample that it could detect and give an appropriate signal. This least detectable amount of sample is referred to as the limit of detection (LOD). The lesser LOD of samples denotes higher efficiency and sensitivity of the sensor.

4.1.5 Phage in biosensors

Phage shows high stability against various extreme conditions such as pH variations, low to high temperatures, enzymes, and even surfactant treatments they do not tend to lose their bacterial infectivity. These properties make the phage as a very good receptor molecule in the sensing platforms. Phage ELISA is the mostly used affinity test for phage-displayed peptides and antibodies. Bringing phage into the sensing platform is more difficult to achieve than any other probe molecules since phage needs a specific strategy to be immobilized on sensing platforms. The application of phage in the sensing platform is emerging due to the development of new strategies and advancements in immobilization techniques [24]. Moreover, the phage-displayed peptides are more stable and have the precise peptide/protein folding when compared to the proteins that are synthesized chemically. So, phage-displayed peptides and antibody fragments are considered as an alternative to monoclonal antibodies, and it allows the manipulation at the molecular level. The phage-based sensing assays were first reported by [25]. In sensing platforms, both lytic and non-lytic phages were used; especially lytic phage was used in the selective killing of bacterial pathogens and releasing intracellular material into the medium. However, the non-lytic filamentous phage is majorly used in displaying a variety of target-specific peptides, antibodies, and proteins and is used as an affinity reagent in therapeutics and in diagnostic biosensors [26]. Since phage could be mass cultivated in bacterial cells, it is very cost-effective to produce compared with other types of probe biomolecules such as antibodies.

4.1.6 Phage-nanoparticle complex in biosensors

Enhancing the performance of phage display-based biosensor is very crucial and certain complex formation of phage with nanostructures has helped in developing advanced sensing platforms. Nanoparticles are mostly used for the immobilization of phage, thereby giving a geometric control but also are used for other purposes like providing charge or magnetic properties which will allow electrochemical detection (Figure 4.2). Microspheres of silica and phage are utilized for sandwich immune assays, and also film nanostructures embedded with phages can be used for the development of colorimetric sensors [11]. Physical and optical properties of a nanoparticle along with the biological properties of a bacteriophage together ensure multifunctional structures with higher potential. These structures have proved to be more stable and help in the higher performance of the biosensor [27].

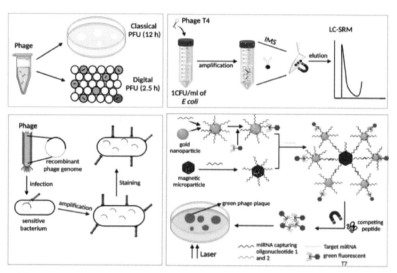

Figure 4.2   Classical phage selection and development of a biosensor. A. Phage is selected for a particular host. B. The selected phage is amplified and eluted using magnetic separation. C. The phage is modified to express the target gene also called as recombinant. D. Development of the fluorescence-based biosensor (adapted from [26].

Yun et al., 2009 explained the development of nanomaterial-based biosensors where the recognition bio-receptor is developed based on nanobiology and is referred to as biomaterial/nanomaterial I, and the second level of nanomaterial used for the binding amplification and signal generation is referred to as the smart material/nanomaterial II [28]. In 2015, Tawil et al. reported the detection of MRSA strains from a heterogenous sample using a phage-nanoparticle (Au) complex; the formation of the complex does not affect the infectivity of the bacteriophage as well as no sample pre-treatment or amplification was required, the detection was solely carried out using dark field microscopy [29]. The technical advancements of phage-nanoparticle composite led Bhardwaj et al., 2017 to develop for the first time a more sensitive detection method where the bacteriophage was interfaced with a metal-organic framework (MOF) and the binding was read by the photoluminescence quenching phenomena with a LOD of 31 CFU/mL [30].

4.1.7 Phage displayed antibodies

McCafferty et al., 1990 first described the display of antibodies on phage surfaces [31]. There are several formats of antibody display including scFv, Fabs,

Fvs, and diabody [32, 33]. Combinatorial phage display antibody libraries were constructed by several groups by screening, selection, and in vitro maturation of antibody fragments [34]. The antibody is selected from a library by phage display, and the most successful selections are those carried out using pure antigens and also minimizing the background binding [33], antibody display-based detectors are suitable for automation and thereby can be used for high throughput screening. Huang et al., 2019 described the expression of an anti-idiotypic nanobody on X27; this acts as a competitive binder to the citrinin mycotoxin and is readily detected using a real-time immuno PCR; this competitive binding assay eliminates the cross-reactions by other mycotoxins [35]. Li et al., in 2016 demonstrated the generation of the anti-testosterone nanobody which was used to detect the presence of testosterone in serum samples using an EIS immunosensor [36]. Kuppevelt et al. generated anti-heparan sulfate antibodies through phage display technology and identified three different types of antibodies that were used to detect heparan sulfate and heparan [37].

4.2 Strategies of Phage Immobilization

The immobilization directly affects the binding of the bacteria or analyte to the phage and influences the efficacy and performance of the biosensor. The immobilization can be random or oriented, out of which oriented phage immobilization ensures maximum binding and therefore orientation of phage during the immobilization process attains much importance. Properly oriented phage can enhance capturing efficiency and can decrease the detection limits (LOD) (Figure 4.3). There are several immobilization techniques and the selection of the immobilization technique depends on the nature of the bioactive material. The phage immobilized on a surface has a certain threshold above which there will be steric hindrances that will affect the binding efficiency.

4.2.1 Physical adsorption

Physical adsorption is the most commonly used immobilization approach. This method is simpler and more economical. Physical adsorption is also known as physisorption, and it generally forms a weak van der Waals force or hydrogen bond or electrostatic interaction which allows the adsorption of certain targets to a surface or substrate [38]. Physical adsorption of bacteriophage on gold surface of quartz crystal microbalance enabled the detection

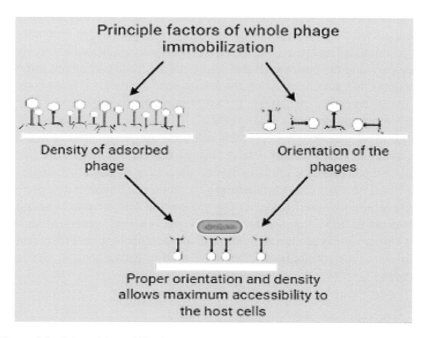

Figure 4.3 Oriented immobilization versus random immobilization: Principal factors of whole phage immobilization. For improving the efficiency of the biosensor, the interaction of the phage bound to the surface should be maximized and proper orientation and density of phage selection ensures maximum interaction thus enhancing the performance.

of galactosidase from *E. coli*. Apart from immobilizing on surfaces, bioprinting on paper can be used which has various advantages like flexibility, low cost, easy fabrication, and recyclable properties. Bacteriophages can be immobilized on silica surfaces via electrostatically facilitated physical adsorption. Similarly, positively charged cellulose membranes as well as glass substrates can be used for the physically adsorbed immobilization of phages [39]. The greatest drawback of physical adsorption is that changes in conditions like pH, shear, ionic concentration, temperature, etc., will result in possible detachment of the phage bound to the surface. Covalent bonding provides a much stronger attachment of phage to the surface [40].

4.2.2 Chemical adsorption

Adsorption can be performed utilizing the affinity of certain substances to another. For instance, the specific affinity of biotin to streptavidin can be exploited for phage immobilization and it is useful for the later binding stages.

During the process of biotinylation, the biotin molecule is covalently attached to the bacteriophage capsid. Otherwise, the phage can be modified to display biotin or biotin-like peptides (Streptavidin binding peptides) on one of its coat proteins [41]. Biotin has a high and specific affinity toward streptavidin. Streptavidin can be bound to the surface of the sensor, so that when the biotin binds to the streptavidin-bound surfaces, the phage can be immobilized on the surface (Figure 4.4). Similarly, other affinity compounds can be used for immobilization. Gold substrate functionalized by sugars and amino acids can also aid in the immobilization of phage onto the sensor surface by means of chemical bonding.

4.2.3 Electric deposition

In presence of an external electric field, the negatively charged phage exhibits certain electric dipole properties and can anchor to the sensor surface. The presence of ionic charges on the sensor surface will promote deposition and

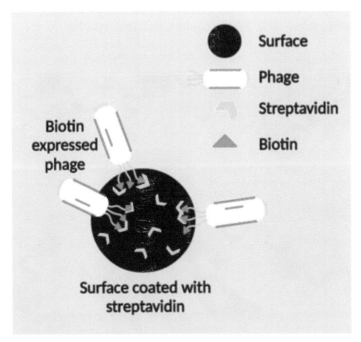

Figure 4.4 Biotin–streptavidin affinity-mediated immobilization of phage. The biotin expressed on the phage surface binds with streptavidin due to its high affinity.

proper orientation of phage [42]. Several materials like silica and cellulose have been modified to surfaces with several ranges of charges, and much higher phage coverage was obtained for modified membranes when compared to unmodified surfaces. Burnham et al., 2014 determined the charge of a mutant species of T4 bacteriophage consisting of only tail and tail fibers as -1.80 ± 0.19 mV whereas the charge of the same with head and tail fibers as -5.31 ± 0.67 [43]. Zhou et al., 2019 modified carbon nanotubes with positively charged polyethyleneimine and an external potential that allowed the deposition of phage via its head to the sensor surface [44].

4.2.4 Phage layer by layer organization

Phage organization can be modified by an alternative layering of phage with polyelectrolytes. The bio-receptor can be bound to a surface that has been functionalized using charged compounds like polylysine, epoxysilane, etc., and once it is bound to the surface it can be fixed by building up a multilayer of polyelectrolytes purely based on the electrostatic interactions (Figure 4.5). The layer-by-layer formation begins with a self-assembled monolayer (SAM) which is spontaneous. The SAM has one end that attaches to the sensor

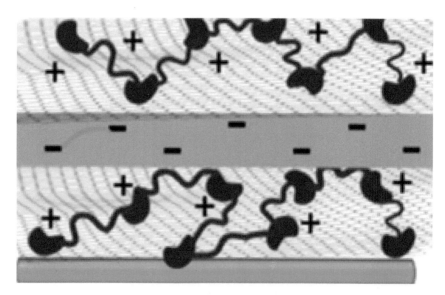

Figure 4.5 Layer by layer organization of chitosan (positively charged) and PSS (polystyrene sulfonate-negatively charged) over the gold electrode.

surface and the other end has functional groups which can interact with other compounds can form a second layer. If the integrity of each monolayer is maintained, then various compounds with opposite charges can be used to achieve layer-by-layer organization [45].

4.3 Types of Phage–bacteria Association and Signal Generation

4.3.1 Reporter phages (RP)

RPs are genetically engineered phages carrying a specific reporter gene that codes for fluorescence, optical markers, or colorimetric components and upon infecting bacterial cells, the expression of these genes helps in the visual identification of specific pathogens [46]. Since phage needs host bacterial cells to replicate and express the reporter gene, the expression of the gene is considered as an indication of the presence of bacteria (Figure 4.6). Phage

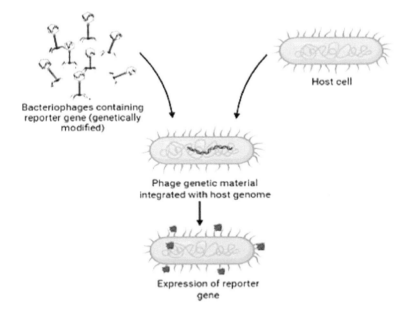

Bacteriophages containing reporter gene (genetically modified)

Host cell

Phage genetic material integrated with host genome

Expression of reporter gene

Figure 4.6 Preparation of reporter phage, detection of a host cell can be achieved by incorporating reporter genes using a recombinant phage.

engineered with reporter genes like inaW, lacZ, celB, and green fluorescent protein (GFP) is mostly applied in the biosensing of pathogens [47]. RPs were very much used in identifying foodborne pathogens, detection of specific strains of *E. coli*, etc. Detection of the signals produced by the reporter gene is done by an electron multiplying charge-coupled device or complementary metal oxide semiconductor imaging technologies which replaced the conventional photomultiplier tube because of the advantages of low read noise and more fidelity. Piuri et al., 2009 incorporated GFP in TM4 phage for the detection of *Mycobacterium* at a count of less than 100 cells [48].

4.3.2 Receptor binding phage

In the biosensing platforms, phage should attach to the target bacterial cells efficiently using host surface receptors such as carbohydrates and proteins by their receptor binding proteins (RBP). This binding initiates the translocation of phage genetic materials into the host cell [11]. Phage RBP could be genetically modified in order to obtain a higher binding affinity or to produce multivalency. Genetically modified RBPs show better stability against stringent conditions such as pH, temperature, and also resistance against certain proteases. Engineered RBPs also allow the addition of appropriate tags without altering their binding affinity against the host bacterial cells [49]. Singh et al., 2010 developed engineered tail-spike proteins of P22 bacteriophage with N-terminal and C-terminal cysteine tags and proved that N-terminal cysteine tags have enhanced bacterial capturing density and were used for highly sensitive and selective detection of *Salmonella enterica* [50].

4.3.3 Stained phage

Biosensors can be developed using phage that are stained with dyes; especially fluorescent dyes. These dyes can discriminate against the target bacterium when they attack and infect the host cells. Several dyes like DAPI (4,6-diamidino-2-phenylindole), biarsenical dyes, and fluorescent dyes like SYBR are used to stain the phage. Detection of *E. coli* in the limit of 20 CFU/mL in water was determined by Yim et al., 2009 by labeling bacteriophages with fluorescent quantum dots within an assay time of 1 hour [51]. Wu et al., 2011 elaborated a sensitive and rapid method in which phage where genetically modified with tetra-cysteine tags and was allowed to amplify in the host cell, followed by staining using biarsenical dye, and the results were obtained through flow cytometry for quantitative and qualitative analysis [52].

4.3.4 Capturing phage

Phage possesses functionally active groups like hydroxyl, aldehyde, and carboxyl groups on their surface which permits them to interact with bacterial surface molecules. For instance, phage 22 which specifically infects *Salmonella* has an amine group in capsid proteins that can covalently bind to other particles. Phage can be immobilized on a solid surface and the binding causes changes like delay in impedance, which can be detected. Streptavidin-coated magnetic beads chemically immobilized with wild-type T7 phage captured 86.2% *E. coli* in 20 minutes [53]. Phage heads are negatively charged, and the tail possess a net positive charge. The immobilizing matrices can be chemically modified to have a positive charge so that the negatively charged phage head will bind to the matrix leaving free tails that can capture the bacteria by specifically binding to the receptors on the host.

4.3.5 Lytic phage

4.3.5.1 Lysed cell contents detection

Phage that is lytic in nature, after infection and replication, causes the lysing of the host cells to release the progenies. The cellular materials are also leaked as a result of the lysis of the cell. They are several components (e.g., ATP) that can be detected, which confirms the presence of bacteria that the phage can specifically infect. T7 phage which infects *E. coli* caused the lysis and release of intracellular components of which β-galactosidase was detected to confirm the presence of the host. β-Galactosidase detection is usually done by colorimetric methods that involve visually observable changes from yellow to red color using chlorophenol red-β-d-galactopyranoside. Similarly, other cellular components can also be detected which ensures the lysis thereby detecting the bacteria present which is specific to the phage.

4.3.5.2 Progeny phage detection

The descendent phage that is released during the lysis of the host cell after infection with a phage can be analyzed to not only detect but also to enumerate the number of target bacterium present in the sample. The number of progenies produced will be directly related to the number of infected bacterial cells. Detection assays like immune or plaque assay and even other molecular approaches like isothermal nucleic acid amplification (ITNAA) and quantitative PCR can be done to determine the progeny phage that is released.

4.3.6 Quantum dot-based biosensors (quantitative analysis)

Quantum dots are synthetic semiconductors that fall within the nanoscale regime. For the development of phage-based quantum dot biosensors, the phage is genetically modified to express a quantum dot binding peptide [54] on its capsid proteins by phage display techniques (for instance, biotin). This modified phage is allowed to infect the bacteria. The progeny phage displays the peptide on its surface which can be allowed to bind to its counterpart (for instance streptavidin for biotin) coated on quantum dots. Exposing this to UV helps to detect the number of progenies which can be related to the number of bacterial cells lysed.

4.4 Types of Signal Output and Reading

4.4.1 Electrochemical biosensors

The electrochemical properties of a biological system are studied in an electrochemical biosensor. Electrochemical sensors are relatively simple, rapid, and real-time readout possible, cost-effective, and allow for miniaturization. In a phage-based electrochemical biosensor, the phage is immobilized onto an electrode surface and serves as the bio-recognition element. The immobilization of phage on the surface of the electrode can be done by physical adsorption using specific interfaces like electrostatic interactions of biotin and avidin, covalent bonding, or by phage entrapment on matrices. Wang et al. developed surfaces of poly-hydroxy-alkanoate on which phages were deposited [55]. Signal generated from the surface due to the binding of bacteria with the phage can be enhanced by the development of electrodes which allows the random orientation of the phage on its surface [56]. This method is limited by the number of phages that are deposited on the surface, if the number of phage that is deposited on the surface exceeds a certain limit it causes steric hindrances between the phage particles that result in inefficiency of the assay.

4.4.2 Impedance-based electrochemical approaches

Impedance is the measure of the opposition of current in a circuit when a voltage is applied and is similar to that of resistance for direct current whereas possesses magnitude and phase for alternating current, while the resistance of an alternative current (AC) possesses only magnitude. Electrochemical impedance is the changes occurring in an electrochemical system when a potential is applied. Electrochemical impedance spectroscopy (EIS) measures the impedance of a system under the influence of an AC potential. For a phage-based biosensor, the change in impedance caused due to the binding of phage to the bacteria or other target molecules is determined. For instance, bacterial detection using modified gold-integrated T4 was carried out by observing the rise in impedance due to the bacterial binding and the fall in impedance due to the lysis of the bacteria. Impedance-based phage sensors can be coupled with many other molecular techniques like loop-mediated isothermal amplification (LAMP) to increase the efficiency of the system. Detection based on an increase in charge transferred resistance with an increase in bacteria was implemented for impedimetric measurements in the presence of a redox couple. EIS is not only used for signal detection but can also be used for electro-chemiluminescence detection [57].

Screen-printed electrodes (SPEs) are disposable electrochemical devices and is reliable in biosensors mainly due to their disposable nature. They are simple, rapid, and cheap. Along with this, SPEs are miniaturized which makes it possible to connect them to portable instrumentation. They can be easily modified into films and also allow the incorporation of biomolecules and nanomaterials.

4.4.2.1 Phage-amperometry-based electrochemical biosensors

Amperometry is a solution-based detection of changes in electric current. Here the lytic phage is allowed to infect the bacteria in the solution. A rise in the conductivity of the solution is noted due to the increase of ions in the vicinity of electrodes which are released from the lysed cells.

4.4.2.2 Resistance-based biosensor

The variations in resistance of the conductor in the presence of a magnetic field can be analyzed. The binding can cause a fall or rise in the resistance of the whole system. For instance, a system using a DNA probe as the biorecognition element, upon hybridization with the analyte; reduces the resistance of the system. These changes in the system are examined to detect the analyte.

4.4.3 Optical biosensors

4.4.3.1 Phage in fluorescence-based bioassays

Relatively probe molecules have one to very few sites for modifications such as providing fluorescence tags or biotinylation, etc. Instead, the probe was expressed on phage, phage offers many sites for the modification or labeling and the sensitivity of the phage-based immunoassays can be amplified to few folds by providing fluorescence tags. For example, Goldman and co-workers in 2010, were developed a sandwich assay in which they immobilized biotinylated single domain antibodies (SdAb) either displayed on phage or free soluble SdAbs covalently to the surface of color-coded microspheres. When the signal was generated using fluorescent phycoerythrin coupled streptavidin, phage-displayed SdAb has shown at least fivefold amplification of signals compared to soluble SdAbs [58]. Semiconducting fluorescence nanoparticles such as quantum dots (QD) can be coupled with phage-based assays as labels and it proved high sensitivity. In a work by Edgar et al., 2006, biotinylated phage was conjugated with streptavidin-coated QDs for bacterial detection, and they have detected 10 CFU/mL sensitivity [59]. Reporter gene green fluorescence protein (GFP)-based phage was developed to detect pathogenic *E. coli* using fluorescence microscopy and flow cytometry [11].

4.4.3.2 Surface plasmon resonance (SPR)-based optical biosensors

In an optical biosensor, the transduced output signal is light, and it is measured as absorption, optical diffraction, refractive index (RI), fluorescence, or electro-chemiluminescence, and it can be with or without labels. SPR was introduced as a label-free optical detection technique in the early 1980s and used in measuring the biomolecule interaction in real time, it is considered as a highly sensitive biosensor technique with a detection limit low to the femtomolar range. SPR techniques are actively being applied in different biological interactions such as surface assembled monolayer (SAM), antibody–antigen binding, protein–DNA, protein–protein interactions, DNA hybridization, etc., SPR is produced when a plane-polarized light target on an electrically conducting layer of noble metals such as gold or silver at the interface of a glass sensor with high RI while at an external liquid medium with low RI. Which internally creates minimum reflected light intensity at a particular angle. At the event of bio-molecular interactions taking place on the surface of the gold layer, the change in the minimum angle observed, and it is highly dependent on the number of molecules interacting on the surface. The change in SPR signal occurs when a very small change (nanoscale) in the interface is detected, and it is accurately measured in real time. Both the lytic and lysogenic phages are applied in SPR-based sensing platforms. Rajaram et al., 2015 have immobilized streptavidin binding peptide (SBP

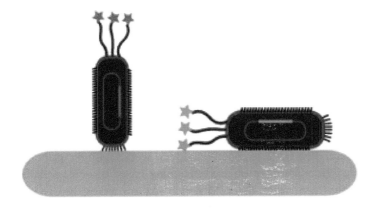

Figure 4.7 Phage orientation.

(biotin-like)) displaying M13 phage on to the streptavidin functionalized chips (SA chips) of SPR Biacore [60]. Since, the SBP is displayed at the pVII of the M13 phage, which is at one end of the filamentous phage, a likely vertical orientation of phage, could be achieved on the sensing platform which allows the attachment of many phage particles as possible and leads to higher sensitivity (Figure 4.7). SBP-displaying phage has provided at least 100-fold higher immobilization compared to the non-SBP displayers.

4.4.3.3 SERS-based biosensor

SERS (surface-enhanced Raman spectroscopy) is an optical technique used to create a molecular fingerprint of a specimen. This fingerprint of a compound can be compared to analyze the changes in the same. SERS-based detection techniques for phage display are limited because the signal is weaker than other fluorescence signals. Lentini et al., 2015 developed a SERS probe with phage and silver nanoparticles for the identification of U937 cells in vitro, this can be used to develop biosensors for hematological cancer cells and for the detection of minimal residual disease [61]. Nguyen et al., 2016 developed phages with plasmonic metallic nanoparticles which contain electromagnetic oscillation on their surfaces which could amplify the electromagnetic fields and enhances SERS [62]. Using SERS peak observation three sepsis-specific markers were determined; this approach could help for diagnostic purposes.

4.4.4 Mechanical biosensors

4.4.4.1 Quartz crystal microbalance

Mass-based bio-sensors:

Mass change based upon the biochemical and immunological interactions on the sensor surface is measured in mass-based sensors. A very good example of a mass-based biosensor is quartz crystal microbalance (QCM). Quartz is one of the abundant and hard materials on the earth which is composed of a network of SiO4. Quartz is known for its piezoelectric effects, meaning that at a given pressure, electricity is generated. QCM works with the principle of the piezoelectric effect and is often used in a mass-based biosensor platform. QCM-D is the next generation of QCM which also measures the dissipation of the sample on the sensor surface. QCM is a label-free approach and measures the biological interactions in real time; very simple sample preparation is needed and is relatively cheaper than many other techniques used in biosensors and provides high resolution and sensitivity. In QCM, the quartz crystal that is used as a sensing surface is coated with gold nanolayer, and at a given voltage, it tends to oscillate in a particular frequency (fundamental frequency); when a mass is adsorbed due to biological interactions on the sensing surface, the frequency changes proportionately and that could be easily measured using Sauerbrey's relationship equation [60]. Other than the fundamental frequency (single harmonic, $n = 1$, 5 MHZ), many different high frequencies are used in readout such as 3rd ($n = 3$, 10 MHZ), 5th ($n = 5$, 15 MHZ) up to 13th ($n = 13$, 65 MHZ). These frequencies are otherwise called overtones and these are very useful for studying the adlayers immobilized on the sensor surface.

Sauerbrey's relationship formula:

$$\Delta m = -C.\Delta f/n$$

Here, C = 17.7 ng Hz^{-1} CM^{-2} for a 5 MHZ quartz crystal,
n is overtone number (1st, 3rd, 5th, ..., 13th)
Δm is the mass change
Δf is the frequency change.

Sauerbrey's relationship equation is applicable only to rigid molecules like antibodies, peptides, proteins, and small molecules. Soft biomolecules such as phage cannot be applied, because it causes dissipation over the sensor and leads to the false calculation of the mass on the sensor surface. The dissipation study helps to understand the viscoelastic properties of the biomolecules on the sensing platform. So, when phage is used as a probe/ligand or receptor molecule, QCM-D systems could be a better choice for the measurement of mass change. Rajaram et al., 2015 have immobilized phage on the biotinylated QCM sensor surface in a likely vertical orientation in a work, while in another work they have prepared a phage complex comprising of phage, autoantibody and magnetic beads attached to either side of the phage displaying autoantigenic target at pVI and SBP at pIX respectively binding to the immobilized anti-human IgG on the QCM-D electrode in a diagnostic method that differentiating rheumatoid arthritis specific autoantibodies in the serum samples [60]. They also measured the dissipation change while measuring the frequency change on the substrate.

4.4.4.2 Magnetoelastic sensors

Magneto-elastic (ME) materials convert or transduce magnetic energy to mechanical energy and mechanical energy to magnetic energy. In a given magnetic field, these sensor surfaces undergo oscillating shape changes (contraction and elongation) thereby inducing a mechanical vibration, and it is directly proportional to the adsorbed mass on the resonator. When the mass increases, the resonant frequency decreases and this change can be measured in real-time. Magneto-elasticity appears when, upon the application of an external magnetic field, the shape of the material changes due to the superposition of its internal magnetic moments by spin–orbital coupling. For a planar material in a given AC field, the fundamental resonance frequency (f) of the longitudinal vibration caused by magnetostriction can be described as follows:

$$f = \frac{\sqrt{E}}{\rho(1-V)2L},$$

where E, ρ, ν, and L represent the elastic modulus, the density, Poisson's ratio, and the length respectively [63]. And the mass adsorbed over the sensor's surface that affects the resonance frequency can be obtained from the following formula

$$\Delta f = \frac{-f_0 \Delta m}{2M},$$

where Δf is the variation in resonance frequency, Δm is the variation in the mass on the sensor surface, and M is the actual mass of the sensor. The negative sign ($-$) designates that the frequency will move to lower values upon the adsorption of mass on the sensor surface. Park et al., 2012 and coworkers (2013) have developed a phage-based ME sensor immobilized using filamentous E2 phage on a gold surface [64]. The phage was genetically modified to specifically detect *Salmonella typhimurium* present on the surface of tomatoes. Using this strategy, they have achieved a higher level of detection than what could be achieved using quantitative PCR and even without culturing or isolating the bacteria from the tomato.

4.5 Applications and Case Studies

4.5.1 Pathogen detection and enumeration

4.5.1.1 Food and environmental samples

Foodborne pathogens and infections are important crisis that should be addressed. Detection of pathogens in the food sample as early as possible helps in the appropriate prognosis of the patient and this can be achieved by advanced highly specific bacteriophage display-based sensors which are highly specific and accurate [65]. Several components or whole organisms can be detected using phage display-based sensor; recognition elements like phage, nucleic acids, and molecular imprinted polymers can be used for the detection of corresponding analyte [66, 67]. Genetically modified T7 phage that can express alkaline phosphatase and upon infecting the host the overproduction of alkaline phosphatase can be detected and quantified. Several beverages including coconut water were used as the sample and the content of *E. coli* was detected. *E. coli* load of as less as 100 organisms per gram was detected in the study. Franche et al., 2017 developed a phage-based sensor to specifically detect enteric bacteria with luminescent reporter genes (*lux* genes) [68]. A novel approach where phage-based magneto elastic biosensors were developed to detect *S. enterica* from soil samples. A comparison was done regarding the detection methods; the limit of detection of the bacteria using a phage-based system is lesser which makes it a better choice than methods like filtration and cation-exchange methods. Minikh et al., 2010 enumerated 1 CFU of *E. coli* in 5 μL of the sample within 2 hour by modifying bacteriophage with either a biotin or cellulose binding molecule affinity tags and immobilizing them in streptavidin or cellulose material; this immobilization makes the attachment strong and irreversible and ensures detection of 70–100% of *E. coli* cells [69].

4.5.1.2 Clinical samples

As per Cao et al. 2016, phages are supramolecule that has various applications including biomedical pathogen detection [70]. Zhao et al., 2017 described the use of phage display to select peptides that could bind to ornithine transcarbamylase that is present in the urine samples of patients having active TB; this assay developed was comparable to the commercially available antibody-based sensors for detection [71]. Phage display-based sensor development for pathogen detection is cost-effective with high selectivity and specificity. Furthermore, Plano et al., 2017 carried out experiments like sequencing and atomic force microscopy (AFM) in order to understand the specificity of binding of random peptides to *Staphylococcus aureus* which have diagnostic purposes [72]. Liu et al., 2016 described the development of a bifunctional nanoprobe that has cysteine-stabilized gold nanoparticles with an *S. aureus*-specific pVIII fusion protein which aggregates in the presence of *S. aureus* with a colorimetric response [73].

Shriver-Lae et al., 2018 selected a single domain antibody from a library of proteins through phage display against DENV NS1 which was effective against all the four serotypes of Dengue and never cross-reacted with the NS1 of other viruses [74]. Similarly, antibody selection against serotype-specific epitopes of Dengue was described by Lebani et al., 2017 [75].

4.5.2 Diagnostic markers detection

Phage display helps to screen sequences and determine and generate high-affinity probes that can bind to targets. Deutscher et al., 2010 explained the usage of phage-displayed peptide probes to detect tumor antigens [76]. Park et al detected peptides that can bind to the tumor antigen CD44 through phage display technique; CD44 is an early tumor antigen and detection of this can aid in the clinical diagnosis and prognosis of certain types of cancers [77]. There have been approaches to develop antibodies that can specifically bind to mycolic acid which can be utilized for the diagnosis of tuberculosis [78]. Identification of epitopes from the phage display library can be used for applications like molecular imaging and the diagnosis of diseases [79].

4.5.3 Vaccine development

Phage display technique enables to express of certain molecules on the surface of the phage especially an epitope to induce immune response and to produce immunity against a specific pathogen. Connor et al., 2016 expressed the entire genome of *Neisseria gonorrhoeae* as oligopeptide on phage to determine the immunogenic sequences; the immunogenicity of full-length proteins was also determined, and this information can be used to develop vaccines against the respective pathogen [80]. In another study, the antigenic regions of *Bacillus anthracis* and *Y. pestis* was expressed simultaneously in T4 phage, thereby generating a dual vaccine (2 antigens for immune elicitation) for anthrax and plague which protected mice, rats and rabbits by specific immune response even when administered with lethal doses of bacteria after immunisation [81].

4.5.4 Material science and industry

There is a diverse application of phage and phage display, especially in the field of material science, for example, M13 was engineered in such a way to express a peptide that binds to graphene. In an aqueous solution, the phage increased the stability of the graphene and allowed nucleation of cations which generated stable electricity which was more capable than ones without phage [82]. Similarly, structures with piezoelectric properties were generated by self-assembling M13-based aerogels with relevant mechanical features like elasticity [5].

4.6 Conclusion

Phage has proven as a better probe molecule than antibodies and other biomolecules, due to its stability over different conditions, ideal functional folding of expressed proteins, etc. Phage-based biosensors are high-throughput sensing methods for rapid and accurate detection of microbial load in a way that is more reliable, economic, and easier. The existing traditional techniques for the detection of bacteria such as culturing, and advanced molecular techniques are time-consuming and require trained personnel for performing the assay. The major drawback of other techniques is that they cannot distinguish between live and dead cells which make the phage-based sensor more accurate and reliable. Phage display technology not only allows the binding of phage to whole bacteria but also can be modified to bind to other proteins or molecules. Different transducers like optical, electrochemical, or mechanical-based detection can be used to analyze the signals generated. Phage display and biopanning help in identifying, selecting, and amplifying a specific phage having the highest affinity toward the analyte. The affinity of the binding can be increased using methods like direct evolution where the genome is modified by introducing point mutations. Selection and amplification of high-affinity phage can result in increased sensitivity of biosensors thus decreasing the limit of detection (LOD). Phage also carries some limitations due to its bigger size, on the sensing platforms/ flow cells, which may occupy more surface area and limit the immobilization of a higher number phage particle. But, the oriented immobilization of phage on sensing surfaces could solve the issue. Advancements in the field of PD-based biosensors like nanoparticle-phage conjugates; antibody/nanobody displaying variants have increased the applicability in various fields like clinical diagnosis, food, environmental, and industrial sampling. Phage-based biosensors are a better alternative in detecting microorganisms and other analytes, and; are very relevant in various fields of research.

References

[1] S. Pavan and F. Berti, "Short peptides as biosensor transducers," *Anal. Bioanal. Chem.*, vol. 402, no. 10, pp. 3055-3070, Apr. 2012, doi: 10.1007/s00216-011-5589-8.

[2] G. Smith, "Filamentous fusion phage: novel expression vectors that display cloned antigens on the virion surface," *Science*, vol. 228, no. 4705, pp. 1315-1317, Jun. 1985, doi: 10.1126/science.4001944.

[3] G. P. Smith and V. A. Petrenko, "Phage Display," *Chem. Rev.*, vol. 97, no. 2, pp. 391-410, Apr. 1997, doi: 10.1021/cr960065d.

[4] W. Ebrahimizadeh and M. Rajabibazl, "Bacteriophage Vehicles for Phage Display: Biology, Mechanism, and Application," *Curr. Microbiol.*, vol. 69, no. 2, pp. 109-120, Aug. 2014, doi: 10.1007/s00284-014-0557-0.

[5] Moon, Choi, Jeong, Sohn, Han, and Oh, "Research Progress of M13 Bacteriophage-Based Biosensors," *Nanomaterials*, vol. 9, no. 10, p. 1448, Oct. 2019, doi: 10.3390/nano9101448.

[6] E. M. Zygiel et al., "Various mutations compensate for a deleterious lacZα insert in the replication enhancer of M13 bacteriophage," *PLoS ONE*, vol. 12, no. 4, p. e0176421, Apr. 2017, doi: 10.1371/journal.pone.0176421.

[7] C. M. Hammers and J. R. Stanley, "Antibody Phage Display: Technique and Applications," *J. Invest. Dermatol.*, vol. 134, no. 2, p. e17, Feb. 2014, doi: 10.1038/jid.2013.521.

[8] A. Demartis et al., "Polypharmacy through Phage Display: Selection of Glucagon and GLP-1 Receptor Co-agonists from a Phage-Displayed Peptide Library," *Sci. Rep.*, vol. 8, no. 1, p. 585, Jan. 2018, doi: 10.1038/s41598-017-18494-5.

[9] L. Rahbarnia et al., "Invert biopanning: A novel method for efficient and rapid isolation of scFvs by phage display technology," *Biologicals*, vol. 44, no. 6, pp. 567-573, Nov. 2016, doi: 10.1016/j.biologicals.2016.07.002.

[10] K. Cung et al., "Rapid, multiplexed microfluidic phage display," *Lab. Chip*, vol. 12, no. 3, pp. 562-565, Jan. 2012, doi: 10.1039/C2LC21129G.

[11] R. Peltomaa, I. López-Perolio, E. Benito-Peña, R. Barderas, and M. C. Moreno-Bondi, "Application of bacteriophages in sensor development," *Anal. Bioanal. Chem.*, vol. 408, no. 7, pp. 1805-1828, Mar. 2016, doi: 10.1007/s00216-015-9087-2.

[12] W. L. Matochko, K. Chu, B. Jin, S. W. Lee, G. M. Whitesides, and R. Derda, "Deep sequencing analysis of phage libraries using Illumina platform," *Methods San Diego Calif*, vol. 58, no. 1, pp. 47-55, Sep. 2012, doi: 10.1016/j.ymeth.2012.07.006.

[13] R. Peltomaa, E. Benito-Peña, R. Barderas, and M. C. Moreno-Bondi, "Phage Display in the Quest for New Selective Recognition Elements for Biosensors," *ACS Omega*, vol. 4, no. 7, pp. 11569-11580, Jul. 2019, doi: 10.1021/acsomega.9b01206.

[14] F. Chen et al., "Screening of Nanobody Specific for Peanut Major Allergen Ara h 3 by Phage Display," *J. Agric. Food Chem.*, vol. 67, no. 40, pp. 11219-11229, Oct. 2019, doi: 10.1021/acs.jafc.9b02388.

[15] D. Hu et al., "Effective Optimization of Antibody Affinity by Phage Display Integrated with High-Throughput DNA Synthesis and Sequencing Technologies," *PLOS ONE*, vol. 10, no. 6, p. e0129125, Jun. 2015, doi: 10.1371/journal.pone.0129125.

[16] M. Park, "Surface Display Technology for Biosensor Applications: A Review," *Sensors*, vol. 20, no. 10, p. 2775, May 2020, doi: 10.3390/s20102775.

[17] K. A. Günay and H.-A. Klok, "Identification of Soft Matter Binding Peptide Ligands Using Phage Display," *Bioconjug. Chem.*, vol. 26, no. 10, pp. 2002-2015, Oct. 2015, doi: 10.1021/acs.bioconjchem.5b00377.

[18] H. Suzuki, "Advances in the Microfabrication of Electrochemical Sensors and Systems," *Electroanalysis*, vol. 12, no. 9, pp. 703–715, May 2000, doi: 10.1002/1521-4109(200005)12:9<703::AID-ELAN703>3.0.CO;2-7.

[19] X. Fan, I. M. White, S. I. Shopova, H. Zhu, J. D. Suter, and Y. Sun, "Sensitive optical biosensors for unlabeled targets: A review,"

Anal. Chim. Acta, vol. 620, no. 1-2, pp. 8-26, Jul. 2008, doi: 10.1016/j.aca.2008.05.022.

[20] N. Zhu, H. Gao, Q. Xu, Y. Lin, L. Su, and L. Mao, "Sensitive impedimetric DNA biosensor with poly(amidoamine) dendrimer covalently attached onto carbon nanotube electronic transducers as the tether for surface confinement of probe DNA," *Biosens. Bioelectron.*, vol. 25, no. 6, pp. 1498-1503, Feb. 2010, doi: 10.1016/j.bios.2009.11.006.

[21] P. Skládal, "Piezoelectric biosensors," *TrAC Trends Anal. Chem.*, vol. 79, pp. 127-133, May 2016, doi: 10.1016/j.trac.2015.12.009.

[22] B. Leca-Bouvier and L. J. Blum, "Biosensors for Protein Detection: A Review," *Anal. Lett.*, vol. 38, no. 10, pp. 1491-1517, Jul. 2005, doi: 10.1081/AL-200065780.

[23] Z. Yu et al., "Integrated Photothermal-Pyroelectric Biosensor for Rapid and Point-of-Care Diagnosis of Acute Myocardial Infarction: A Convergence of Theoretical Research and Commercialization," *Small*, vol. 18, no. 30, p. 2202564, 2022, doi: 10.1002/smll.202202564.

[24] L. K. Harada et al., "Biotechnological applications of bacteriophages: State of the art," *Microbiol. Res.*, vol. 212-213, pp. 38-58, Jul. 2018, doi: 10.1016/j.micres.2018.04.007.

[25] V. A. Petrenko, "Landscape Phage as a Molecular Recognition Interface for Detection Devices," *Microelectron. J.*, vol. 39, no. 2, pp. 202-207, Feb. 2008, doi: 10.1016/j.mejo.2006.11.007.

[26] Y. Tan, T. Tian, W. Liu, Z. Zhu, and C. J. Yang, "Advance in phage display technology for bioanalysis," *Biotechnol. J.*, vol. 11, no. 6, pp. 732-745, Jun. 2016, doi: 10.1002/biot.201500458.

[27] N. Tawil, E. Sacher, E. Boulais, R. Mandeville, and M. Meunier, "X-ray Photoelectron Spectroscopic and Transmission Electron Microscopic Characterizations of Bacteriophage-Nanoparticle Complexes for Pathogen Detection," *J. Phys. Chem. C*, vol. 117, no. 40, pp. 20656-20665, Oct. 2013, doi: 10.1021/jp406148h.

[28] Y.-H. Yun et al., "Tiny Medicine: Nanomaterial-Based Biosensors," *Sensors*, vol. 9, no. 11, pp. 9275-9299, Nov. 2009, doi: 10.3390/s91109275.

[29] N. Tawil, E. Sacher, D. Rioux, R. Mandeville, and M. Meunier, "Surface Chemistry of Bacteriophage and Laser Ablated Nanoparticle Complexes for Pathogen Detection," *J. Phys. Chem. C*, p. 150610143229008, Jun. 2015, doi: 10.1021/acs.jpcc.5b02169.

[30] N. Bhardwaj, S. K. Bhardwaj, J. Mehta, K.-H. Kim, and A. Deep, "MOF-Bacteriophage Biosensor for Highly Sensitive and Specific

Detection of Staphylococcus aureus," *ACS Appl. Mater. Interfaces*, vol. 9, no. 39, pp. 33589-33598, Oct. 2017, doi: 10.1021/acsami.7b07818.

[31] J. McCafferty, A. D. Griffiths, G. Winter, and D. J. Chiswell, "Phage antibodies: filamentous phage displaying antibody variable domains," *Nature*, vol. 348, no. 6301, pp. 552-554, Dec. 1990, doi: 10.1038/348552a0.

[32] L. Ledsgaard, M. Kilstrup, A. Karatt-Vellatt, J. McCafferty, and A. Laustsen, "Basics of Antibody Phage Display Technology," *Toxins*, vol. 10, no. 6, p. 236, Jun. 2018, doi: 10.3390/toxins10060236.

[33] H. R. Hoogenboom, "Overview of Antibody Phage-Display Technology and Its Applications," in *Antibody Phage Display*, New Jersey: Humana Press, 2001, pp. 001-037. doi: 10.1385/1-59259-240-6:001.

[34] R. Barderas and E. Benito-Peña, "The 2018 Nobel Prize in Chemistry: phage display of peptides and antibodies," *Anal. Bioanal. Chem.*, vol. 411, no. 12, pp. 2475-2479, May 2019, doi: 10.1007/s00216-019-01714-4.

[35] Huang, Tu, Ning, He, and Li, "Development of Real-Time Immuno-PCR Based on Phage Displayed an Anti-Idiotypic Nanobody for Quantitative Determination of Citrinin in Monascus," *Toxins*, vol. 11, no. 10, p. 572, Sep. 2019, doi: 10.3390/toxins11100572.

[36] G. Li et al., "Generation of Small Single Domain Nanobody Binders for Sensitive Detection of Testosterone by Electrochemical Impedance Spectroscopy," *ACS Appl. Mater. Interfaces*, vol. 8, no. 22, pp. 13830-13839, Jun. 2016, doi: 10.1021/acsami.6b04658.

[37] T. H. van Kuppevelt, M. A. B. A. Dennissen, W. J. van Venrooij, R. M. A. Hoet, and J. H. Veerkamp, "Generation and Application of Type-specific Anti-Heparan Sulfate Antibodies Using Phage Display Technology," *J. Biol. Chem.*, vol. 273, no. 21, pp. 12960-12966, May 1998, doi: 10.1074/jbc.273.21.12960.

[38] P. Pourhakkak, A. Taghizadeh, M. Taghizadeh, M. Ghaedi, and S. Haghdoust, "Chapter 1 – Fundamentals of adsorption technology," in *Interface Science and Technology*, M. Ghaedi, Ed., in *Adsorption: Fundamental Processes and Applications*, vol. 33. Elsevier, 2021, pp. 1-70. doi: 10.1016/B978-0-12-818805-7.00001-1.

[39] E. Vonasek, P. Lu, Y.-L. Hsieh, and N. Nitin, "Bacteriophages immobilized on electrospun cellulose microfibers by non-specific adsorption, protein–ligand binding, and electrostatic interactions," *Cellulose*, vol. 24, no. 10, pp. 4581-4589, Oct. 2017, doi: 10.1007/s10570-017-1442-3.

[40] S. Horikawa et al., "Effects of surface functionalization on the surface phage coverage and the subsequent performance of phage-immobilized magnetoelastic biosensors," *Biosens. Bioelectron.*, vol. 26, no. 5, pp. 2361-2367, Jan. 2011, doi: 10.1016/j.bios.2010.10.012.

[41] K. Rajaram, V. Vermeeren, K. Somers, V. Somers, and L. Michiels, "Construction of helper plasmid-mediated dual-display phage for autoantibody screening in serum," *Appl. Microbiol. Biotechnol.*, vol. 98, no. 14, pp. 6365-6373, Jul. 2014, doi: 10.1007/s00253-014-5713-8.

[42] M. Janczuk-Richter, I. Marinović, J. Niedziółka-Jönsson, and K. Szot-Karpińska, "Recent applications of bacteriophage-based electrodes: A mini-review," *Electrochem. Commun.*, vol. 99, pp. 11-15, Feb. 2019, doi: 10.1016/j.elecom.2018.12.011.

[43] S. Burnham et al., "Towards rapid on-site phage-mediated detection of generic Escherichia coli in water using luminescent and visual readout," *Anal. Bioanal. Chem.*, vol. 406, no. 23, pp. 5685-5693, Sep. 2014, doi: 10.1007/s00216-014-7985-3.

[44] Y. Zhou, Y. Fang, and R. Ramasamy, "Non-Covalent Functionalization of Carbon Nanotubes for Electrochemical Biosensor Development," *Sensors*, vol. 19, no. 2, p. 392, Jan. 2019, doi: 10.3390/s19020392.

[45] C. M. A. Brett, "Perspectives and challenges for self-assembled layer-by-layer biosensor and biomaterial architectures," *Curr. Opin. Electrochem.*, vol. 12, pp. 21-26, Dec. 2018, doi: 10.1016/j.coelec.2018.11.004.

[46] A. E. Smartt and S. Ripp, "Bacteriophage reporter technology for sensing and detecting microbial targets," *Anal. Bioanal. Chem.*, vol. 400, no. 4, pp. 991-1007, May 2011, doi: 10.1007/s00216-010-4561-3.

[47] M. Schmelcher and M. J. Loessner, "Application of bacteriophages for detection of foodborne pathogens," *Bacteriophage*, vol. 4, no. 2, p. e28137, Apr. 2014, doi: 10.4161/bact.28137.

[48] M. Piuri, L. Rondón, E. Urdániz, and G. F. Hatfull, "Generation of Affinity-Tagged Fluoromycobacteriophages by Mixed Assembly of Phage Capsids," *Appl. Environ. Microbiol.*, vol. 79, no. 18, pp. 5608-5615, Sep. 2013, doi: 10.1128/AEM.01016-13.

[49] E. Stone, K. Campbell, I. Grant, and O. McAuliffe, "Understanding and Exploiting Phage–Host Interactions," *Viruses*, vol. 11, no. 6, p. 567, Jun. 2019, doi: 10.3390/v11060567.

[50] A. Singh et al., "Bacteriophage tailspike proteins as molecular probes for sensitive and selective bacterial detection," *Biosens. Bioelectron.*, vol. 26, no. 1, pp. 131-138, Sep. 2010, doi: 10.1016/j.bios.2010.05.024.

[51] P. B. Yim et al., "Quantitative characterization of quantum dot-labeled lambda phage for Escherichia coli detection," *Biotechnol. Bioeng.*, vol. 104, no. 6, pp. 1059-1067, Dec. 2009, doi: 10.1002/bit.22488.

[52] L. Wu, T. Huang, L. Yang, J. Pan, S. Zhu, and X. Yan, "Sensitive and Selective Bacterial Detection Using Tetracysteine-Tagged Phages in Conjunction with Biarsenical Dye," *Angew. Chem. Int. Ed.*, vol. 50, no. 26, pp. 5873-5877, Jun. 2011, doi: 10.1002/anie.201100334.

[53] Z. Wang, D. Wang, J. Chen, D. A. Sela, and S. R. Nugen, "Development of a novel bacteriophage based biomagnetic separation method as an aid for sensitive detection of viable Escherichia coli," *The Analyst*, vol. 141, no. 3, pp. 1009-1016, Feb. 2016, doi: 10.1039/c5an01769f.

[54] K. H. Lee, J.-W. Lee, J. Song, and M. Hwang, "Nanoscale bacteriophage biosensors beyond phage display," *Int. J. Nanomedicine*, p. 3917, Oct. 2013, doi: 10.2147/IJN.S51894.

[55] C. Wang, D. Sauvageau, and A. Elias, "Immobilization of Active Bacteriophages on Polyhydroxyalkanoate Surfaces," *ACS Appl. Mater. Interfaces*, vol. 8, no. 2, pp. 1128-1138, Jan. 2016, doi: 10.1021/acsami.5b08664.

[56] J. Xu, Y. Chau, and Y. Lee, "Phage-based Electrochemical Sensors: A Review," *Micromachines*, vol. 10, no. 12, p. 855, Dec. 2019, doi: 10.3390/mi10120855.

[57] H. Yue, Y. He, E. Fan, L. Wang, S. Lu, and Z. Fu, "Label-free electrochemiluminescent biosensor for rapid and sensitive detection of pseudomonas aeruginosa using phage as highly specific recognition agent," *Biosens. Bioelectron.*, vol. 94, pp. 429-432, Aug. 2017, doi: 10.1016/j.bios.2017.03.033.

[58] E. R. Goldman, G. P. Anderson, R. D. Bernstein, and M. D. Swain, "Amplification of immunoassays using phage-displayed single domain antibodies," *J. Immunol. Methods*, vol. 352, no. 1-2, pp. 182-185, Jan. 2010, doi: 10.1016/j.jim.2009.10.014.

[59] R. Edgar et al., "High-sensitivity bacterial detection using biotin-tagged phage and quantum-dot nanocomplexes," *Proc. Natl. Acad. Sci. U. S. A.*, vol. 103, no. 13, pp. 4841-4845, Mar. 2006, doi: 10.1073/pnas.0601211103.

[60] K. Rajaram et al., "Real-time analysis of dual-display phage immobilization and autoantibody screening using quartz crystal microbalance with dissipation monitoring," *Int. J. Nanomedicine*, p. 5237, Aug. 2015, doi: 10.2147/IJN.S84800.

[61] G. Lentini et al., "Phage–AgNPs complex as SERS probe for U937 cell identification," *Biosens. Bioelectron.*, vol. 74, pp. 398-405, Dec. 2015, doi: 10.1016/j.bios.2015.05.073.

[62] A. H. Nguyen, Y. Shin, and S. J. Sim, "Development of SERS substrate using phage-based magnetic template for triplex assay in sepsis diagnosis," *Biosens. Bioelectron.*, vol. 85, pp. 522-528, Nov. 2016, doi: 10.1016/j.bios.2016.05.043.

[63] M. Johnson et al., "A wireless biosensor using microfabricated phage-interfaced magnetoelastic particles," *Sens. Actuators Phys.*, vol. 144, pp. 38-47, May 2008, doi: 10.1016/j.sna.2007.12.028.

[64] M.-K. Park, H. C. Wikle, Y. Chai, S. Horikawa, W. Shen, and B. A. Chin, "The effect of incubation time for Salmonella Typhimurium binding to phage-based magnetoelastic biosensors," *Food Control*, vol. 26, no. 2, pp. 539-545, Aug. 2012, doi: 10.1016/j.foodcont.2012.01.061.

[65] A. Singh, S. Poshtiban, and S. Evoy, "Recent Advances in Bacteriophage Based Biosensors for Food-Borne Pathogen Detection," *Sensors*, vol. 13, no. 2, pp. 1763-1786, Jan. 2013, doi: 10.3390/s130201763.

[66] B. Van Dorst et al., "Recent advances in recognition elements of food and environmental biosensors: A review," *Biosens. Bioelectron.*, vol. 26, no. 4, pp. 1178-1194, Dec. 2010, doi: 10.1016/j.bios.2010.07.033.

[67] N. Wisuthiphaet, X. Yang, G. M. Young, and N. Nitin, "Rapid detection of Escherichia coli in beverages using genetically engineered bacteriophage T7," *AMB Express*, vol. 9, no. 1, p. 55, Dec. 2019, doi: 10.1186/s13568-019-0776-7.

[68] N. Franche, M. Vinay, and M. Ansaldi, "Substrate-independent luminescent phage-based biosensor to specifically detect enteric bacteria such as E. coli," *Environ. Sci. Pollut. Res.*, vol. 24, no. 1, pp. 42-51, Jan. 2017, doi: 10.1007/s11356-016-6288-y.

[69] O. Minikh, M. Tolba, L. Y. Brovko, and M. W. Griffiths, "Bacteriophage-based biosorbents coupled with bioluminescent ATP assay for rapid concentration and detection of Escherichia coli," *J. Microbiol. Methods*, vol. 82, no. 2, pp. 177-183, Aug. 2010, doi: 10.1016/j.mimet.2010.05.013.

[70] B. Cao, M. Yang, and C. Mao, "Phage as a Genetically Modifiable Supramacromolecule in Chemistry, Materials and Medicine," *Acc. Chem. Res.*, vol. 49, no. 6, pp. 1111-1120, Jun. 2016, doi: 10.1021/acs.accounts.5b00557.

[71] N. Zhao, J. Spencer, M. A. Schmitt, and J. D. Fisk, "Hyperthermostable binding molecules on phage: Assay components for point-of-care diagnostics for active tuberculosis infection," *Anal. Biochem.*, vol. 521, pp. 59-71, Mar. 2017, doi: 10.1016/j.ab.2016.12.021.

[72] L. M. De Plano, S. Carnazza, G. M. L. Messina, M. G. Rizzo, G. Marletta, and S. P. P. Guglielmino, "Specific and selective probes for Staphylococcus aureus from phage-displayed random peptide libraries," *Colloids Surf. B Biointerfaces*, vol. 157, pp. 473-480, Sep. 2017, doi: 10.1016/j.colsurfb.2017.05.081.

[73] P. Liu, L. Han, F. Wang, V. A. Petrenko, and A. Liu, "Gold nanoprobe functionalized with specific fusion protein selection from phage display and its application in rapid, selective and sensitive colorimetric biosensing of Staphylococcus aureus," *Biosens. Bioelectron.*, vol. 82, pp. 195-203, Aug. 2016, doi: 10.1016/j.bios.2016.03.075.

[74] L. C. Shriver-Lake et al., "Selection and Characterization of Anti-Dengue NS1 Single Domain Antibodies," *Sci. Rep.*, vol. 8, no. 1, p. 18086, Dec. 2018, doi: 10.1038/s41598-018-35923-1.

[75] K. Lebani et al., "Isolation of serotype-specific antibodies against dengue virus non-structural protein 1 using phage display and application in a multiplexed serotyping assay," *PLOS ONE*, vol. 12, no. 7, p. e0180669, Jul. 2017, doi: 10.1371/journal.pone.0180669.

[76] S. L. Deutscher, "Phage Display in Molecular Imaging and Diagnosis of Cancer," *Chem. Rev.*, vol. 110, no. 5, pp. 3196-3211, May 2010, doi: 10.1021/cr900317f.

[77] H.-Y. Park, K.-J. Lee, S.-J. Lee, and M.-Y. Yoon, "Screening of Peptides Bound to Breast Cancer Stem Cell Specific Surface Marker CD44 by Phage Display," *Mol. Biotechnol.*, vol. 51, no. 3, pp. 212-220, Jul. 2012, doi: 10.1007/s12033-011-9458-7.

[78] C. E. Chan et al., "Novel phage display-derived mycolic acid-specific antibodies with potential for tuberculosis diagnosis," *J. Lipid Res.*, vol. 54, no. 10, pp. 2924-2932, Oct. 2013, doi: 10.1194/jlr.D036137.

[79] L.-F. Wang and M. Yu, "Epitope Identification and Discovery Using Phage Display Libraries: Applications in Vaccine Development and Diagnostics," *Curr. Drug Targets*, vol. 5, no. 1, pp. 1-15, Jan. 2004, doi: 10.2174/1389450043490668.

[80] D. O. Connor, J. Zantow, M. Hust, F. F. Bier, and M. von Nickisch-Rosenegk, "Identification of Novel Immunogenic Proteins of Neisseria gonorrhoeae by Phage Display," *PLOS ONE*, vol. 11, no. 2, p. e0148986, Feb. 2016, doi: 10.1371/journal.pone.0148986.

[81] P. Tao et al., "A Bacteriophage T4 Nanoparticle-Based Dual Vaccine against Anthrax and Plague," *mBio*, vol. 9, no. 5, pp. e01926-18, /mbio/9/5/mBio.01926-18.atom, Oct. 2018, doi: 10.1128/mBio.01926-18.

[82] D. Oh et al., "Graphene Sheets Stabilized on Genetically Engineered M13 Viral Templates as Conducting Frameworks for Hybrid Energy-Storage Materials," *Small*, vol. 8, no. 7, pp. 1006-1011, Apr. 2012, doi: 10.1002/smll.201102036.

5

Methods and Challenges of Tissue Clearing Engineering for Tissue Complexity

Anh H. Nguyen[1,3], Donna H. Tran[2], and Hung Cao[1,2]

[1]Department of Electrical and Computer Engineering, UC Irvine, Irvine, CA 92697, USA
[2]Department of Biomedical Engineering, UC Irvine, Irvine, CA 92697, USA
[3]Sensoriis Inc, Edmonds, WA, 98026-7617, USA

Abstract

The properties of opaque tissues include high scattering and absorption which restrict the passage of light into deep native tissues. Tissue-clearing methods allow optics-based analysis at subcellular levels from specific organs to the entire body of small animals. When paired with light-sheet microscopy, tissue-clearing methods can expedite sample imaging and analysis in clinical settings compared to conventional immunohistochemistry by several orders of magnitude. Labeling chemistry allows a specific and robust antibody label for whole organ and thick tissues. In addition, advances in image and data processing allow scientists to extract comprehensive data from structural and functional tissue networks at cellular and subcellular levels. In this Chapter, we discuss how tissue-clearing methods provide insight on unbiased, system-level tissue patterns. We also cover future opportunities for use of computational approaches and imaging processing integrated with tissue-clearing methods in tissue engineering.

Keywords: Tissue-clearing methods, light-sheet microscopy, immunohisto-chemistry, labeling chemistry, imaging process, functional tissue networks, tissue engineering

5.1 Introduction

Advancements in optical physics and chemical engineering promote innovative discoveries in robust and reproducible clearing techniques and their applications. Visualization through tissue clearing and fluorescence volumetric imaging of whole tissues, organs, and subcellular levels can provide insights on tumor complexity [1, 2]. The complexity of tumor tissues is attributed to the heterogenous refractive index (RI) which is reflected in differences in components of connective tissue fibers throughout the tissue. In tumors, signaling networks and cell interactions offer critical data to determine key milestones in tumor progression and become potential biomarkers for cancer diagnostics [3, 4]. However, cancer tissues have varying properties that are caused by higher blood vessels density, heterogeneity refractive index (RI) as an intrinsic imaging contract, enriched lipid profiles, and rich pigment epithelium-derived factors (PEDF) which cause light scattering and contribute to tissue opacity [5, 6]. This obstacle reduces the sufficient resolution for cell and tissue profiles in three-dimensional (3D) resolution, which can result in loss of information. Although 3D visualization techniques for large volume tissues and anatomical structures such as computer tomography (CT) and magnetic resonance imaging (MRI) do not require image reassembly, they do not provide sufficient resolution for cellular profiling. Tissue-clearing techniques enable the preparation of high transparent whole-organ cell profiling through imaging with a high-speed imaging system.

Current methods of tissue analysis involve two-dimensional (2D) histopathology where samples are sectioned into thin slices for imaging (Figure 5.1). Thin slides are a few micrometers thick, allowing for the profiling of individual cells. However, only a few 2D sections are analyzed practically, leaving most of the sample unexposed. Visualizing whole samples through 3D imaging enables comprehending entire specimens through mapping of spatial relationships and architectures of cancer tissues by the standard immunohistochemistry after tissue clearing [7].

Tissue-clearing technology targets this obstacle by homogenizing the RI to produce clear tissue samples. A simple breakdown of the methodology involves the following: tissue fixation, delipidation through permeabilization, decolorization, and RI matching [8]. The fixation ensures that the tissue remains as close to its natural state for examination and analysis. Lipid extraction and decolorization minimize light scattering and facilitate light absorption only to the observed region. Finally, RI matching homogenizes

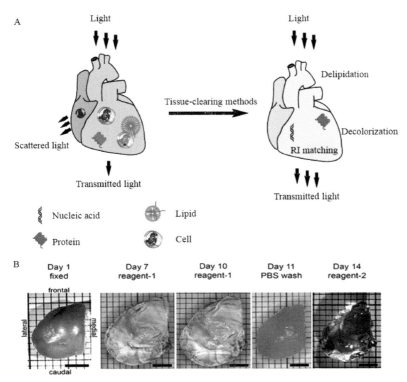

Figure 5.1 (A) Concepts of tissue clearing methods. Tissue opacity inhibit light transmission due to light scattering and light absorption. Lipids, proteins, hemoglobin, and melanin in tissues are key components that cause light scattering and absorption. After clearing, tissue components have a match refractive index which minimizes light scattering and absorption to make tissue optically transparent. (B) An example of a cleared tissue is through several steps. A hemisphere of the songbird brain was cleared by removing lipid components (7 days), increasing tissue transparency (10 days), and creating refractive index matching (13 days). Part (B) was adapted from Ref [10].

light scattering in the tissue to make it transparent. While there is an expanding quantity of published tissue-clearing protocols which go into further detail on these techniques, most are categorized into one of three categories: hydrogel-based clearing, hydrophobic or organic solvent-based clearing, and aqueous-based clearing. Within the three subgroups, the most prominent methods for tissue clearing are clear lipid-exchanged acrylamide-hybridized rigid imaging/immunostaining/in situ hybridization-compatible tissue-hydrogel (CLARITY); DISCO; and clear, unobstructed brain imaging cocktails and computational analysis (CUBIC), respectively.

Table 5.1 Safety information of some commercial reagents for CUBIC tissue-clearing methods.

Product name	Physical state	Specification	Safety and Regulations
Tissue-Clearing Reagent CUBIC-L [for delipidation and decoloring]	Liquid	Colorless to almost colorless clear liquid	Causes serious eye damage.
Tissue-Clearing Reagent CUBIC-R+(N) [for RI matching]	Liquid	Colorless to light yellow clear liquid	Causes skin irritation. Causes serious eye irritation. Causes damage to organs.
Tissue-Clearing Reagent CUBIC-R+(M) [for RI matching]	Liquid	Colorless to light orange to Yellow clear liquid	Causes skin irritation.Causes serious eye irritation.Causes damage to organs.
Tissue-Clearing Reagent CUBIC-B [for decalcification]	Liquid	Colorless to almost colorless clear liquid	Causes severe skin burns and eye damage.
Tissue-Clearing Reagent CUBIC-HL [for highly fatty tissue and quenching autofluorescence]	Liquid	Colorless to light yellow to light orange clear liquid	Causes severe skin burns and eye damage. May cause damage to organs. Toxic to aquatic life.
Tissue-Clearing Reagent CUBIC-P [efficiently aids perfusion fixation]	Liquid	Colorless to almost colorless clear liquid	Causes severe skin burns and eye damage. May damage fertility or the unborn child. May cause damage to organs. Causes damage to organs through prolonged or repeated exposure.
Tissue-Clearing Reagent CUBIC-X1 [for tissue expansion]	Liquid	Colorless to almost colorless clear liquid	Harmful if swallowed. Causes severe skin burns and eye damage. Suspected of damaging fertility or the unborn child. May cause damage to organs.

Table 5.1   (Continued).

Product name	Physical state	Specification	Safety and Regulations
Tissue-Clearing Reagent CUBIC-X2 [for RI matching while keeping the expanded size]	Liquid	Colorless to light yellow to light orange clear liquid	Causes skin irritation. Causes serious eye irritation. Suspected of damaging fertility or the unborn child. Causes damage to organs.
Mounting Solution (RI 1.520) [for CUBIC-R+]	Liquid	Colorless to light yellow clear liquid	
Mounting Solution (RI 1.467) [for CUBIC-X2]	Liquid	Colorless to light yellow clear liquid	
CUBIC-HV^{TM}1 3D nuclear staining kit	Liquid (Heat Sensitive)		Heat sensitive Causes serious eye irritation.
CUBIC-HV^{TM}1 3D immunostaining kit (Casein separately)	Liquid (Heat Sensitive)		Causes serious eye irritation.

The process of selecting the most appropriate clearing protocol to use is highly dependent on the experimental objectives and requires careful consideration of the strengths and weaknesses of each technique. For instance, considering factors such as tissue size, changes to morphology, and time for clearing can aid in determining which method is most ideal on a case-by-case basis. Moreover, after reducing tissue opacity there is also a selection of imaging technologies for cleared tissues including light-sheet fluorescence, confocal, and two-photon microscopy [9]. Employing tissue clearing procedures (note to its chemical safety, Table 5.1) followed by microscopic imaging facilitates subcellular comprehension of cellular behavior and can aid in tumor detection and diagnosis, applicable to clinical settings. In this Chapter, we describe the current scene of clearing methods and their application including clearing and labeling chemicals, light-sheet microscopy, and data analysis. We also discuss the challenges and opportunities related to transferring the technology to clinical settings.

Safety notes. Always use personal protective equipment (a lab coat, nitrile gloves, and splash goggles), proper respiratory protection, and a chemical hood during conducting clearing tissues. Glass and secondary containers should be used to store all clearing reagents in the dark. This information is adapted from TCI chemicals.

5.2 Tissue Clearing Methods

5.2.1 Hydrogel tissue clearing methods

CLARITY which stands for "cleared lipid-extracted acryl-hybridized rigid immunostaining/in situ hybridization-compatible tissue hydrogel" is a typical clearing technique enabling immunofluorescent labeling and imaging of large-volume three-dimensional tissue. This technique provides greater insight of tissue morphology and spatial interactions between distinct cell types. The basic steps of a typical CLARITY are as follows: tissue fixation, hydrogel formation, lipid extraction, staining, and imaging [11]. Through crosslinking with polymers, for example, acrylamide and others, the process secures intact tissues in a reticular hydrogel structure where biomolecules are positioned at their cellular locations through their covalent bonds to acryl-based hydrogel [12]. After biomolecules are immobilized, lipid profiles are removed by applying an electrical current (active CLARITY) or passive thermal diffusion (passive CLARITY) (Figure 5.2).

This process leaves the tissue transparent by removing cellular components like lipids which strongly impact the refractive index and homogenizing tissue refractive indices while retaining proteins and nucleic acids in the hydrogel matrix [13]. In addition, CLARITY preserves the physical structure and permittivity of tissue and turns it into optically transparent tissue without severely damaging biomolecules [14]. Therefore, it can allow for multiple staining and destaining trials without sectioning for the observation of 3D-structured tissue. However, due to the complete removal of light-scattering lipid structures, the network complexity of tissue morphology can be significantly altered and significantly transforms tissue histochemistry due to loss of antigenicity and immunolabeling in conjunction with CLARITY. In addition, CLARITY also induces damage to tissues due to the high current applied throughout the whole process, facing a significant risk of sample loss when working with valuable tissue samples [15]. Therefore, some modifications

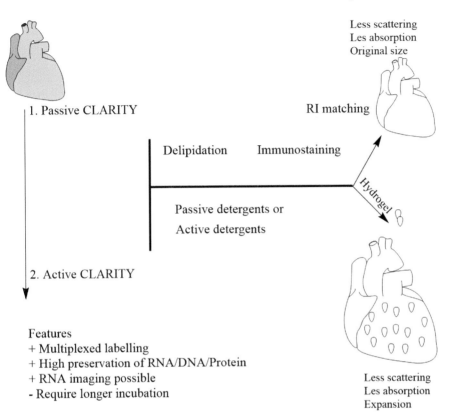

Figure 5.2 Two formats of hydrogel-based methods. Tissues are cross-linked to a hydrogel-based solution containing 4% paraformaldehyde, 4% acrylamide, and 0.05% bis-acrylamide. Sodium dodecyl sulfate (SDS) is used to remove lipids in either passive (top) or active (bottom) CLARITY. In passive CLARITY, lipid-embedded SDS micelles are diffused out of tissues and leave a transparent hydrogel-tissue hybrid behind. In active CLARITY, lipid-embedded SDS micelles are electrophorized to enhance the extraction rate and leave a transparent hydrogel-tissue hybrid behind. In both methods, a matched refractive index of the medium makes tissues complete transparency. During these processes, tissues absorb water and expand their volumes, which improves the permeability of antibodies and immunohistochemistry. Polymers used in the method are present in the format of synthetic gel or reinforced tissue gel.

to CLARITY have been developed to alleviate the challenge of low tissue clearing efficiency, reduce tissue damage, and produce low-cost equipment for CLARITY techniques [16]. Correspondingly, derivative CLARITY techniques are classified into three subgroups: (i) active, (ii) passive, and (ii) a hybrid method between active and passive CLARITY (see Table 5.2).

Table 5.2 Comparison of active and passive CLARITY.

Methods	Active CLARITY	Passive CLARITY
Force	Electrophoresis	Passive thermal diffusion
Clearing time	Hours to days	Days to weeks
Tissue morphology	Less retaining	Good
Cellular biology	Less retaining	Good
Operation	Costly and complicated	Cheap and easy
Circulation system	Need to maintain temperature and flow rate	Needless
Clearing buffer	4% SDS in PBS, pH 8.5	8% SDS in PBS buffer, pH 7.5

Technical notes: One of the disadvantages of the passive CLARITY method is the time compared to that of the active CLARITY. The table is adapted from [7],

These advancements provide affordable clearing methods to overcome the current limitations of CLARITY techniques by increasing the removal speed, tissue transparency, and retaining tissue morphology. For example, using poly(acrylamide-*co*-styrenesulfonate) gels as a polyelectrolyte gel for the passive CLARITY method could reduce the clearing time and increase three-dimensional fluorescence imaging, especially for DNA staining in intact tissues [17]. By combining this with immunofluorescent labeling, mitochondrial proteins in intact cerebellum tissues have been observed at different neural subregions and blood vessels [18]. Interesting insight into mechanisms of neurodegenerative diseases relating to mitochondrial disease has been discovered by revealing protein respiratory chain deficiency associated with mitochondrial disease. By retaining permittivity, the cleared tissue can be used to strip and re-probe new antibodies for large volumes of intact cerebellum to further discoveries of the neuronal changes in neurological disorders such as Alzheimer's [19]. Recently, CLARITY has been used to study three-dimensional imaging and quantitative analysis using quadruple immunofluorescence for staining multiple markers and confocal microscopy imaging of breast cancer tissues [20].

5.2.2 Hydrophobic tissue clearing

In this method, organic solvents (see chemical list in Table 5.3) are used to quickly render intact tissue completely transparent. A typical protocol of this method is 3D imaging of solvent-cleared organs (3DISCO) which was developed by Ertürk who cleared a whole adult mouse brain in 1–2 days [21]. Compared to hydrophilic tissue-clearing methods, 3DISCO is a straightforward method using hydrophobic solvents to sequentially incubate

Table 5.3 Chemical agents for dehydration in tissue clearing.

Chemical	Function in tissue clearing	Protocol	Functional groups
Methanol	Dehydration and permeabilization	IDISCO, iDISCO(+)	Alcohol
Ethanol	Dehydration	BABB	Alcohol
Ter-Butanol	Dehydration	FluoClearBABB, uDISCO, PEGASOS	Alcohol
Tetrahydrofuran	Dehydration	3DISCO, iDISCO, vDISCO	Ether
Hexane	Dehydration and delipidation?	BABB	Alkane

Technical note: Dehydration is an important step of the tissue-clearing process for collagenous and cellular tissue.

the specimens. 3DISCO and its derivatives have been widely explored in studies of neural circuits, cardiac damaging tissue, in vivo stem cell differentiation, and metastasis in unsectioned organ and human biopsy specimens [22]. By retaining tissue permittivity, deep tissue immunolabeling approaches can be used to stain specific tissues, embryonic samples, and whole body of specimens (Figure 5.3).

The major light scatters in tissues is attributed to the variance in the refractive index of water (1.22) which is lower than that of soft tissue (1.44–1.56). Thus, the first step of 3DISCO method is to remove water from tissues. The next step is to extract most of the lipid with organic solvents in order to increase the RI to emulate the average RI of biological tissue. For

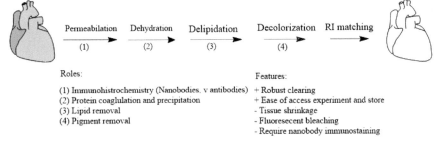

Roles:
(1) Immunohistrochemistry (Nanobodies. v antibodies)
(2) Protein coaglulation and precipitation
(3) Lipid removal
(4) Pigment removal

Features:
+ Robust clearing
+ Ease of access experiment and store
- Tissue shrinkage
- Fluoresecent bleaching
- Require nanobody immunostaining

Figure 5.3 Hydrophobic methods. A typical process of hydrophobic methods is to use organic solvents (Table 5.3) to completely dehydrate, extract lipids, and create refractive index (RI) matching of tissues. Although hydrophobic methods allow for conducting tissue-clearing processes fast and fully, the methods can cause photobleaching for fluorescent protein rapidly.

example, the RI for both organic solvents and cleared brain tissue is 1.56. However, dehydration (see chemical lists in Table 5.3) leads to shrinking the specimen size facilitating the application of light-sheet microscopy to depict whole bodies at cellular resolution [23]. Typically, the DISCO method isotropically shrinks mouse bodies to about a third of their actual size to visualize neural networks in the central nervous system of mice [24]. In DISCO methods, cleared specimens preserve tissue morphology, allowing numerous imaging sessions and analysis of the same samples, especially for endogenous fluorescent signal regenerated from immunolabeling methods in larger samples [25]. However, other difficulties arise due to changes in tissue permeability and epitope modifications that make it difficult to label large tissues with explicit dyes and antibodies during the clearing process. [26]. An improved method, iDISCO, developed by Renier and Tessier-Lavigne is used as a rapid method to immunolabel large tissue samples like large mouse embryos and adult organs for volume imaging [27]. Immunolabeled imaging provides insight on neurodevelopment in embryos and neurodegeneration in adults mice [27, 28]. Compared to 3DISCO, iDISCO requires pretreatment of the sample with H_2O_2 and methanol to increase the porosity of the mouse brain. This process may be damaging to the tissue sample as it also removes most of the epitope on the tissue for further immunostaining. This becomes a challenge for hydrophobic tissue clearing methods in current deep tissue labeling approaches. Therefore, full epitope preservation and retaining tissue permittivity are crucial factors to increase application in whole organ clearing. Recently, a new method entitled vDISCO (v refers to variable chains) was developed [29]. In this method, small variable domain of heavy chain antibodies combined with high-pressure delivery was used to complete the far-red fluorescent (bright Atto dyes) immuno-labeling of the whole organ and body. The far-red fluorescent molecules, compared to blue-green regions, surpass the low signal intensity and aut-ofluorescence of many tissues. Nanobodies can permeate deeply in intact clear bodies through bones and muscles and allow subcellular imaging and single-cell quantification with amplified fluorescent signal by two orders of magnitude. vDISCO allows the analysis of whole-body neuronal proteome of adult mice, which supports studies of neurodegeneration and inflam-mation in CNS lesions and skull-meninges connections [30]. In addition, nanobodies used in vDISCO remarkably amplify signal contrast, which provides high-quality imaging for data processing of transgenically labeled mammalian brains.

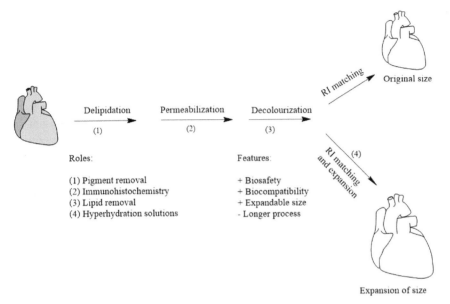

Figure 5.4 Hydrophilic methods include serial steps of delipidation, permeabilization, decolorization, RI matching, and expansion (as an option). In these methods, water-soluble solutions allow the tissue-clearing process with higher biocompatibility and biosafety than hydrophobic methods, but some hydrophilic methods need longer time to complete the process.

5.2.3 Hydrophilic tissue clearing

In this method, water-soluble, biocompatible, and biosafe materials are used for tissue clearing (Figure 5.4). This results in the preservation of protein function and epitope specificity. Hydrophilic reagents form hydrogen bonds with functional groups of proteins in tissues and surrounding water molecules, possibly preserving 3D-structured tissues and signal intensity of fluorescent proteins [31]. Hydrophilic reagents can be used at a high concentration as a RI-matching medium to provide a transparent medium for imaging. Hydrophilic tissue clearing has some derivatives. FocusClear method, developed by Liu et al., used an X-ray contrast agent (diatrizoate acid) and a detergent (Tween 20) for whole cockroach brain imaging [32]. Scale, developed by Hama et al., used urea to increase hyperhydration and reduce tissue RI, as a key tissue-clearing component, to develop Scale, a hydrophilic derivate for efficient preservation of fluorescent proteins, for broader samples of whole mouse embryos, infant mouse brains, and adult mouse brains slide [33]. This method is consistent with multiple fluorescent staining in different tissue specimens. For example, it allowed imaging yellow, green fluorescent-conjugated antibodies bound to neuron biomarkers in infant and adult mice, and neural stem cells in the adult mouse brain [33, 34]. Recently, sorbitol and fructose, key RI-matching components, have been used to increase the properties of the Scale method for imaging old and neurodegenerative mouse brain hemispheres and tracing neural circuits in mouse olfactory bulbs [35, 36]. This is a key advancement because it prevents the resizing of biological samples and therefore maintains their morphology. To obtain super-resolution imaging, the updated method (SeeDB2) integrated X-ray contrast agent with a higher RI to analyze chemical profiling in neural circuits in mouse brains [37]. The profiling of at least 1,600 hydrophilic chemicals has been analyzed with imaging cocktails and computation analysis (CUBIC) by using a series of amino alcohols [38]. The amino alcohols (e.g., N-butyl diethanolamine) function as delipidation and decolorization

Table 5.4 Chemical agents for decolorization in tissue clearing.

Chemical	Function in tissue clearing	Protocol	Functional groups
1-Methylimidazole	Decolorization	CUBIC-P	Imidazole
1,3-Bis(aminomethyl) cyclohexane	Delipidation, decolorization, and pH adjustment	CUBIC-HL	Aliphatic amine
N, N, N′,N′-Tetrakis(2-hydroxypropyl) Ethylenediamine (Quadrol)	Delipidation, decolorization, and RI matching	ScaleCUBIC-1, PACT-deCAL– Bone CLARITY, PEGASOS, vDISCO	Amino alcohol
Triethanolamine	Delipidation, decolorization, and RI matching	ScaleCUBIC-2, RTF	Amino alcohol
N-Butyldiethanolamine	Delipidation, decolorization, and pH adjustment	CUBIC-L, CUBIC-R+	Amino alcohol

Technical note: Decolorization is an important step to reduce background signal and increase the quality of fluorescent imaging.

Table 5.5 Chemical agents for RI matching and delipidation in tissue clearing.

Chemical	Function in tissue clearing	Protocol	Functional groups
Benzyl alcohol	RI matching and delipidation (putative)	BABB, vDISCO	Aromatic alcohol
Benzyl benzoate	RI matching and delipidation (putative)	BABB, PEGASOS, vDISCO	Aromatic ester
Ethyl-3-phenylprop-2-enoate (ethyl cinnamate)	RI matching	Ethyl cinnamate method	Aromatic ester
Dibenzyl ether	RI matching and delipidation (putative)	3DISCO, iDISCO, iDISCO+	Aromatic ether
Diphenyl ether	RI matching and delipidation (putative)	uDISCO	Aromatic ether
Dichloromethane	Delipidation and dehydration	3DISCO, uDISCO, iDISCO, iDISCO+	Alkyl halide
Glycerol	RI matching	Scale reagents, CLARITY, LUCID	Polyol
2,2′-Thiodiethanol	RI matching	Single treatment, CLARITY, LUCID, TOMEI	Thioalcohol
Sucrose	RI matching	ScaleCUBIC2, UbasM-2	Oligosaccharide
d(−)-Fructose	RI matching	SeeDB, FRUIT	Monosaccharide
d(−)-Sorbitol	RI matching	sRIMS, ScaleS, STP tomography	Monosaccharide
Xylitol	Dehydration and RI matching	ClearSee	Monosaccharide
Methyl-β-cyclodextrin	Delipidation (extracting cholesterol)	ScaleS, vDISCO	Oligosaccharides
γ-Cyclodextrin	Delipidation (extracting cholesterol)	ScaleS	Oligosaccharides
Polyethylene glycol	RI matching	ClearT2, PEGASOS	Polyether
Dimethyl sulfoxide	RI matching	ScaleS, STP tomography	Sulfoxide

Table 5.5 (*Continued*).

Chemical	Function in tissue clearing	Protocol	Functional groups
Urea	Hydration (putatively contributes to expansion and hence RI matching) and chemical penetration	Scale reagents, ScaleCUBIC reagents UbasM, FRUIT, ClearSee, vDISCO	Urea
1,3-Dimethyl-2-imidazolidinone	Hydration (putatively contributes to expansion and hence RI matching) and chemical penetration	UbasM-1–UbasM-2	Urea
Formamide	Putatively similar mechanism to urea	ClearT, ClearT2, RTF	Amide
N-Methylacetamide	RI matching	Ce3D	Amide
Nicotinamide	RI matching	CUBIC-R, CUBIC-R+	Aromatic amide
N-Methylnicotinamide	RI matching	CUBIC-R, CUBIC-R+	Aromatic amide
Antipyrine	RI matching and expansion	CUBIC-X, CUBIC-R, CUBIC-R+	Phenyl pyrazolone
5-(N-2,3-Dihydroxypropyla-cetamido)-2,4,6-triiodo-N, N*l*-bis(2,3-dihydroxypropyl) isophthalamide (iohexol)	RI matching	RIMS–SeeDB2–Ce3D	Contrast reagent
Diatrizoate acid	RI matching	RI matching reagent in SWITCH protocol, FocusClear	Contrast agent
Iodixanol	RI matching	RI matching reagent in SWITCH protocol	Contrast reagent
N-Acetyl-l-hydroxyproline (trans-1-acetyl-4-hydroxy-l-proline)	Relaxing collagen structure	ScaleS, vDISCO	Amino acid

Table 5.5 (*Continued*).

Chemical	Function in tissue clearing	Protocol	Functional groups
Meglumine or N-methyl-d-glucamine	Delipidation and RI matching	UbasM-1, RI matching reagent in SWITCH protocol	Amino alcohol
Imidazole	Expansion and decalcification (with EDTA)	CUBIC-X, CUBIC-B	Imidazole
Sodium dodecyl sulfate	Delipidation	CLARITY and related protocols (PACT, PARS, ACT, FACT, FASTClearand simplified CLARITY), SWITCH	Ionic detergent
Sodium dodecyl-benzenesulfonate	Delipidation	CUBIC-HL	Ionic detergent
Sodium deoxycholate	Delipidation	ClearSee	Ionic detergent
Triton X-100	Delipidation	Scale reagents, ScaleCUBIC-1, UbasM-1 (0.2%), Ce3D (0.1%), SUT, CUBIC-L, CUBIC-HL, vDISCO	Non-ionic detergent
Tween 20	Delipidation	FocusClear	Non-ionic detergent
EDTA	Decalcification	PACT-deCal, Bone CLARITY, CUBIC-B, PEGASOS, vDISCO	EDTA

Technical note: Refractive index (RI) matching allows facilitating light to pass unhindered through the 3D cleared samples during imaging. BABB, benzyl alcohol and benzyl benzoate; Ce3D, clearing-enhanced 3D; EDTA, ethylenediaminetetraacetic acid; FACT, fast free-of-acrylamide clearing tissue; LUCID, illuminate cleared organs to identify target molecules; PEGASOS, polyethylene glycol-associated solvent system; RTF, rapid clearing method based on triethanolamine and formamide; RI, refractive index; RIMS, refractive index-matching solution; STP tomography, serial two-photon tomography; TOMEI, transparent plant organ method for imaging; UbasM, urea-based amino-sugar mixture. A portion of the table is adapted from [7] agents (see chemical lists in Tables 5.3 and 5.4) and are potential potent-RI matching reagents (see chemical lists in Table 5.5) for tissue samples [39]. By using amino alcohols for advanced delipidation and RI matching, the hydrophilic methods become like hydrophobic methods in terms of delipidation and decolorization. However, they still preserve biocompatibility, biosafety, and protein characteristics. For this reason, CUBIC has been used to monitor of instant early gene expression in whole organ or body imaging induced by optogenetics platform and small molecules [2, 40, 41]. Namely, the hydrophilic method has succeeded in whole-brain imaging of cancer metastasis, individual neurons, and taste-sensing neural circuits of some species [42–44]. The method also showed detailed imaging performance of 3D imaging of hematopoietic stem cells in bone marrow [45], glutaminergic synaptic connections, carcinoma in the lung, hypothalamic neural subtypes, mammary gland single cell lineage tracing, layer-specific astrocyte morphology, depictions of retinal ganglion cells, and development of the heart and central neuron system [46].

As described in hydrophobic tissue-clearing methods, delipidations in hydrophilic tissue-clearing methods increase the permeability of tissue structures for large biomolecules, for example, antibodies and DNA probes, to penetrate more rapidly and deeply into tissues [47]. Combined with conjugated nanobodies, the hydrophilic CUBIC method allows for conducting 3D immunohistochemistry in large samples such as heart, brain, lung, stomach, and intestine during their development [48, 49]. Recently, in combination with whole-brain 3D immunohistochemistry, two derivative techniques (CUBIC-L–CUBIC-R) of the CUBIC method (Table 5.6) show high resolution of blood vessels in entire mouse brain with neurons expressing vesicular acetylcholine transporter [50]. Other versions of CUBIC, for

Table 5.6   Tissue-clearing reagent for CUBIC methods.

Name	Role
CUBIC-L	For delipidation and decoloring
CUBIC-R+(N)	For RI matching
CUBIC-R+(M)	For RI matching
CUBIC-B	For decalcification
CUBIC-HL	For delipidation strongly and quenching autofluorescence
CUBIC-P	With perfusion before tissue excision
CUBIC-X1	For expansion
CUBIC-X2	For RI matching with expansion

Note. This information is adapted from [62].

example, AbScale, can be implemented with brain-wide immunohistochemistry to image amyloid- β plaques formation, a milestone pathological feature of Alzheimer's disease [51, 52]. In order to expand the size of tissue samples, a derivate of the hydrophilic method, CUBIC-X (X is X-fold expansion), was designed to use an imidazole and antipyrine as hyperhydrating reagents [53]. With this method, the adult mouse brain can be expanded 10-fold in volume for whole-brain cell profiling and creating a single-cell resolution 3D mouse brain atlas, which allows mapping gene expression, cell differentiation, and communication [53].

5.3 Labeling in Tissue-clearing Techniques

Three typical methods of tissue-clearing are hydrophobic, hydrophilic, and hydrogel-based tissue-clearing techniques that are suited for variable domain antibodies and nanobodies staining of intact organs (Figure 5.5 A). The staining step supports tissue-clearing methods and is fully capable of integrating information from post-mortem samples with functional biomarkers [54]. For example, the staining allows tracking transcriptional, translational, and biochemical changes in response to heart (Figure 5.5 B), kidney (Figure 5.5 C), and brain-wide neuronal activity after death. After staining with specific probes, single-molecule fluorescence in situ hybridization is employed to detect changes in RNA transcripts in large-scale tissue clearing (PACT) [55]. The staining also allows conducting multiple biomarkers by using multiple tagged probes in deep tissue to understand pathophysiological processes while preserving the spatial relationships of the microenvironment [56]. In combination with light sheet microscopy, 3D single-molecule super-resolution imaging of neuronal synapses can be detected in brain slices in which their thickness is on a millimeter scale, which allows reducing fluorescence background, photobleaching, and photodamage during imaging of biological samples [57]. At this thickness, microbial identification after Passive CLARITY (MiPACT), MiPACT-cleared samples showed the identity and growth rate of pathogens in cleared sputum samples in cystic fibrosis patients using rRNA probes [58]. Therefore, labeling methods are considered a crucial step in tissue-clearing methods. To maximize the performance of tissue-clearing methods, especially in long-range projections of the central and peripheral nervous systems, the clearing methods need to be paired with specific labeling methods while preserving their tissue identity. Another factor that should be considered in labeling steps is that densely labeled neurons like those in neuronal circuits, morphology-filling markers, and dopaminergic projections in the brain output large data sets, making it difficult to recover and reconstruct tissue morphology. [59]. The fraction of cells from labeling individual cells in brain has been obtained by multiple engineering systemic adeno-associated viruses. The engineered virus can cross the blood-brain barrier, peripheral nervous system, and other organs and emit a variety of hues of modified encoded fluorescent proteins. The advantage of using virus-assisted spectral monitoring is that it enables customizing multicolor labeling system and controlling labeling density for gene expression, protein profiles, and cell communications in 3D morphology of cells in whole, thick, and cleared tissues [60, 61]. Therefore, the combination of tissue-clearing and labeling techniques substantially enhances the understanding of molecular pathogenesis of multiple factors-related diseases.

5.4 Applications in Clinical Samples and Data Processing

Structure-function relationships have been investigated by traditional anatomical methods that require the sectioning of the target tissues such as the heart or the brain (Figure 5.6). The nervous system is a high-structured organ, with hundreds of individual cell types forming complicated circuits that are difficult to analyze on 2D slides. The advances in tissue clearing chemistry and optical microscopy (e.g., light sheet microscopy) provide a great tool for the in-depth study of tissue structure of clinical samples, for example in imaging entire clinical samples of neurodegenerative and cardiac diseases [63, 64]. Figure 5.6A shows a typical scheme of the tissue-clearing process for clinical samples.

The next challenge involves translating analyzed data of the nervous system to applications in clinical research for post-mortem pathological analysis of neurodegenerative diseases. Recently, many methods (CLARITY, iDISCO, and Scale) have been developed to visualize both samples: healthy and pathological human samples [65–67]. Cortical pyramidal neurons and interneurons, α-synuclein inclusions, dopaminergic axons, mitochondria, and amyloid plaques have been imaged and analyzed in intact animal and human brain tissue using chemical probes [68] and FxClear-based label-free imaging [69]. These techniques can detect diffused plaques and links between microglia and amyloid plaques in Alzheimer's pathology [19, 70]. In cancer metastasis, CUBIC integrated with whole-brain immunohistochemistry is a procedure used to see metastatic cancerous cells of blood vessels. Moreover, active CLARITY technique-pressure-related efficient and stable transfer of macromolecules into organs, ACT-PRESTO) allows imaging of GABAergic

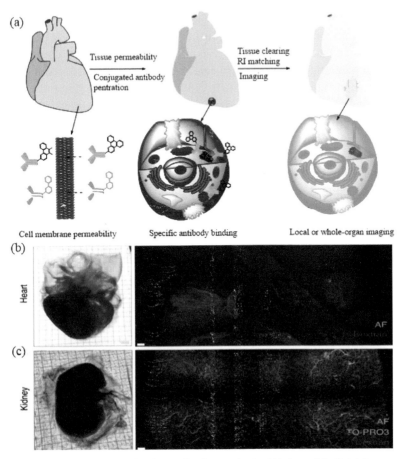

Figure 5.5 Labeling systems for tissue-clearing methods. (A) A model of staining process. (B) SHANEL protocol for heart clearing and staining, the red (Dextran, Texas red) shows blood vessels of the coronary artery, and autofluorescence (AF, 488 nm) shows myocardium. Scale bars, 1 cm and 2 mm, respectively. (C) SHANEL protocol for kidney clearing and staining. The pink (Dextran, Cy5.5) shows the structure of glomeruli. TO-PRO3 shows nuclei in blue. Scale bars, 1 cm and 2 mm, respectively. Parts (B, C) are adapted from [63].

interneurons in slices of hemimegalencephalic cortex [71, 72]. Some clearing reagents (see Table 5.3) are compatible with lipophilic dyes and minimize distortion and destruction during tissue delipidation, which has been applied to image mossy fiber axons in adult human cerebellum.

The entire section (uppermost, left); details of the entire gut wall (uppermost, right); Paneth cells in the crypts (PC); enterocytes, goblet cells (GC);

myenteric (MN) and submucous neurons (SM). In both preparations, the PACT solution obviously protects the tips of the villi better than 4% PFA (B, uppermost). Scale bars: entire gut cross sections 500 μm; gut wall overviews 200 μm; high magnification images 20 μm. Parts A, B are adapted from [73].

However, some challenges of human brain clearing persist. First, the size and dimensions of tissue blocks are less than a few hundred cubic micrometers, the best ratio is about 1/1,500 of the total human brain volume [74]. To obtain sufficient RI matching and transparency, human tissues need more clearing time. Second, larger and more complex structures such as human brain tissue need to be fixed by formalin, but it generates higher autofluorescence [75]. Some lipofuscin-grouped pigments and neuromelanin in blood are difficult to render transparent [76, 77]. Third, the permittivity of conjugated antibodies or nanobodies is limited to clear samples of less than a few hundred micrometers, and many conjugated antibodies failed to bind to biomarkers of specific deep tissues like in the 3D imaging of transparent adult human CNS [78–80]. Regarding lower structures of human tissues, tissue clearing has successfully provided interesting data on 3D cytoarchitecture human embryonic development and few millimeter-sized brain organoids [81]. DISCO methods conjugated with immunolabeling successfully made the gestational

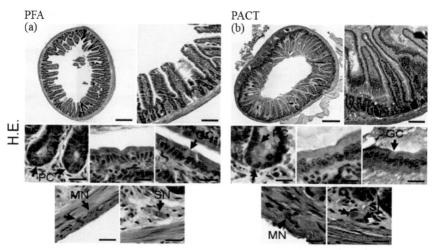

Figure 5.6   Classical histology of cleared gut tissue. Transversal paraffin sections of the gut were fixed with 4% para formaldehyde (PFA, (A)) and cleared with large-scale tissue clearing (PACT, (B)) solution and stained by hematoxylin–eosin (HE) subsequently.

development from human embryos to fetuses transparent. By using polyi-nosinic: polycytidylic acid (poly(I:C)), 3D reconstructions of whole lung airway structures and airway subepithelial collagen were observed by using large-scale, label-free multiphoton microscopy (MPM) and second harmonic generation microscopy (SHGM) (Figure 5.7 E, F) [82]. By using a large panel of antibodies, 3D imaging of DISCO-cleared human embryos detected neurons secreting gonadotropin-releasing hormones involving two biochem-ical pathways in neural subsections outside the hypothalamus [83]. 3DISCO methods also show heterogeneous and unsystematic sensory nerve branching pattern formation detected in the hands, limbs, and head [74, 78].

The clearing-tissue methods have also been used to investigate molecular correlations (combination between protein and nucleic acid imaging) between tumor tissues and virus infections in in situ 3D tissue microenvironments [84–86]. Some recent reports of tissue-clearing methods present a significant outperformance in the cellular resolution for HIV-infected human T lympho-cytes in mouse lymphoid organs, which is better than that of bioluminescent assay regarding quantitative assays [87, 88]. Therefore, tissue-clearing meth-ods are expected to contribute to the analysis of 3D imaging of multiple kinds of tissues in different stages of pathological and healthy tissues while maintaining their native context during the process of tissue clearing and imaging.

Together with the development of innovative chemicals for clearing-tissue methods, optical engineering currently becomes one of the interesting elements in clearing-tissue techniques. Exact structure information at the subcellular level is required to capture the images at rapid, precise, high reso-lution, and comprehensive assessment of 3D large volume of complex tissues, for example in experimental nerve regeneration and tumor tissue complexity. Multicolor two-photon light-sheet microscopy with both spatial and temporal specifications becomes a key solution for remarkable imaging advances [89]. The key characteristics of light-sheet microscopy lie in the illumination of the specimen by a thin sheet of laser from the side and the acquisition of a representation of the illuminated plane by a CCD camera-based detection system. The most common operation of light-sheet microscopy involves a wide-field detection arm that is targeted at a specific angle onto the light-sheet illumination axis with varying objectives used for tissue illumination and feature detection.

With the use of modern sCMOS detectors (option: nanophotonic inte-gration), light-sheet microscopy offers faster data acquisition rates (up to several hundred million voxels per second) for the imaging of large tissue

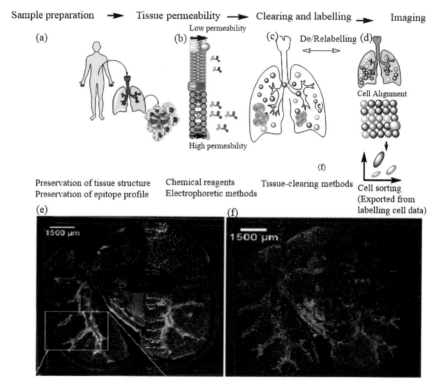

Figure 5.7 A model of tissue clearing and staining of human tissues. (A) samples preparation; (B) treatments for enhancing tissue permeability; (C) tissue clearing and labeling with different fluorescent tags (destaining and relabeling is an option); (D) Imaging and imaging processing; (E) treated by poly(I:C), large-scale, label-free multiphoton microscopy (MPM) and second harmonic generation microscopy (SHGM) for 3D reconstructions of whole lung airway structures and airway subepithelial collagen throughout the full lung. (F) A poly(I:C)-treated lungs are represented in a 3D rendering of whole lung collagen. Parts (E, F) are adapted from [82].

volumes, compared to conventional microscopy [90, 91]. Photobleaching and phototoxic effects on tissue samples are negligible since the emitted laser light is only directed onto the focus plane of samples of the detection system. Therefore, the total emitted energy that the samples are exposed to is constant and independent from sample size and volume and used for data processing.

High-speed and high-resolution imaging of large specimens create large-scale data that needs to be analyzed by a new computational algorithm. In only a single mouse brain analysis, we can record thousands of raw data of individual 3D subvolumes recorded in millions of terabytes [92]. Due to the complexity of interconnected structures of the nervous system and tumor tissues, comprehensive analysis can only be retrieved from thousands of cells in networks of large volumes. To that end, the tiling of image acquisition is an important step to extract only a fragment of the specimen volume at a time and ultimately render an image of the whole specimen, with larger amount of resulting image data. To exactly extract connectomically biological data, an order of powerful computational methods is conducted as follows: large-scale data regulation; multi-tile image profiling and fusion; image processing and mapping of image data to existing brain atlases; and building interactive

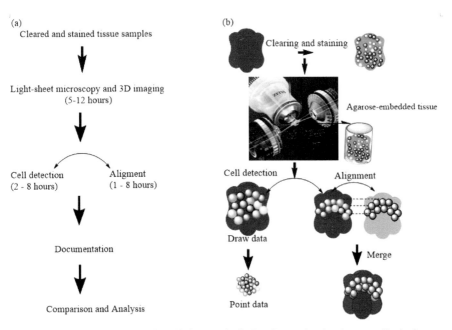

Figure 5.8 A model of single-cell data analysis for tissue-clearing images. Typical steps to analyze whole-brain profiling of single cells. The data obtained from cleared tissues can be registered in the 3D single-cell resolution respiratory (e.g., CUBIC-Atlas). First, tissue specimens are cleared and stained with a chosen method. The cleared tissues are imaged using high-resolution light-sheet microscopy integrated with a high-speed cell counting unit to determine all cells from the acquired images, transferring the whole tissue into the assembled cellular point. Subsequently, individual tissue samples are registered and documented onto the assigned common tissue coordinates (such as heart, lung, and brain) to conduct quantitative comparison and analysis.

visualization of the 3D image data for single cell analysis (Figure 5.8) [93]. Recently, processes for rapid compression and decompression of large-scale image data have been developed [94]. For example, a high writing and reading speech file, the Keller Lab Block (KLB) file format, is a block-based, lossless compression format that enables recording at 500 and 1,000 MB per second, respectively [95, 96]. At such speech, the block-based approach offers rapid access to micro-regions in large 3D images, toward explaining connectomics in tissue specimens. However, big data processing is still a challenge in connectomics. Other approaches include the Big-DataViewer and ImageJ ecosystem which also allows for the manipulation of many hboxblock-based data formats including hierarchical data format version 5 (HDF5), hierarchical data format version 5 (HDF5), and KLB with ease [97]. These latest breakthroughs including high-speech, high-resolution imaging with new large-aperture, and long-working-distance detection optics used in data processing are integrated into light-sheet microscopy for imaging cleared tissues and multiple strategies of imaging large tissue volumes.

5.5 Perspectives and Conclusions

Although 2D traditional histology remains useful in tissue analysis, this technique only operates on a few isolated thin slices of tissue samples, making it susceptible to unavoidable biases because key details can be omitted from the isolated sample. Important information on pathological and healthy tissue could be located elsewhere outside the thin slices. By contrast, tissue-clearing techniques allow the analysis of 3D histology on whole transparent specimens, which provides more insight into molecular and cellular interactions that can be used to explain anatomy and pathology at a lower cost than that of histology analysis. However, the tissue-clearing process needs to be improved considerably to understand the biological and chemical interactions in deep tissues. This can be solved if labeling, imaging, and data analysis tools are made more robust when implanted in combination. A series of techniques can be considered: immunohistochemistry for whole organs, especially for RNA labeling; new algorithms for light-sheet microscopy to image tissues as large as organs and bodies with less than 1- μm isotropic subcellular resolution; accurate data management for data sets extending beyond tens of terabytes. Furthermore, transcriptional and translational profiles containing extremely informative data on health and pathogenesis of tissues should be analyzed in situ in 3D tissue imaging of intact organs. This allows a quantitative comparison of expression levels between mRNA and proteins in different individuals that is more informative than quantitative PCR, western blotting, or fluorescence in situ hybridization (FISH) in 2D sliced tissues. It would be an informative approach if innovative methods such as multiplexed robust FISH can image a panel of mRNAs on tissue sections alongside tissue-clearing methods. Interestingly, tissue-clearing methods can be integrated with other informative techniques such as single-cell RNA sequencing and mass spectrometry. The combination allows for the investigation of spatial and temporal data of RNA and protein expression over entire organs and bodies. This approach can contribute to drug development assays for monitoring molecular changes in single cells in entire organs and bodies versus control groups. For example, some drug candidates have been developed for neurodegenerative diseases and small molecules that cross the blood-brain barriers to bind neuronal receptors or dysfunctional proteins (beta amyloids) observed underlying 3D intact tissues. Up to date, many attempts have targeted amyloid peptides in the plaques by nanobodies for Alzheimer's disease treatment, but clinical trials have not shown convincible data of their efficacy. It is as if there is lacking the technology to provide informative data on therapeutics, for example, epitope and isotype specificities of antibodies to beta-amyloid peptide, at single-molecule resolution in intact 3D tissue specimens. For example, single-molecule resolution facilitates the analysis of single plaque clearance under the treatment of specific antibodies in an animal model of Alzheimer's disease. Moreover, single-molecule resolution also allows for the monitoring of biodistributions and actions of specific antibodies in cellular and biochemical mechanisms of plaque clearance. In this regard, nanoparticle-based therapeutics also become a potential candidate to use in tandem with tissue-clearing techniques to image targets in 3D intact tissues. Bioluminescence, fluorescence, surface-enhanced Raman scattering, and quantum dots can be conducted with plasmonic nanoparticles-based theragnostics. Recently, nanoparticle-chelator conjugates are explored as a potential novel therapeutic candidate for Alzheimer's disease. In combination with functional nanoparticles, the performance of tissue-clearing methods will improve the identification of molecular drug interactions in whole rodents at single-cell resolution in 3D intact tissues. Taken together, tissue-clearing methods provide informative data of 3D histology of healthy and pathological tissues at a scale-up level. These support basic, preclinical, and clinical investigations of human tissues and organs. Tissue-clearing methods are being combined with immunolabeling and imaging to investigate whole human biopsy specimens at single-cell resolution. This combination will deliver a much more rapid and accurate diagnosis and staging of the pathology on the order of centimeters, becoming a key technology to investigate the molecular network structure and interaction circuits of whole organs using the following strategies: clearing, labeling, and imaging patterns, toward connectomic data collection and processing.

References

[1] Almagro, J.; Messal, H. A.; Zaw Thin, M.; van Rheenen, J.; Behrens, A., Tissue clearing to examine tumour complexity in three dimensions. *Nature Reviews Cancer* 2021, 21, (11), 718-730.

[2] Yang, B.; Treweek, Jennifer B.; Kulkarni, Rajan P.; Deverman, Benjamin E.; Chen, C.-K.; Lubeck, E.; Shah, S.; Cai, L.; Gradinaru, V., Single-Cell Phenotyping within Transparent Intact Tissue through Whole-Body Clearing. *Cell* 2014, 158, (4), 945-958.

[3] Kamińska, K.; Szczylik, C.; Bielecka, Z. F.; Bartnik, E.; Porta, C.; Lian, F.; Czarnecka, A. M., The role of the cell-cell interactions in cancer progression. *J Cell Mol Med* 2015, 19, (2), 283-296.

[4] Kürten, C. H. L.; Kulkarni, A.; Cillo, A. R.; Santos, P. M.; Roble, A. K.; Onkar, S.; Reeder, C.; Lang, S.; Chen, X.; Duvvuri, U.; Kim, S.; Liu, A.; Tabib, T.; Lafyatis, R.; Feng, J.; Gao, S.-J.; Bruno, T. C.; Vignali, D. A. A.; Lu, X.; Bao, R.; Vujanovic, L.; Ferris, R. L., Investigating immune and non-immune cell interactions in head and neck tumors by single-cell RNA sequencing. *Nature Communications* 2021, 12, (1), 7338.

[5] Marusyk, A.; Polyak, K., Tumor heterogeneity: causes and consequences. *Biochim Biophys Acta* 2010, 1805, (1), 105-117.

[6] Vickman, R. E.; Faget, D. V.; Beachy, P.; Beebe, D.; Bhowmick, N. A.; Cukierman, E.; Deng, W.-M.; Granneman, J. G.; Hildesheim, J.; Kalluri, R.; Lau, K. S.; Lengyel, E.; Lundeberg, J.; Moscat, J.; Nelson, P. S.; Pietras, K.; Politi, K.; Puré, E.; Scherz-Shouval, R.; Sherman, M. H.; Tuveson, D.; Weeraratna, A. T.; White, R. M.; Wong, M. H.; Woodhouse, E. C.; Zheng, Y.; Hayward, S. W.; Stewart, S. A., Deconstructing tumor heterogeneity: the stromal perspective. *Oncotarget* 2020, 11, (40), 3621-3632.

[7] Ueda, H. R.; Ertürk, A.; Chung, K.; Gradinaru, V.; Chédotal, A.; Tomancak, P.; Keller, P. J., Tissue clearing and its applications in neuroscience. *Nature Reviews Neuroscience* 2020, 21, (2), 61-79.

[8] Ariel, P., A beginner's guide to tissue clearing. *Int J Biochem Cell Biol* 2017, 84, 35-39.

[9] Santi, P. A., Light sheet fluorescence microscopy: a review. *J Histochem Cytochem* 2011, 59, (2), 129-138.

[10] Rocha, M. D.; Düring, D. N.; Bethge, P.; Voigt, F. F.; Hildebrand, S.; Helmchen, F.; Pfeifer, A.; Hahnloser, R. H. R.; Gahr, M., Tissue Clearing and Light Sheet Microscopy: Imaging the Unsectioned Adult Zebra Finch Brain at Cellular Resolution. *Front Neuroanat* 2019, 13, 13.

[11] Tomer, R.; Ye, L.; Hsueh, B.; Deisseroth, K., Advanced CLARITY for rapid and high-resolution imaging of intact tissues. *Nature Protocols* 2014, 9, (7), 1682-1697.

[12] Choi, S. W.; Guan, W.; Chung, K., Basic principles of hydrogel-based tissue transformation technologies and their applications. *Cell* 2021, 184, (16), 4115-4136.

[13] Phillips, J.; Laude, A.; Lightowlers, R.; Morris, C. M.; Turnbull, D. M.; Lax, N. Z., Development of passive CLARITY and immunofluorescent labelling of multiple proteins in human cerebellum: understanding mechanisms of neurodegeneration in mitochondrial disease. *Scientific Reports* 2016, 6, (1), 26013.

[14] Tomer, R.; Ye, L.; Hsueh, B.; Deisseroth, K., Advanced CLARITY for rapid and high-resolution imaging of intact tissues. *Nat Protoc* 2014, 9, (7), 1682-97.

[15] Pavlova, I. P.; Shipley, S. C.; Lanio, M.; Hen, R.; Denny, C. A., Optimization of immunolabeling and clearing techniques for indelibly labeled memory traces. *Hippocampus* 2018, 28, (7), 523-535.

[16] Guo, Z.; Zheng, Y.; Zhang, Y., CLARITY techniques based tissue clearing: types and differences. *Folia Morphol (Warsz)* 2022, 81, (1), 1-12.

[17] Ono, Y.; Nakase, I.; Matsumoto, A.; Kojima, C., Rapid optical tissue clearing using poly(acrylamide-co-styrenesulfonate) hydrogels for three-dimensional imaging. *J Biomed Mater Res B Appl Biomater* 2019, 107, (7), 2297-2304.

[18] Phillips, J.; Laude, A.; Lightowlers, R.; Morris, C. M.; Turnbull, D. M.; Lax, N. Z., Development of passive CLARITY and immunofluorescent labelling of multiple proteins in human cerebellum: understanding mechanisms of neurodegeneration in mitochondrial disease. *Sci Rep* 2016, 6, 26013.

[19] Liebmann, T.; Renier, N.; Bettayeb, K.; Greengard, P.; Tessier-Lavigne, M.; Flajolet, M., Three-Dimensional Study of Alzheimer's Disease

Hallmarks Using the iDISCO Clearing Method. *Cell Rep* 2016, 16, (4), 1138-1152.

[20] Chen, Y.; Shen, Q.; White, S. L.; Gokmen-Polar, Y.; Badve, S.; Goodman, L. J., Three-dimensional imaging and quantitative analysis in CLARITY processed breast cancer tissues. *Scientific Reports* 2019, 9, (1), 5624.

[21] Ertürk, A.; Becker, K.; Jährling, N.; Mauch, C. P.; Hojer, C. D.; Egen, J. G.; Hellal, F.; Bradke, F.; Sheng, M.; Dodt, H.-U., Three-dimensional imaging of solvent-cleared organs using 3DISCO. *Nature Protocols* 2012, 7, (11), 1983-1995.

[22] Qi, Y.; Yu, T.; Xu, J.; Wan, P.; Ma, Y.; Zhu, J.; Li, Y.; Gong, H.; Luo, Q.; Zhu, D., FDISCO: Advanced solvent-based clearing method for imaging whole organs. *Science Advances* 2019, 5, (1), eaau8355.

[23] Fang, C.; Yu, T.; Chu, T.; Feng, W.; Zhao, F.; Wang, X.; Huang, Y.; Li, Y.; Wan, P.; Mei, W.; Zhu, D.; Fei, P., Minutes-timescale 3D isotropic imaging of entire organs at subcellular resolution by content-aware compressed-sensing light-sheet microscopy. *Nature Communications* 2021, 12, (1), 107.

[24] Wan, P.; Zhu, J.; Xu, J.; Li, Y.; Yu, T.; Zhu, D., Evaluation of seven optical clearing methods in mouse brain. *Neurophotonics* 2018, 5, (3), 035007.

[25] Li, W.; Germain, R. N.; Gerner, M. Y., Multiplex, quantitative cellular analysis in large tissue volumes with clearing-enhanced 3D microscopy (C_e3D). *Proceedings of the National Academy of Sciences* 2017, 114, (35), E7321-E7330.

[26] Molbay, M.; Kolabas, Z. I.; Todorov, M. I.; Ohn, T. L.; Ertürk, A., A guidebook for DISCO tissue clearing. *Mol Syst Biol* 2021, 17, (3), e9807.

[27] Renier, N.; Wu, Z.; Simon, D. J.; Yang, J.; Ariel, P.; Tessier-Lavigne, M., iDISCO: a simple, rapid method to immunolabel large tissue samples for volume imaging. *Cell* 2014, 159, (4), 896-910.

[28] Liang, X.; Luo, H., Optical Tissue Clearing: Illuminating Brain Function and Dysfunction. *Theranostics* 2021, 11, (7), 3035-3051.

[29] Cai, R.; Pan, C.; Ghasemigharagoz, A.; Todorov, M. I.; Förstera, B.; Zhao, S.; Bhatia, H. S.; Parra-Damas, A.; Mrowka, L.; Theodorou, D.; Rempfler, M.; Xavier, A. L. R.; Kress, B. T.; Benakis, C.; Steinke, H.; Liebscher, S.; Bechmann, I.; Liesz, A.; Menze, B.; Kerschensteiner, M.; Nedergaard, M.; Ertürk, A., Panoptic imaging of transparent mice

reveals whole-body neuronal projections and skull–meninges connections. *Nature Neuroscience* 2019, 22, (2), 317-327.

[30] Cai, R.; Pan, C.; Ghasemigharagoz, A.; Todorov, M. I.; Förstera, B.; Zhao, S.; Bhatia, H. S.; Parra-Damas, A.; Mrowka, L.; Theodorou, D.; Rempfler, M.; Xavier, A. L. R.; Kress, B. T.; Benakis, C.; Steinke, H.; Liebscher, S.; Bechmann, I.; Liesz, A.; Menze, B.; Kerschensteiner, M.; Nedergaard, M.; Ertürk, A., Panoptic imaging of transparent mice reveals whole-body neuronal projections and skull-meninges connections. *Nat Neurosci* 2019, 22, (2), 317-327.

[31] Tainaka, K.; Murakami, T. C.; Susaki, E. A.; Shimizu, C.; Saito, R.; Takahashi, K.; Hayashi-Takagi, A.; Sekiya, H.; Arima, Y.; Nojima, S.; Ikemura, M.; Ushiku, T.; Shimizu, Y.; Murakami, M.; Tanaka, K. F.; Iino, M.; Kasai, H.; Sasaoka, T.; Kobayashi, K.; Miyazono, K.; Morii, E.; Isa, T.; Fukayama, M.; Kakita, A.; Ueda, H. R., Chemical Landscape for Tissue Clearing Based on Hydrophilic Reagents. *Cell Rep* 2018, 24, (8), 2196-2210.e9.

[32] Liu, Y. C.; Chiang, A. S., High-resolution confocal imaging and three-dimensional rendering. *Methods* 2003, 30, (1), 86-93.

[33] Hama, H.; Kurokawa, H.; Kawano, H.; Ando, R.; Shimogori, T.; Noda, H.; Fukami, K.; Sakaue-Sawano, A.; Miyawaki, A., Scale: a chemical approach for fluorescence imaging and reconstruction of transparent mouse brain. *Nat Neurosci* 2011, 14, (11), 1481-8.

[34] Parra-Damas, A.; Saura, C. A., Tissue Clearing and Expansion Methods for Imaging Brain Pathology in Neurodegeneration: From Circuits to Synapses and Beyond. *Frontiers in Neuroscience* 2020, 14.

[35] Yu, T.; Zhu, J.; Li, D.; Zhu, D., Physical and chemical mechanisms of tissue optical clearing. *iScience* 2021, 24, (3), 102178.

[36] Zhu, X.; Huang, L.; Zheng, Y.; Song, Y.; Xu, Q.; Wang, J.; Si, K.; Duan, S.; Gong, W., Ultrafast optical clearing method for three-dimensional imaging with cellular resolution. *Proceedings of the National Academy of Sciences* 2019, 116, (23), 11480-11489.

[37] Ke, M.-T.; Nakai, Y.; Fujimoto, S.; Takayama, R.; Yoshida, S.; Kitajima, Tomoya S.; Sato, M.; Imai, T., Super-Resolution Mapping of Neuronal Circuitry With an Index-Optimized Clearing Agent. *Cell Reports* 2016, 14, (11), 2718-2732.

[38] Tainaka, K.; Murakami, T. C.; Susaki, E. A.; Shimizu, C.; Saito, R.; Takahashi, K.; Hayashi-Takagi, A.; Sekiya, H.; Arima, Y.; Nojima, S.; Ikemura, M.; Ushiku, T.; Shimizu, Y.; Murakami, M.; Tanaka, K. F.; Iino, M.; Kasai, H.; Sasaoka, T.; Kobayashi, K.; Miyazono, K.; Morii,

E.; Isa, T.; Fukayama, M.; Kakita, A.; Ueda, H. R., Chemical Landscape for Tissue Clearing Based on Hydrophilic Reagents. *Cell Reports* 2018, 24, (8), 2196-2210.e9.

[39] Richardson, D. S.; Lichtman, J. W., Clarifying Tissue Clearing. *Cell* 2015, 162, (2), 246-257.

[40] Franceschini, A.; Costantini, I.; Pavone, F. S.; Silvestri, L., Dissecting Neuronal Activation on a Brain-Wide Scale With Immediate Early Genes. *Front Neurosci* 2020, 14, 569517.

[41] Simpson, S.; Chen, Y.; Wellmeyer, E.; Smith, L. C.; Aragon Montes, B.; George, O.; Kimbrough, A., The Hidden Brain: Uncovering Previously Overlooked Brain Regions by Employing Novel Preclinical Unbiased Network Approaches. *Front Syst Neurosci* 2021, 15, 595507.

[42] Mano, T.; Albanese, A.; Dodt, H. U.; Erturk, A.; Gradinaru, V.; Treweek, J. B.; Miyawaki, A.; Chung, K.; Ueda, H. R., Whole-Brain Analysis of Cells and Circuits by Tissue Clearing and Light-Sheet Microscopy. *J Neurosci* 2018, 38, (44), 9330-9337.

[43] Cong, L.; Wang, Z.; Chai, Y.; Hang, W.; Shang, C.; Yang, W.; Bai, L.; Du, J.; Wang, K.; Wen, Q., Rapid whole brain imaging of neural activity in freely behaving larval zebrafish (Danio rerio). *eLife* 2017, 6, e28158.

[44] Amato, S. P.; Pan, F.; Schwartz, J.; Ragan, T. M., Whole Brain Imaging with Serial Two-Photon Tomography. *Frontiers in Neuroanatomy* 2016, 10.

[45] Nowzari, F.; Wang, H.; Khoradmehr, A.; Baghban, M.; Baghban, N.; Arandian, A.; Muhaddesi, M.; Nabipour, I.; Zibaii, M. I.; Najarasl, M.; Taheri, P.; Latifi, H.; Tamadon, A., Three-Dimensional Imaging in Stem Cell-Based Researches. *Frontiers in Veterinary Science* 2021, 8.

[46] Furuta, T.; Yamauchi, K.; Okamoto, S.; Takahashi, M.; Kakuta, S.; Ishida, Y.; Takenaka, A.; Yoshida, A.; Uchiyama, Y.; Koike, M.; Isa, K.; Isa, T.; Hioki, H., Multi-scale light microscopy/electron microscopy neuronal imaging from brain to synapse with a tissue clearing method, ScaleSF. *iScience* 2022, 25, (1), 103601.

[47] Sylwestrak, E. L.; Rajasethupathy, P.; Wright, M. A.; Jaffe, A.; Deisseroth, K., Multiplexed Intact-Tissue Transcriptional Analysis at Cellular Resolution. *Cell* 2016, 164, (4), 792-804.

[48] Ueda, H. R.; Dodt, H.-U.; Osten, P.; Economo, M. N.; Chandrashekar, J.; Keller, P. J., Whole-Brain Profiling of Cells and Circuits in Mammals by Tissue Clearing and Light-Sheet Microscopy. *Neuron* 2020, 106, (3), 369-387.

[49] Messal, H. A.; Almagro, J.; Zaw Thin, M.; Tedeschi, A.; Ciccarelli, A.; Blackie, L.; Anderson, K. I.; Miguel-Aliaga, I.; van Rheenen, J.; Behrens, A., Antigen retrieval and clearing for whole-organ immunofluorescence by FLASH. *Nature Protocols* 2021, 16, (1), 239-262.

[50] Mano, T.; Albanese, A.; Dodt, H.-U.; Erturk, A.; Gradinaru, V.; Treweek, J. B.; Miyawaki, A.; Chung, K.; Ueda, H. R., Whole-Brain Analysis of Cells and Circuits by Tissue Clearing and Light-Sheet Microscopy. *The Journal of Neuroscience* 2018, 38, (44), 9330-9337.

[51] Bhatia, H. S.; Brunner, A.-D.; Rong, Z.; Mai, H.; Thielert, M.; Al-Maskari, R.; Paetzold, J. C.; Kofler, F.; Todorov, M. I.; Ali, M.; Molbay, M.; Kolabas, Z. I.; Kaltenecker, D.; Müller, S.; Lichtenthaler, S. F.; Menze, B. H.; Theis, F. J.; Mann, M.; Ertürk, A., Proteomics of spatially identified tissues in whole organs. *bioRxiv* 2021, 2021.11.02.466753.

[52] Hama, H.; Hioki, H.; Namiki, K.; Hoshida, T.; Kurokawa, H.; Ishidate, F.; Kaneko, T.; Akagi, T.; Saito, T.; Saido, T.; Miyawaki, A., ScaleS: an optical clearing palette for biological imaging. *Nature Neuroscience* 2015, 18, (10), 1518-1529.

[53] Murakami, T. C.; Mano, T.; Saikawa, S.; Horiguchi, S. A.; Shigeta, D.; Baba, K.; Sekiya, H.; Shimizu, Y.; Tanaka, K. F.; Kiyonari, H.; Iino, M.; Mochizuki, H.; Tainaka, K.; Ueda, H. R., A three-dimensional single-cell-resolution whole-brain atlas using CUBIC-X expansion microscopy and tissue clearing. *Nature Neuroscience* 2018, 21, (4), 625-637.

[54] Susaki, E. A.; Shimizu, C.; Kuno, A.; Tainaka, K.; Li, X.; Nishi, K.; Morishima, K.; Ono, H.; Ode, K. L.; Saeki, Y.; Miyamichi, K.; Isa, K.; Yokoyama, C.; Kitaura, H.; Ikemura, M.; Ushiku, T.; Shimizu, Y.; Saito, T.; Saido, T. C.; Fukayama, M.; Onoe, H.; Touhara, K.; Isa, T.; Kakita, A.; Shibayama, M.; Ueda, H. R., Versatile whole-organ/body staining and imaging based on electrolyte-gel properties of biological tissues. *Nature Communications* 2020, 11, (1), 1982.

[55] Yang, B.; Treweek, J. B.; Kulkarni, R. P.; Deverman, B. E.; Chen, C. K.; Lubeck, E.; Shah, S.; Cai, L.; Gradinaru, V., Single-cell phenotyping within transparent intact tissue through whole-body clearing. *Cell* 2014, 158, (4), 945-958.

[56] Hofmann, J.; Gadjalova, I.; Mishra, R.; Ruland, J.; Keppler, S. J., Efficient Tissue Clearing and Multi-Organ Volumetric Imaging Enable Quantitative Visualization of Sparse Immune Cell Populations During Inflammation. *Front Immunol* 2021, 11.

[57] Gagliano, G.; Nelson, T.; Saliba, N.; Vargas-Hernández, S.; Gustavs-son, A.-K., Light Sheet Illumination for 3D Single-Molecule Super-Resolution Imaging of Neuronal Synapses. *Frontiers in Synaptic Neuroscience* 2021, 13.

[58] DePas, W. H.; Starwalt-Lee, R.; Sambeek, L. V.; Kumar, S. R.; Gradinaru, V.; Newman, D. K., Exposing the Three-Dimensional Biogeography and Metabolic States of Pathogens in Cystic Fibrosis Sputum via Hydrogel Embedding, Clearing, and rRNA Labeling. *mBio* 2016, 7, (5), e00796-16.

[59] Lin, R.; Wang, R.; Yuan, J.; Feng, Q.; Zhou, Y.; Zeng, S.; Ren, M.; Jiang, S.; Ni, H.; Zhou, C.; Gong, H.; Luo, M., Cell-type-specific and projection-specific brain-wide reconstruction of single neurons. *Nature Methods* 2018, 15, (12), 1033-1036.

[60] Wang, L.; Challis, C.; Li, S.; Fowlkes, C. C.; Ravindra Kumar, S.; Yuan, P. Q.; Taché, Y. F., Multicolor sparse viral labeling and 3D digital tracing of enteric plexus in mouse proximal colon using a novel adeno-associated virus capsid. *Neurogastroenterol Motil* 2021, 33, (8), e14014.

[61] Chan, K. Y.; Jang, M. J.; Yoo, B. B.; Greenbaum, A.; Ravi, N.; Wu, W. L.; Sánchez-Guardado, L.; Lois, C.; Mazmanian, S. K.; Deverman, B. E.; Gradinaru, V., Engineered AAVs for efficient noninvasive gene delivery to the central and peripheral nervous systems. *Nat Neurosci* 2017, 20, (8), 1172-1179.

[62] Mai, H.; Rong, Z.; Zhao, S.; Cai, R.; Steinke, H.; Bechmann, I.; Ertürk, A., Scalable tissue labeling and clearing of intact human organs. *Nature Protocols* 2022.

[63] Rocha, M. D.; Düring, D. N.; Bethge, P.; Voigt, F. F.; Hildebrand, S.; Helmchen, F.; Pfeifer, A.; Hahnloser, R. H. R.; Gahr, M., Tissue Clearing and Light Sheet Microscopy: Imaging the Unsectioned Adult Zebra Finch Brain at Cellular Resolution. *Frontiers in Neuroanatomy* 2019, 13.

[64] Chakraborty, T.; Driscoll, M. K.; Jeffery, E.; Murphy, M. M.; Roudot, P.; Chang, B.-J.; Vora, S.; Wong, W. M.; Nielson, C. D.; Zhang, H.; Zhemkov, V.; Hiremath, C.; De La Cruz, E. D.; Yi, Y.; Bezprozvanny, I.; Zhao, H.; Tomer, R.; Heintzmann, R.; Meeks, J. P.; Marciano, D. K.; Morrison, S. J.; Danuser, G.; Dean, K. M.; Fiolka, R., Light-sheet microscopy of cleared tissues with isotropic, subcellular resolution. *Nature Methods* 2019, 16, (11), 1109-1113.

[65] Liu, A. K.; Hurry, M. E.; Ng, O. T.; DeFelice, J.; Lai, H. M.; Pearce, R. K.; Wong, G. T.; Chang, R. C.; Gentleman, S. M., Bringing CLARITY to the human brain: visualization of Lewy pathology in three dimensions. *Neuropathol Appl Neurobiol* 2016, 42, (6), 573-87.

[66] Rusch, H.; Brammerloh, M.; Stieler, J.; Sonntag, M.; Mohammadi, S.; Weiskopf, N.; Arendt, T.; Kirilina, E.; Morawski, M., Finding the best clearing approach - Towards 3D wide-scale multimodal imaging of aged human brain tissue. *NeuroImage* 2022, 247, 118832.

[67] Vigouroux, R. J.; Belle, M.; Chédotal, A., Neuroscience in the third dimension: shedding new light on the brain with tissue clearing. *Mol Brain* 2017, 10, (1), 33.

[68] Lai, H. M.; Ng, W. L.; Gentleman, S. M.; Wu, W., Chemical Probes for Visualizing Intact Animal and Human Brain Tissue. *Cell Chem Biol* 2017, 24, (6), 659-672.

[69] Lee, B.; Lee, E.; Kim, J. H.; Kim, H.-J.; Kang, Y. G.; Kim, H. J.; Shim, J.-K.; Kang, S.-G.; Kim, B.-M.; Kim, K.; Kim, Y.; Cho, K.; Sun, W., Sensitive label-free imaging of brain samples using FxClear-based tissue clearing technique. *iScience* 2021, 24, (4), 102267.

[70] Hansen, D. V.; Hanson, J. E.; Sheng, M., Microglia in Alzheimer's disease. *J Cell Biol* 2018, 217, (2), 459-472.

[71] Yoneda, T.; Sakai, S.; Maruoka, H.; Hosoya, T., Large-scale Three-dimensional Imaging of Cellular Organization in the Mouse Neocortex. *J Vis Exp* 2018, (139).

[72] Tremblay, R.; Lee, S.; Rudy, B., GABAergic Interneurons in the Neocortex: From Cellular Properties to Circuits. *Neuron* 2016, 91, (2), 260-92.

[73] Neckel, P. H.; Mattheus, U.; Hirt, B.; Just, L.; Mack, A. F., Large-scale tissue clearing (PACT): Technical evaluation and new perspectives in immunofluorescence, histology, and ultrastructure. *Sci Rep* 2016, 6, 34331.

[74] Ueda, H. R.; Ertürk, A.; Chung, K.; Gradinaru, V.; Chédotal, A.; Tomancak, P.; Keller, P. J., Tissue clearing and its applications in neuroscience. *Nat Rev Neurosci* 2020, 21, (2), 61-79.

[75] Davis, A. S.; Richter, A.; Becker, S.; Moyer, J. E.; Sandouk, A.; Skinner, J.; Taubenberger, J. K., Characterizing and Diminishing Autofluorescence in Formalin-fixed Paraffin-embedded Human Respiratory Tissue. *J Histochem Cytochem* 2014, 62, (6), 405-423.

[76] Double, K. L.; Dedov, V. N.; Fedorow, H.; Kettle, E.; Halliday, G. M.; Garner, B.; Brunk, U. T., The comparative biology of neuromelanin and lipofuscin in the human brain. *Cell Mol Life Sci* 2008, 65, (11), 1669-82.

[77] Zucca, F. A.; Vanna, R.; Cupaioli, F. A.; Bellei, C.; De Palma, A.; Di Silvestre, D.; Mauri, P.; Grassi, S.; Prinetti, A.; Casella, L.; Sulzer, D.; Zecca, L., Neuromelanin organelles are specialized autolysosomes that accumulate undegraded proteins and lipids in aging human brain and are likely involved in Parkinson's disease. *npj Parkinson's Disease* 2018, 4, (1), 17.

[78] Hansmeier, N. R.; Büschlen, I. S.; Behncke, R. Y.; Ulferts, S.; Bisoendial, R.; Hägerling, R., 3D Visualization of Human Blood Vascular Networks Using Single-Domain Antibodies Directed against Endothelial Cell-Selective Adhesion Molecule (ESAM). *Int J Mol Sci* 2022, 23, (8), 4369.

[79] Park, T. E.; Mustafaoglu, N.; Herland, A.; Hasselkus, R.; Mannix, R.; FitzGerald, E. A.; Prantil-Baun, R.; Watters, A.; Henry, O.; Benz, M.; Sanchez, H.; McCrea, H. J.; Goumnerova, L. C.; Song, H. W.; Palecek, S. P.; Shusta, E.; Ingber, D. E., Hypoxia-enhanced Blood-Brain Barrier Chip recapitulates human barrier function and shuttling of drugs and antibodies. *Nat Commun* 2019, 10, (1), 2621.

[80] Shin, Y.; Choi, S. H.; Kim, E.; Bylykbashi, E.; Kim, J. A.; Chung, S.; Kim, D. Y.; Kamm, R. D.; Tanzi, R. E., Blood-Brain Barrier Dysfunction in a 3D In Vitro Model of Alzheimer's Disease. *Adv Sci (Weinh)* 2019, 6, (20), 1900962.

[81] de Bakker, B. S.; de Jong, K. H.; Hagoort, J.; Oostra, R. J.; Moorman, A. F., Towards a 3-dimensional atlas of the developing human embryo: the Amsterdam experience. *Reprod Toxicol* 2012, 34, (2), 225-36.

[82] Ochoa, L. F.; Kholodnykh, A.; Villarreal, P.; Tian, B.; Pal, R.; Freiberg, A. N.; Brasier, A. R.; Motamedi, M.; Vargas, G., Imaging of Murine Whole Lung Fibrosis by Large Scale 3D Microscopy aided by Tissue Optical Clearing. *Scientific Reports* 2018, 8, (1), 13348.

[83] Belle, M.; Godefroy, D.; Couly, G.; Malone, S. A.; Collier, F.; Giacobini, P.; Chédotal, A., Tridimensional Visualization and Analysis of Early Human Development. *Cell* 2017, 169, (1), 161-173.e12.

[84] Jiang, S.; Chan, C. N.; Rovira-Clavé, X.; Chen, H.; Bai, Y.; Zhu, B.; McCaffrey, E.; Greenwald, N. F.; Liu, C.; Barlow, G. L.; Weirather, J. L.; Oliveria, J. P.; Nakayama, T.; Lee, I. T.; Matter, M. S.; Carlisle,

A. E.; Philips, D.; Vazquez, G.; Mukherjee, N.; Busman-Sahay, K.; Nekorchuk, M.; Terry, M.; Younger, S.; Bosse, M.; Demeter, J.; Rodig, S. J.; Tzankov, A.; Goltsev, Y.; McIlwain, D. R.; Angelo, M.; Estes, J. D.; Nolan, G. P., Combined protein and nucleic acid imaging reveals virus-dependent B cell and macrophage immunosuppression of tissue microenvironments. *Immunity* 2022, 55, (6), 1118-1134.e8.

[85] Anders, M.; Hansen, R.; Ding, R.-X.; Rauen, K. A.; Bissell, M. J.; Korn, W. M., Disruption of 3D tissue integrity facilitates adenovirus infection by deregulating the coxsackievirus and adenovirus receptor. *Proceedings of the National Academy of Sciences* 2003, 100, (4), 1943-1948.

[86] Moysi, E.; Del Rio Estrada, P. M.; Torres-Ruiz, F.; Reyes-Terán, G.; Koup, R. A.; Petrovas, C., In Situ Characterization of Human Lymphoid Tissue Immune Cells by Multispectral Confocal Imaging and Quantitative Image Analysis; Implications for HIV Reservoir Characterization. *Front Immunol* 2021, 12.

[87] Au - Zhang, T.; Au - Gupta, A.; Au - Frederick, D.; Au - Layman, L.; Au - Smith, D. M.; Au - Gianella, S.; Au - Kieffer, C., 3D Visualization of Immune Cell Populations in HIV-Infected Tissues via Clearing, Immunostaining, Confocal, and Light Sheet Fluorescence Microscopy. *JoVE* 2021, (171), e62441.

[88] Ventura, J. D.; Beloor, J.; Allen, E.; Zhang, T.; Haugh, K. A.; Uchil, P. D.; Ochsenbauer, C.; Kieffer, C.; Kumar, P.; Hope, T. J.; Mothes, W., Longitudinal bioluminescent imaging of HIV-1 infection during antiretroviral therapy and treatment interruption in humanized mice. *PLoS Pathog* 2019, 15, (12), e1008161.

[89] Mahou, P.; Vermot, J.; Beaurepaire, E.; Supatto, W., Multicolor two-photon light-sheet microscopy. *Nature Methods* 2014, 11, (6), 600-601.

[90] Orcutt, J. S.; Khilo, A.; Holzwarth, C. W.; Popović, M. A.; Li, H.; Sun, J.; Bonifield, T.; Hollingsworth, R.; Kärtner, F. X.; Smith, H. I.; Stojanović, V.; Ram, R. J., Nanophotonic integration in state-of-the-art CMOS foundries. *Opt. Express* 2011, 19, (3), 2335-2346.

[91] Kosmidis, S.; Negrean, A.; Dranovsky, A.; Losonczy, A.; Kandel, E. R., A fast, aqueous, reversible three-day tissue clearing method for adult and embryonic mouse brain and whole body. *Cell Reports Methods* 2021, 1, (7), 100090.

[92] Foxley, S.; Sampathkumar, V.; De Andrade, V.; Trinkle, S.; Sorokina, A.; Norwood, K.; La Riviere, P.; Kasthuri, N., Multi-modal imaging

of a single mouse brain over five orders of magnitude of resolution. *NeuroImage* 2021, 238, 118250.

[93] Lichtman, J. W.; Pfister, H.; Shavit, N., The big data challenges of connectomics. *Nat Neurosci* 2014, 17, (11), 1448-54.

[94] Balázs, B.; Deschamps, J.; Albert, M.; Ries, J.; Hufnagel, L., A real-time compression library for microscopy images. *bioRxiv* 2017, 164624.

[95] Lemon, W. C.; Pulver, S. R.; Höckendorf, B.; McDole, K.; Branson, K.; Freeman, J.; Keller, P. J., Whole-central nervous system functional imaging in larval Drosophila. *Nature Communications* 2015, 6, (1), 7924.

[96] Amat, F.; Höckendorf, B.; Wan, Y.; Lemon, W. C.; McDole, K.; Keller, P. J., Efficient processing and analysis of large-scale light-sheet microscopy data. *Nature Protocols* 2015, 10, (11), 1679-1696.

[97] Pietzsch, T.; Saalfeld, S.; Preibisch, S.; Tomancak, P., BigDataViewer: visualization and processing for large image data sets. *Nature Methods* 2015, 12, (6), 481-483.

6

Novel Imaging Contrast Agents and Systems for Biomedical Sensing

Shuai Yu[1] and J.-C. Chiao[2]

[1]Senior Optical R&D Engineer, Rockley Photonics, Irvine, CA 92612;
Department of Bioengineering, The University of Texas at Arlington,
Arlington, TX 76019, USA
[2]Department of Electrical and Computer Engineering, Southern Methodist
University, Dallas, Texas, USA

Abstract

In the past decades, imaging has become more important in biomedical
sensing applications for healthcare because it provides non-invasive nature,
wide ranges of anatomical structure and physiological sensing, and potential for real-time continuous monitoring. Conventional biomedical imaging
techniques include X-ray and X-ray-based computed tomography (CT), ultrasound, magnetic resonance imaging (MRI), optical imaging, positron emission tomography (PET), and single-photon emission computerized tomography (SPECT). Combined with these modalities, imaging contrast agents
play a key role in building a bridge between external detectors and internal
biological/pathological processes. In recent years, novel imaging techniques
such as multi-modal imaging have been fast developed to fulfill the gaps
between needs and limitations in the conventional modalities. Meanwhile,
new imaging contrast agents are quickly emerged and covered in broader
bio-sensing applications. This chapter covers current cutting-edge imaging
techniques, particularly, current imaging contrast agents and systems for a
variety of biomedical sensing.

Keywords: Imaging contrast agents, multi-modality imaging, photoacoustic
imaging, ultrasound-switchable fluorescence imaging

6.1 Introduction

Over the past decades, biomedical imaging has been playing a critical role in biological sensing and disease diagnosis. Biomedical imaging has many advantages including non-invasive detection, real-time monitoring, and wide ranges of spatial resolution (10^{-9} to 10^{-1} m; i.e., nanometers to tens of centimeters) and temporal resolution (10^{-12} to 10^{9} s; picoseconds to years) for biological and pathological processes. Currently, the most common biomedical imaging modalities include: X-ray[1] and X-ray-based computed tomography (CT),[2] ultrasound,[1, 3] magnetic resonance imaging (MRI),[4, 5] positron emission tomography (PET)[6, 7] and single-photon emission computerized tomography (SPECT),[8] and optical imaging.[9–16] In many of them, contrast agents are adopted to ensure and/or improve the image representation. For example, (1) fluorescence imaging requires fluorescent probes attached to the specimen for fluorescent optical signal representation; (2) microbubbles are intravenously injected for ultrasound imaging with a better imaging contrast on vascular tissue.

An important consideration in selecting an imaging modality is its imaging quality and performance on pathological representation, including (1) imaging depth, (2) imaging contrast and sensitivity, (3) spatial resolution, and (4) temporal resolution and acquisition.[17] At the same time, other considerations such as cost and bio-safety are essential too. While each modality carries advantages over others, there are inherent trade-off relationships between its imaging qualities. For example, CT, MRI, and ultrasound have excellent imaging depth (up to tens of centimeters) as well as good spatial resolution. However, their sensitivity to biological tissue is relatively low. On the contrary, optics provides the highest sensitivity in biological tissue and is capable of tracking many interesting physiological and pathological micro-phenomena such as microcirculation, angiogenesis, and cancer metastasis.[18–20] Optics also carries many other advantages such as multicolor imaging [21, 22] because of its wide range of wavelengths (from 400 to 1300 nm). However, due to the high scattering property is tissue, its spatial resolution decreases significantly with the increase of imaging depth.[20]

One solution to overcome the barriers in the trade-offs is to develop multimodality imaging techniques. That is, by combining two sensing mechanisms from the existing modalities, the image will provide biological features with advantageous qualities from both. In general, multi-modality imaging is classified into two cases: (1) there is no physical (energy) interaction between the two sensing modalities, the two separate modalities are spatially registered and temporally synchronized for analysis; (2) there is a physical interaction between the two energy sources so that it forms a new sensing mechanism. An example of case (1) is a PET-CT scan.[23] The CT image provides high spatial resolution and structural information of the tissue (i.e., structural imaging), while the PET image provides the high sensitivity of molecular information of the tissue (functional imaging) yet with low spatial resolution. CT and PET images are spatially registered so that it provides both high resolution of the structural information and high sensitivity of the molecular information. An example of case (2) is photoacoustic (PA) imaging, which is a biomedical imaging modality based on photoacoustic effect.[24] Optical pulses are delivered into the biological tissue and the energy is partially absorbed and converted into heat, leading to a transient thermo-elastic expansion and thus ultrasound emission. The ultrasound signal is detected by the ultrasound transducer array and analyzed to produce the images. Because the signal represents the information of optical absorption properties, it provides physiological properties of biological tissue with high sensitivity. Meanwhile, because ultrasound provides a higher spatial resolution due to a less scattering coefficient of two or three orders of magnitude in the biological tissue,[25] photoacoustic imaging provides a high spatial resolution and high sensitivity in centimeter-deep tissue.

Besides the two examples of multi-modality imaging, many others have been well developed for better sensing quality in biomedical imaging, such as ultrasound-modulated fluorescence (UMF) and ultrasound switchable fluorescence (USF) imaging. Among many of them, the contrast agents play a key role in connecting the external detectors and the internal biological/pathological sensing processes. In this chapter, we will discuss the novel multi-modality imaging techniques, along with their contrast agents and systems, for a variety of biomedical sensing applications.

6.2 Photoacoustic Imaging

Photoacoustic imaging is a multimodal biomedical imaging technique that takes advantage of both high imaging contrast of optical absorption in biological tissue and imaging resolution of acoustic signal in centimeter tissue depth. It breaks the traditional depth limits of ballistic optical imaging and the resolution limits of diffuse optical imaging. By detecting the acoustic waves generated from thermos-elastic expansion in response to the absorption of short-pulsed light, photoacoustic imaging provides an image of optical absorption coefficients at depths from a few millimeters to several centimeters with acoustic resolution capability. In the past two decades, photoacoustic imaging has been fast developed and widely applied in many bio-sensing applications such as quantifying angiogenesis and hyper-metabolism in early-stage cancer detection.

6.2.1 Principle of photoacoustic imaging

Electromagnetic (EM) energy is the optical (from visible to NIR) and radio-frequency (RF) regions used for photoacoustic excitation in soft tissues because these regions are safe (e.g., non-ionizing) and also provide high contrast as well as adequate penetration depth.[25] The former is also called optoacoustic (OA) imaging and the latter is also called thermoacoustic (TA) imaging. Typically, a short low-fluence EM pulse is illuminated on the target to excite ultrasonic waves through the thermos-elastic mechanism when the EM pulse triggers a slight temperature rise (millikelvin) by energy absorption inside the biological tissue.

Two confinements, that is, thermal confinement and stress confinement must be met in order to generate PA signals efficiently.[25, 26] First, the EM pulse duration t_p must be shorter than the time scale for the heat dissipation of the absorbed energy t_{th}, where $t_{th} =\sim L_p{}^2/4D_T$ (L_p is the linear dimension of the tissue volume being heated, and D_T is the thermal diffusivity of the sample), to ensure heat diffusion is negligible during the excitation pulse, so-called thermal confinement. Second, the time for the stress to transit the heated volume $t_s = L_p/c$ (c is the speed of sound in the volume) should be shorter than t_p so that the thermos-elastic pressure in the volume can build up high fast, so called stress confinement. When both confinements are met, an initial pressure rise p_0 by thermal expansion is estimated $p_0= (\beta c^2/C_p)\mu_a F$, where β is the isobaric volume expansion coefficient, C_p is the specific heat, μ_a is the absorption coefficient and F is the local light fluence.

The initial pressure p_0 acts as the acoustic source and wave propagate in three-dimensional space. The wave carries a frequency range and the ultrasound in low-megahertz is selected due to its sufficient resolution, low scattering, and deep penetration, and thus captured by an ultrasound transducer when it reaches the tissue surface. The photoacoustic signal reconstruction is similar to ultrasonography, where its lateral resolution is determined by the focal diameter of the transducer and the central frequency of the received photoacoustic signals; and its axial resolution is determined by the impulse response of the transducer and the EM pulse width.[25] Each detected time-resolved signal after an EM pulse excitation provides a 1D image along the acoustic axis of the transducer (i.e., A line), and by scanning of the transducer, a 2D or 3D image is formed.

A reconstruction-based photoacoustic tomography provides more flexibility in dealing with the received photoacoustic signals than the image-forming methods with focused transducers. By combining both the temporal and spatial information of signals received at various transducer locations, it affords sufficient information of a complete 3D reconstruction. 3D reconstruction tomography involves both frequency domain and time domain algorithms.

6.2.2 Multiscale photoacoustic microscopy and tomography

Photoacoustic imaging can be classified into two major implementations, one is photoacoustic computed tomography (PACT) which focuses on revealing the centimeter deep tissue optical contrast through a reconstruction algorithm,[24, 25] and the other is so call photoacoustic microscopy (PAM) which adopts a raster-scanning of optical and acoustic foci and forms images from the depth-resolved signals (i.e., A-lines).[24, 27] The PACT provides a comparably larger imaging depth and a faster acquisition speed by adopting ultrasound transducer arrays (or linear/planar transducers), yet a lower imaging contrast and a mathematically complicated reconstruction algorithm. The PAM, in comparison, maximizes the detection sensitivity by confocally aligning its optical illumination and ultrasound detection. The PAM can be further classified into three groups based on its multiscale imaging depth and resolution27: (1) optical resolution PAM (OR-PAM); (2) acoustic resolution PAM (AR-PAM); and (3) photoacoustic macroscopy. The OR-PAM has its optical focusing much tighter than its acoustic focusing so that its spatial resolution is higher, but it suffers from less penetration depth due to optical diffusion; while the AR-PAM has its acoustic focusing tighter (e.g., with a central frequency at 50 MHz) and its penetration depth is greater at the sacrifice of resolution to some extent. In photoacoustic microscopy, a further-lower ultrasound transducer (e.g., 5 MHz) can be used along with a NIR light source (e.g., wavelength at 804 nm) to further scale the penetration depth to 30 mm and its axial resolution 144 μm and later 560 μm, respectively. Hence, photoacoustic microscopy becomes macroscopy.

The imaging speed is another consideration when selecting a PAM or PACT. Due to a mechanical raster scanning method, PAM has a relatively lower frame rate than PACT which adopts a transducer array and a reconstruction algorithm. Great efforts have been made to improve the scanning

Table 6.1   Multiscale photoacoustic imaging.[27]

Rank	OR-PAM	AR-PAM	Photoacoustic macroscopy	PACT
Imaging depth	$\sim 10^{-4}$ to 10^{-3} m	$\sim 10^{-3}$ m	$\sim 10^{-2}$ m	$\sim 10^{-2}$ m
Lateral resolution	$\sim 10^{-7}$ to 10^{-6} m	$\sim 10^{-5}$ m	$\sim 10^{-4}$ m	$\sim 10^{-4}$ m
Axial resolution	$\sim 10^{-5}$ m	$\sim 10^{-5}$ m	$\sim 10^{-4}$ m	$\sim 10^{-4}$ m
Imaging contrast	Optical absorption	Optical absorption	Optical absorption	Optical absorption

speed in OR-PAM, through an optical scanning of the laser beam within the focal spot of the static ultrasound transducer.[28, 29] Also, its imaging speed can be further improved through parallel acoustic detection by ultrasound transducer arrays (i.e., a multi-focal OR-PAM system).[30]

6.2.3 Photoacoustic contrast agents

There are various *endogenous* and *exogenous* contrast agents in photoacoustic imaging. The endogenous contrast agents are non-toxic natural biomarkers in biological tissue and do not induce an external perturbation (e.g., injection) thus do not require costly and time-consuming regulatory approval. The endogenous contrast agents have broad absorption wavelengths from ultraviolet (UV, 180–400 nm) and visible (400–700 nm), to NIR (700–1400 nm). DNA/RNA are the absorbers mainly in UV range. Cytochrome c, myoglobin, hemoglobin, and melanin are the primary absorbers in UV to visible wavelength range.[31] For example, hemoglobin, the major oxygen carrier, has been extensively in photoacoustic imaging. Because of the identically different absorption properties of its two states: oxygenated (HbO_2) and deoxygenated (HbR) hemoglobin in red-to-NIR wavelengths, multiple excitation wavelengths (e.g., 660 nm and 940 nm) can be adopted for functional photoacoustic imaging, including oxygen saturation and total hemoglobin concentration. Functional photoacoustic imaging enables a comprehensive understanding of vascularization such as angiogenesis and hyper-metabolism, especially in early-stage cancer diagnosis. In the NIR wavelengths (700–1400 nm), lipids and glucose are the main absorbers for photoacoustic imaging. A main limit in adopting NIR wavelengths is its insufficient optical energy at a high repetition-rate laser required for photoacoustic imaging.

The exogenous contrast agents can be classified into three groups: organic dyes, nanoparticles, and genetic-encoded fluorescent proteins or reporters.[24, 27] The exogenous contrast agents, unlike the endogenous ones, can be specially designed for different imaging purposes. They can be modified with targeting molecules for disease tracking. The organic dyes have small sizes (i.e., a few nanometers) and quickly circulate in the vascular system with a relatively short circulation time. For example, indocyanine green (ICG) is an FDA-approved NIR dye that shows good absorption at $\sim$ 800 nm for improving the photoacoustic imaging contrast. However, due to its short half-lifetime in bloodstream (i.e., several minutes), it is preferably conjugated with polyethylene glycol (ICG-PEG) or encapsulated in nanoparticles to improve its circulation time significantly for sustained imaging. The conjugation or encapsulation will further protect these dyes from in vivo biodegradation. The nanoparticle-based contrast agents have another advantage: it has a relatively larger size resulting in better optical absorbance. For example, gold nanoparticles have been extensively studied for many applications in photoacoustic sensing. One major drawback is, most nanoparticles are still under toxicity and long-term side effects of accumulation and chronic exposure and need more thorough safety investigation before clinical trials. The genetic-encoded contrast agents are expressed in living cells and do not require exogenous delivery. For example, PAM imaging of LacZ gene activity has been demonstrated by using the blue product as the contrast agent.[32]

6.2.4 Summary

Photoacoustic imaging is a fast-growing area of biomedical sensing technology over the past years, based on its high detection sensitivity through optical absorption contrast as well as its high spatial resolution by acoustic confinement. In addition to the 2D or 3D structural imaging, many have adopted photoacoustic sensing in functional or molecular tracking of in vivo physiological and pathological conditions, such as oxygen saturation of hemoglobin (SO_2) for angiogenesis and cancer diagnosis, blood flow measurement through photoacoustic Doppler flowmetry [33], and single-wavelength SO_2 measurement based on the relaxation times of HbO_2 and HbR [32]. As a novel multimodal imaging technique, photoacoustic imaging has many unique advantages over other modalities based on the considerations of imaging qualities, safety, and cost. It promises in vivo imaging at multiple depth/resolution scales, as well as many functional and molecular imaging applications based on its unique optical-acoustic sensing mechanism.

6.3 Ultrasound Switchable Fluorescence Imaging

Over the past years, researchers have been investigating fluorescence imaging in centimeter-deep tissue with high spatial resolution. Several technologies have been proposed and developed to overcome the limitation of low spatial resolution. One general idea is to use focused ultrasound (FU) to confine fluorescence emission into a small volume. Thus, a fluorescence image of a centimeter deep tissue can be represented with acoustic resolution (10^{-4} m, depending on the frequency). Among them, one technology that confines fluorescence imaging by FU, so-called ultrasound switchable fluorescence (USF), is quickly developed in the past years.[35–50]

6.3.1 Principle of USF imaging

In USF imaging, there are two key components: (1) an excellent USF contrast agent; (2) a highly sensitive USF imaging system.[37] The USF contrast agent is temperature-sensitive and its fluorescence intensity and lifetime change as a function of temperature. When the temperature is below a threshold, the fluorescence is highly quenched and stays in "switched-off" state; when the temperature rises over the threshold, the contrast agent will be "switched-on" and release strong fluorescence (Figure 6.1). The contrast agents can be repetitively "switched-on/off" many times in USF imaging. In a USF imaging system, the key components are an excitation light, a focused ultrasound (FU) transducer, and a sensitive optical detector. An ultrasound pulse is delivered in the tissue and generates a temperature rise at its focus.

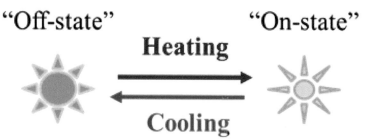

Figure 6.1   The USF contrast agent's fluorescence is quenched (i.e., "off-state") when the temperature is cool and becomes "on-state" (i.e., strong emission) when the temperature rises over a threshold. This "switching on and off" property can be repeated many times when the temperature repetitively cools and heats. Note that the excitation illumination should present the contrast agent in this phenomenon.

When the USF contrast agent is the ultrasound focus and encounters the temperature increase over its threshold, with the excitation light on, it releases strong fluorescence emission (i.e., termed as USF signal). The USF photons are scattered out of the tissue and captured by the optical detector in the imaging system. By scanning the FU pulses over the tissue, a USF image is acquired which provides the distribution of the contrast agents.

USF imaging has many unique advantages. First, because both the delivery of excitation light and the collection of emission light depend on the highly scattered photons, USF can acquire the signal at centimeter-deep biological tissue. Second, because the USF signal only comes from the FU focus (i.e., acoustic confinement), the USF imaging provides a high spatial resolution (10^{-4} m). Third, USF has high detection sensitivity because (a) the USF contrast agents are designed highly sensitive to a small temperature change (only a few Celsius degrees) and have a high quantum efficiency after being switched on[36–40]; and (b) several sensitive USF imaging systems are developed [35, 37, 38, 41, 43, 45] to acquire the weak fluorescence emission that comes from deep tissue to ensure a desired signal-to-noise ratio (SNR). Fourth, USF is safe for in vivo imaging because (a) ultrasound is safe with an appropriate power; (b) the contrast agent is safe when it is designed bio-degradable and non-toxic; and (c) the USF contrast agent's temperature threshold can be well controlled slighted above the body temperature ($\sim$37°C) with a narrow temperature transition bandwidth (usually 3–5°C) so that it only requires a temperature increase of a few Celsius degrees in biological tissue. In comparison to (1) fluorescence diffuse optical tomography (FDOT) which collects highly scattered photons with reconstruction algorithms for centimeter deep tissue imaging and (2) ultrasound imaging, USF imaging shows its desired imaging properties (Table 6.2).

6.3.2 USF contrast agents

A general idea of designing a USF contrast agent is to encapsulate a polarity- and viscosity-sensitive fluorophore into temperature-sensitive polymers so that its fluorescence intensity and/or lifetime will change as a function of temperature. This is because when the temperature rises over the lower critical solution temperature (LCST) of the polymers, the polymers will have a phase transition from hydrophilic to hydrophobic, leading to a microenvironment change from water-rich to polymer-rich where the fluorophores are encapsulated. Because the fluorophores are sensitive to polarity and viscosity, their quantum efficiency will increase significantly. As a result, the

Table 6.2   Comparison between ultrasound, FDOT, and USF.

Rank	Ultrasound	FDOT	USF
Sensitivity	Relatively poor	Excellent	Excellent (=FDOT)
Imaging Depth	Good (a few centimeters)	Good (a few centimeters)	Good (=ultrasound)
Resolution	Good ($\sim 10^1$ to 10^2 microns)	Poor (>1 mm)	Good (=ultrasound)
Imaging Contrast	Acoustic contrast (structural information)	Optical contrast (functional information)	Optical contrast (=FDOT)
Multiplex Imaging	Not possible	Possible	Possible (=FDOT)

fluorescence emission will stay quenched (i.e., "switched off" state) when the temperature is below the LCST, and the fluorescence will be "switched on" and release strong emission when the temperature is above the LCST. In addition, repeatability is essential in designing a USF contrast agent. That is, the fluorescence should be able to "switch on and off" repeatability when the temperature is repetitively above and below the LCST. In acquiring a USF signal, the USF contrast agents are switched "on" by the FU beam in a small focused volume and switched "off" by thermal diffusion. Thus, through a point-to-point scan method, and the agents are switched "on" and "off" repetitively, a USF image is acquired.

This "switching" performance of a USF contrast agent is characterized by five parameters (Figure 6.2): (1) the fluorescence peak excitation and emission wavelengths (λ_{ex} and λ_{em}); (2) the fluorescence intensity ratio between its on- and off- states (I_{on}/I_{off}); (3) the fluorescence lifetime ratio between its on- and off-states (τ_{on}/τ_{off}); (4) the temperature threshold to switch on the fluorescence (T_{th}); and (5) the temperature transition bandwidth (T_{BW}). An ideal USF contrast agent should have (a) near-infrared (NIR) λ_{ex} and λ_{em}; (b) a high I_{on}/I_{off} and a high τ_{on}/τ_{off}; (c) an adjustable T_{th}; and (d) a narrow T_{BW}. This is because (a) NIR wavelengths provide good penetration to biological tissue; (b) a high I_{on}/I_{off} and a high τ_{on}/τ_{off} provide high sensitivity to detect a USF signal from background fluorescence; (c) an adjustable T_{th} provides USF contrast agents appropriate for both ex vivo and in vivo scenarios; and (d) a narrow T_{BW} provides fluorescence-switching efficiency by ultrasound heating (usually a few Celsius degrees).

The most frequently used USF contrast agents are listed here: (1) ICG-encapsulated PNIPAM-based nanoparticles [37, 39]; (2) ADP

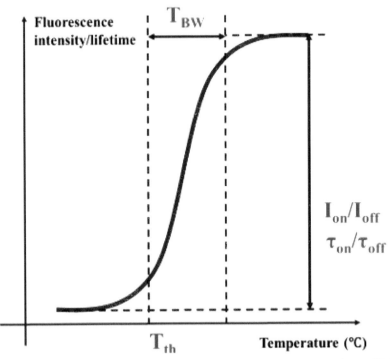

Figure 6.2 Characterization of a USF contrast agent's fluorescence intensity or lifetime as a function of temperature.36,48 I_{on}/I_{off}: the fluorescence intensity ratio between its on- and off-states; τ_{on}/τ_{off}: the fluorescence lifetime ratio between its on- and off-states; T_{th}: the temperature threshold to switch on the fluorescence; T_{BW}: the temperature transition bandwidth.

or ZnPc-encapsulated pluronic-based nanocapsules [40, 48]; (3) ICG-encapsulated liposomes.[46] These contrast agents have their advantages and disadvantages. All the contrast agents adopt near-infrared (NIR) fluorophores and they have good imaging depth with high detection sensitivity. ICG-PNIPAM nanoparticles have long-term stability (up to years), a precisely-controllable T_{th} and a narrow T_{BW} appropriate for in vivo USF imaging and can be modified with functional groups for in vivo targeting (e.g., tracking a cancer). The particle sizes are also controllable so they can bio-distribute to various in vivo targets (e.g., organs and tumor) depending on the size. Its main drawback is in vivo cytotoxicity. ADP or ZnPc-encapsulated nanocapsules have an extreme high I_{on}/I_{off} ratio and some of them also have a high τ_{on}/τ_{off} so that it provides a USF image with a higher SNR especially in a time-domain USF imaging system [45]. Also, its T_{th} is controllable and its T_{BW} is narrow for USF imaging. However, a main drawback is its bio-instability thus make it not suitable for in vivo USF imaging. The ICG liposomes have overcome the bio-toxicity issue and are relatively safe (i.e., bio-degradable) and bio-stable in vivo so they are promising for clinical use. Its USF properties such as I_{on}/I_{off} ratio (only a few times) can be further improved for a better USF imaging quality in the future development.

6.3.3 USF imaging systems

In a USF imaging system, there are three key components: (1) an excitation light; (2) an FU transducer; and (3) a sensitive optical detector. The excitation light is typically NIR and illuminates tissue up to centimeter thickness and excites the USF contrast agents in the tissue. The FU transducer sends an ultrasound pulse into the tissue and at its acoustic focus increases the tissue temperature in the small volume by a few Celsius degrees. The contrast agents in the volume are "switched on" and emit USF photons. The USF photons scatter out of the tissue and are captured by the optical detector. By scanning the ultrasound focus across the tissue, a USF imaging is acquired representing the distribution of contrast agents in the tissue. Based on the principle of USF imaging, several types of USF imaging systems were developed and characterized the performances.

6.3.4 PMT-based and camera-based USF imaging system

When a photo-multiplier tube is adopted as the optical detector in the USF imaging system, it has several advantages. First, PMT is very sensitive to a weak optical signal. Second, it has a fast detection speed (<1 ns response time) and thus is capable of fast continuous acquisition. Figure 6.3 (A) shows a frequency-domain USF imaging system [44] that adopts the PMT as the optical detector. A diode laser is adopted to illuminate the sample at a modulated frequency. When an ultrasound pulse is sent to the object and a USF signal is generated, the photons are scattered out and collected in an optical fiber which is coupled to the PMT. Following the PMT and its amplifier, a lock-in amplifier is to further amplify the USF signal with the reference signal (same as the laser modulation frequency). The output USF signal is finally acquired by a data-acquisition card (DAC) and recorded in the computer. One main limitation of using the PMT is that it is a single sensing element and is not capable of providing the spatial information of the scattered USF photons. In comparison, when a camera is adopted, it

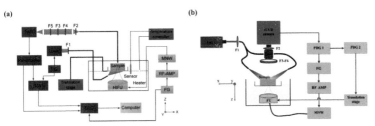

Figure 6.3 (A) PMT-based frequency domain USF imaging system versus an (B) ICCD camera-based USF imaging system.

overcomes the limitation. Figure 6.3 (B) shows an ICCD camera-based USF imaging system [47]. The main difference is that it adopts a camera as the optical detector placed above the sample and the image plane is focused on the sample's surface through a lens system. Since the camera (CCD) does not support a high acquisition rate (i.e., with an exposure time = hundreds of milliseconds and a frame rate = a few frames per second), a continuous wave (CW) laser should be in direct current (DC) mode for steady illumination, or, a high-repetition-rate (e.g., 20 MHz) pulsed pico-second laser can be adopted when the ICCD camera's intensifier is set at the "time-gated" mode at the same repetition frequency, so-called a time-domain USF imaging system [45]. The camera not only provides two-dimensional (2D) spatial information of scattered USF photons, but it also ensures a capture of sufficient USF photons scattered with a desired field of view (FOV) so that it provides a desired image quality based on its collection efficiency.

6.3.5 Direct current, frequency-domain, and time-domain USF imaging system

In a direct current (DC) USF imaging system, a direct current continuous wave (CW) laser is adopted as the exication light. A diagram of acquiring a USF signal from a DC USF imaging system is shown in Figure 6.4 (A). The first line represents that the FU transducer sends a ultrasound signal to the tissue with a fixed exposure time. The second line represents that the temperature at the FU focus rises up after the start of ultrasound exposure. At the end of ultrasound exposure, the temperature at the ultrasound focus starts to decrease. The third line represents the DC laser output. The output laser intensity/power is fixed at a constant. The fourth line represents the acquired USF signal. Because the USF signal is extracted from a background fluorescence only by a small intensity increase at a low-frequency range, and also the fact that many noises are added during signal acquisition (such as dark current noise in the optical sensor), the main limitation of a DC USF imaging system is a low signal-to-noise ratio (SNR). A frequency-domain USF imaging system overcomes this limitation (Figure 6.4 (B)) by modulating the laser at a specific frequency (e.g., 1 kHz) and that the fluorescence is modulated at the same frequency. Consequently, the USF signal is also modulated at this frequency and thus could be specifically identified by a lock-in amplifier. By lock-in detection at a specific frequency, the SNR of a USF signal can be significantly improved. In contrast to the two systems, a time-domain USF imaging system adopts a pulsed laser instead of a CW laser, and a time-gated detection of the USF signal (Figure 6.4 (C)). In the fourth line of Figure 6.4 (C), there is one input the laser pulse (the red dash arrow), and two output pulses (two green solid curves), of which the first is relatively short lifetime and represents the background (i.e., the laser leakage, tissue auto-fluorescence, and non-100%-off background fluorescence from the agents), and the second is relative long life-time and represents the USF signal. The plot represents a scenario that which a USF contrast agent with a high τ_{on}/τ_{off} is adopted in the system and the USF signal (i.e., the green shadow area in the plot) is acquired in a time-gated window with a desired delay, most of the background signal can be removed and thus the SNR will be significantly improved by a significant increase of signal-to-background ratio (SBR). Note that the pulsed USF signals can be at a high repetition rate and acquired over an integrated time. The time-domain USF imaging system was first developed in a PMT detector-based USF imaging system[35] and then developed in an ICCD camera-based system [45]. Overall, the time-domain system is preferable because it can achieve the highest SNR by reducing most kinds of noises through a time-gated acquisition mechanism. In the meantime, however, a time-domain system is the most expensive because of the use of the picosecond pulsed laser and the fast detector (e.g., an ICCD camera), which makes a frequency-domain system preferable when considering the cost.

6.3.6 In vivo USF imaging

In vivo USF imaging has been successfully achieved in several mice models (BALB/cJ, NU/J). Two types of contrast agents are successfully adopted for in vivo USF imaging: ICG-encapsulated PNIPAM-based nanoparticles and ICG-encapsulated liposomes because of their bio-stability. The in vivo USF

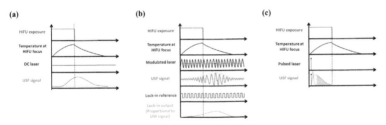

Figure 6.4 Sequence diagram of USF signal acquisition through (A) direct current system; (B) frequency-domain system; and (C) time-domain system.

imaging have been successfully acquired in both PMT-based and camera-based system. Figure 6.5 shows an example of in vivo USF imaging in the ICCD camera-based USF imaging system. [47] In this example, large-size ICG PNIPAM nanoparticles (particle size = ~330 nm) are intravenously (IV) injected and accumulated in the spleen of the mouse after a few hours. In the work, it is known that the large nanoparticles (particle size = ~330 nm) will accumulate in the spleen while the smaller nanoparticles (particle size = ~70 nm) will bio-distribute to the liver after IV injection. Both T_{th}s are designed at ~36–39°C, slightly above the mouse's body temperature in an anesthetized condition. It is also proven that the nanoparticles are bio-stable and retain USF properties when they accumulate in these organs. Through FU scanning across the spleen area, a USF image of its contrast agent distribution in the spleen was acquired. Also, an X-ray CT contrast agent (ExiTron nano 12000) is IV injected at the same time and accumulates in the spleen. Thus, the USF image is co-registered with the CT image for comparison (Figure 6.5 (e)–(i)).

6.3.7 Summary

USF imaging has a high detection sensitivity and also a high image resolution at centimeter tissue-depth, because of its unique acoustic-optical sensing mechanism. Currently, the maximum USF imaging depth can go up to ~5 cm.[50] Because contrast agents can be modified with functional groups that can track molecular and cells in vivo, and also because of its excellent imaging properties, it provides a promising method to the existing challenges in biomedical imaging and sensing. One application is to adopt USF imaging in detecting and tracking early-stage cancers and provide early treatment and reduce the mortality rate.

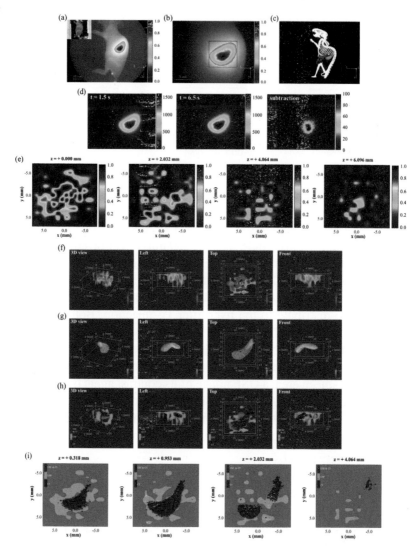

Figure 6.5 (A) Normalized 2D planar fluorescence image of the mouse on its left side in FOV 1. The fluorescence mainly accumulated in the mouse's spleen. (B) Zoomed-in normalized 2D planar fluorescence image of the mouse's spleen area in FOV 2. (C) Top view of the 3D CT image of the mouse's left side. (D) An example of acquired USF signals from the spleen area of the mouse. (E) Acquired USF images on x–y plane at different z locations. (F) The USF image of the USF contrasts volume from the 3D view, left side view, top view, and front view. (G) The CT image of the CT contrasts volume from a 3D view, left side view, top view, and front view. (H) 3D co-registration results of the USF and CT image. (I) The co-registration results on x–y plane at different z locations.

6.4 Other Novel Multi-modal Imaging Techniques

Besides photoacoustic imaging and ultrasound-switchable fluorescence imaging, there exist many new multi-modal imaging modalities which have been fast developed in the past years, including X-ray luminescence/fluorescence computed tomography (XLCT and XFCT) [51], ultrasound modulated optical/fluorescence imaging [52–55], and magnetic-fluorescent nanoparticles for multi-modal imaging [56], etc.

X-ray luminescence and X-ray fluorescence computed tomography are two fast-developing multi-modal imaging techniques that uses X-ray illumination and X-ray sensitive contrast agents that convert the X-ray energy into optical or secondary X-ray emission signals. It provides functional and molecular imaging capability which tracks the bio-distribution of its contrast agents. It provides high spatial resolution through spatial excitation confinement by adopting a collimated X-ray beam and a scanning method. One existing problem is its long scanning time, but various efforts have been made to improve the acquisition speed such as adopting a cone-beam scan method [57, 58] in XLCT imaging. Various contrast agents have been developed for this imaging technology. In XLCT, scintillation is the conversion of X-ray energy into optical light energy, and thus microscopic scintillating nanoparticles are used as imaging probes in XLCT, such as the rare-earth phosphor nanoparticles [59]. In XFCT, stabilized gold nanoparticles have been particularly used and demonstrated their biocompatibility, low toxicity, and surface modification for biological targeting [60]. Similar to conventional X-ray CT imaging, a major drawback of this multi-modal technique, is its bio-safety concerns due to the use of ionizing EM radiation (i.e., X-ray).

Considering safety and cost, optics and ultrasound are the most two favorable sensing energy sources. Meanwhile, optics usually provides the highest detection sensitivity and thus a high imaging contrast, while ultrasound provides a relatively high penetration depth and a good spatial resolution. Thus, multi-modal imaging techniques based on ultrasound and optics have been widely researched, proposed, and developed over the past decades. Among them, photoacoustic imaging and ultrasound switchable fluorescence imaging are the two outstanding multimodal imaging techniques as introduced in this chapter. Besides the two techniques, there have been several other ultrasound-optical multi-modal imaging methods that are well-developed. When the acoustic signal is used for excitation and the optical signal for detection, a general method is to adopt the focused ultrasound to confine the optical emission into a small volume. Time-reversed ultrasonically encoded (TRUE) optical focusing aims to focus the excitation light into deep tissue with acoustic confinement.[54, 55] Its advantage is that the fluorophores can be excited directly without any modification. Its disadvantage is that it requires a sophisticated system because of its phase conjugation method and thus limits its application when there is motion. On the contrary, ultrasound-modulated fluorescence (UMF) imaging aims to confine the fluorescence emission into the ultrasound focal volume by specifically designing the fluorescent probes that are sensitive to the acoustic signals,[52, 53] which is a similar idea to USF imaging. Unlike the fluorescence-thermal-sensitive contrast agents in USF imaging, UMF adopts fluorophore-conjugated microbubbles whose fluorescent quantum efficiency is correlated to the surface density of the conjugated fluorophores so that when an acoustic signal is driving the oscillation of microbubbles, its fluorescence intensity will be modulated at the same acoustic frequency. UMF only needs a low driving power for microbubble oscillation and does not require a high acoustic energy to increase the microenvironment temperature. Its disadvantage is a relatively low SNR due to a low modulation intensity from its contrast agent and thus has a limited imaging depth (i.e., a few millimeters).

6.5 Conclusion

In this chapter, we have discussed several novel multi-modal imaging techniques. Their bio-sensing mechanisms, as well as the systems and imaging contrast agents, are introduced. When selecting a specific imaging modality, there are many considerations including different perspectives of imaging qualities, cost and safety, the contrast agents, as well as indicated use. The advantages and disadvantages of each modality are carefully discussed and their capabilities for sensing in multiscale applications are revealed. There exist natural trade-off relationships in achieving a desired image, such as the trade-off between imaging depth and detection sensitivity, not only within a single modality with multiscale depths but also among the various imaging modalities. The contrast agents should also be thoroughly taken into consideration about their desirable features as well as limitations. Thus, this chapter provides a guide when considering a multimodal imaging technique for various bio-sensing applications. Overall, these novel imaging technologies have provided advantageous sensing methods in many biomedical applications. Like many other bio-sensing methods, these imaging technologies can provide non-invasive and real-time monitoring, as well as functional and molecular tracking of physiological and pathological conditions, and thus can benefit human beings in medical prevention, diagnosis, and treatment.

References

[1] Jalalian, A. et al. Computer-aided detection/diagnosis of breast cancer in mammography and ultrasound: a review. *Clin. Imaging* 37, 420-426 (2013).

[2] Cnudde, V. & Boone, M. N. High-resolution X-ray computed tomography in geosciences: A review of the current technology and applications. *Earth-Science Reviews* 123, 1-17 (2013).

[3] Fenster, A., Downey, D. B. & Cardinal, H. N. Three-dimensional ultrasound imaging. *Phys. Med. Biol.* 46, R67 (2001).

[4] Warner, E. et al. Systematic review: using magnetic resonance imaging to screen women at high risk for breast cancer. *Ann. Intern. Med.* 148, 671-679 (2008).

[5] Houssami, N. & Hayes, D. F. Review of preoperative magnetic resonance imaging (MRI) in breast cancer: should MRI be performed on all women with newly diagnosed, early stage breast cancer? *CA Cancer J. Clin.* 59, 290-302 (2009).

[6] Yang, W. T. et al. Inflammatory breast cancer: PET/CT, MRI, mammography, and sonography findings. *Breast Cancer Res. Treat.* 109, 417-426 (2008).

[7] Pennant, M. et al. A systematic review of positron emission tomography (PET) and positron emission tomography/computed tomography (PET/CT) for the diagnosis of breast cancer recurrence. *Health Technol. Assess.* (2010).

[8] Mariani, G. et al. A review on the clinical uses of SPECT/CT. *European journal of nuclear medicine molecular Imaging* 37, 1959-1985 (2010).

[9] Schmitt, J. M. Optical coherence tomography (OCT): a review. *IEEE Journal of selected topics in quantum electronics* 5, 1205-1215 (1999).

[10] Lichtman, J. W. & Conchello, J.-A. Fluorescence microscopy. *Nature methods* 2, 910 (2005).

[11] Corlu, A. et al. Three-dimensional in vivo fluorescence diffuse optical tomography of breast cancer in humans. *Opt. Express* 15, 6696-6716 (2007).

[12] Andresen, V. et al. Infrared multiphoton microscopy: subcellular-resolved deep tissue imaging. *Curr. Opin. Biotechnol.* 20, 54-62 (2009).

[13] Hilderbrand, S. A. & Weissleder, R. Near-infrared fluorescence: application to in vivo molecular imaging. *Curr. Opin. Chem. Biol.* 14, 71-79 (2010).

[14] Van Dam, G. M. et al. Intraoperative tumor-specific fluorescence imaging in ovarian cancer by folate receptor-α targeting: first in-human results. *Nat. Med.* 17, 1315 (2011).

[15] Xu, C. T. et al. High-resolution fluorescence diffuse optical tomography developed with nonlinear upconverting nanoparticles. *ACS nano* 6, 4788-4795 (2012).

[16] Pifferi, A. et al. New frontiers in time-domain diffuse optics, a review. *J. Biomed. Opt.* 21, 091310 (2016).

[17] Webb, A. & Kagadis, G. C. Introduction to biomedical imaging. *Med. Phys.* 30, 2267-2267 (2003).

[18] Weissleder, R., Tung, C.-H., Mahmood, U. & Bogdanov Jr, A. In vivo imaging of tumors with protease-activated near-infrared fluorescent probes. *Nat. Biotechnol.* 17, 375 (1999).

[19] Hoffman, R. M. Green fluorescent protein imaging of tumour growth, metastasis, and angiogenesis in mouse models. *The lancet oncology* 3, 546-556 (2002).

[20] Frangioni, J. V. In vivo near-infrared fluorescence imaging. *Curr. Opin. Chem. Biol.* 7, 626-634 (2003).

[21] Hu, C.-D. & Kerppola, T. K. Simultaneous visualization of multiple protein interactions in living cells using multicolor fluorescence complementation analysis. *Nat. Biotechnol.* 21, 539 (2003).

[22] Yamamoto, N., Tsuchiya, H. & Hoffman, R. M. Tumor imaging with multicolor fluorescent protein expression. *Int. J. Clin. Oncol.* 16, 84-91 (2011).

[23] Goertzen, A. L., Meadors, A. K., Silverman, R. W. & Cherry, S. R. Simultaneous molecular and anatomical imaging of the mouse in vivo. *J Physics in Medicine & Biology* 47, 4315 (2002).

[24] Wang, L. V. Multiscale photoacoustic microscopy and computed tomography. *Nature photonics* 3, 503-509 (2009).

[25] Xu, M. & Wang, L. V. Photoacoustic imaging in biomedicine. *Rev. Sci. Instrum.* 77, 041101 (2006).

[26] Gusev, V. & Karabutov, A. Laser Optoacoustics AIP. *J New York*, 304 (1993).

[27] Yao, J. & Wang, L. V. Photoacoustic microscopy. *Laser photonics reviews* 7, 758-778 (2013).

[28] Hajireza, P., Shi, W. & Zemp, R. Label-free in vivo fiber-based optical-resolution photoacoustic microscopy. *Optics letters* 36, 4107-4109 (2011).

[29] Rao, B. et al. Real-time four-dimensional optical-resolution photoacoustic microscopy with Au nanoparticle-assisted subdiffraction-limit resolution. *Optics letters* 36, 1137-1139 (2011).

[30] Song, L., Maslov, K. & Wang, L. V. Multifocal optical-resolution photoacoustic microscopy in vivo. *Optics letters* 36, 1236-1238 (2011).

[31] Yao, J. & Wang, L. V. Photoacoustic tomography: fundamentals, advances and prospects. *Contrast media molecular imaging* 6, 332-345 (2011).

[32] Li, L., Zemp, R. J., Lungu, G. F., Stoica, G. & Wang, L. V. Photoacoustic imaging of lacZ gene expression in vivo. *J. Biomed. Opt.* 12, 020504 (2007).

[33] Yao, J., Maslov, K. I., Zhang, Y., Xia, Y. & Wang, L. V. Label-free oxygen-metabolic photoacoustic microscopy in vivo. *J. Biomed. Opt.* 16, 076003 (2011).

[34] Danielli, A., Favazza, C. P., Maslov, K. & Wang, L. V. Single-wavelength functional photoacoustic microscopy in biological tissue. *Optics letters* 36, 769-771 (2011).

[35] Yuan, B., Uchiyama, S., Liu, Y., Nguyen, K. T. & Alexandrakis, G. High-resolution imaging in a deep turbid medium based on an ultrasound-switchable fluorescence technique. *Applied physics letters* 101, 033703 (2012).

[36] Cheng, B. et al. Development of ultrasound-switchable fluorescence imaging contrast agents based on thermosensitive polymers and nanoparticles. *IEEE Journal of Selected Topics in Quantum Electronics* 20, 67-80 (2014).

[37] Pei, Y. et al. High resolution imaging beyond the acoustic diffraction limit in deep tissue via ultrasound-switchable NIR fluorescence. *Sci. Rep.* 4, 4690 (2014).

[38] Cheng, B. et al. High-resolution ultrasound-switchable fluorescence imaging in centimeter-deep tissue phantoms with high signal-to-noise ratio and high sensitivity via novel contrast agents. *PLoS One* 11, e0165963 (2016).

[39] Yu, S. et al. New generation ICG-based contrast agents for ultrasound-switchable fluorescence imaging. *Sci. Rep.* 6, 35942 (2016).

[40] Cheng, B. et al. The mechanisms and biomedical applications of an NIR BODIPY-based switchable fluorescent probe. *Int. J. Mol. Sci.* 18, 384 (2017).

[41] Kandukuri, J. et al. A dual-modality system for both multi-color ultrasound-switchable fluorescence and ultrasound imaging. *Int. J. Mol. Sci.* 18, 323 (2017).

[42] Kandukuri, J., Yu, S., Yao, T. & Yuan, B. Modulation of ultrasound-switchable fluorescence for improving signal-to-noise ratio. *J. Biomed. Opt.* 22, 076021 (2017).

[43] Yao, T., Yu, S., Liu, Y. & Yuan, B. Ultrasound-Switchable Fluorescence Imaging via an EMCCD Camera and a Z-Scan Method. *IEEE Journal of Selected Topics in Quantum Electronics* 25, 1-8 (2019).

[44] Yao, T., Yu, S., Liu, Y. & Yuan, B. In vivo ultrasound-switchable fluorescence imaging. *Sci. Rep.* 9, 9855 (2019).

[45] Yu, S., Yao, T. & Yuan, B. An ICCD camera-based time-domain ultrasound-switchable fluorescence imaging system. *Sci. Rep.* 9, 10552 (2019).

[46] Liu, Y. et al. A Biocompatible and Near-Infrared Liposome for In Vivo Ultrasound-Switchable Fluorescence Imaging. *Advanced Healthcare Materials* 9, 1901457 (2020).

[47] Yu, S., Yao, T., Liu, Y. & Yuan, B. In vivo ultrasound-switchable fluorescence imaging using a camera-based system. *Biomedical Optics Express* 11, 1517-1538 (2020).

[48] Yu, S., Wang, Z., Yao, T., Yuan, B. & Surgery. Near-infrared temperature-switchable fluorescence nanoparticles. *Quantitative Imaging in Medicine* 11, 1010 (2021).

[49] Liu, R. et al. Temperature-sensitive polymeric nanogels encapsulating with β-cyclodextrin and ICG complex for high-resolution deep-tissue ultrasound-switchable fluorescence imaging. *Nano Res* 13, 1100-1110 (2020).

[50] Yao, T., Liu, Y., Ren, L. & Yuan, B. Improving sensitivity and imaging depth of ultrasound-switchable fluorescence via an EMCCD-gain-controlled system and a liposome-based contrast agent. *Quantitative Imaging in Medicine Surgery* 11, 957 (2021).

[51] Ahmad, M., Pratx, G., Bazalova, M. & Xing, L. X-ray luminescence and x-ray fluorescence computed tomography: new molecular imaging modalities. *Ieee Access* 2, 1051-1061 (2014).

[52] Liu, Y., Feshitan, J. A., Wei, M.-Y., Borden, M. A. & Yuan, B. Ultrasound-modulated fluorescence based on fluorescent microbubbles. *J. Biomed. Opt.* 19, 085005 (2014).

[53] Liu, Y., Feshitan, J. A., Wei, M.-Y., Borden, M. A. & Yuan, B. Ultrasound-modulated fluorescence based on donor-acceptor-labeled microbubbles. *J. Biomed. Opt.* 20, 036012 (2015).

[54] Wang, Y. M., Judkewitz, B., DiMarzio, C. A. & Yang, C. Deep-tissue focal fluorescence imaging with digitally time-reversed ultrasound-encoded light. *Nature communications* 3, 928 (2012).

[55] Judkewitz, B., Wang, Y. M., Horstmeyer, R., Mathy, A. & Yang, C. Speckle-scale focusing in the diffusive regime with time reversal of variance-encoded light (TROVE). *Nature photonics* 7, 300 (2013).

[56] Mulder, W. J. et al. Magnetic and fluorescent nanoparticles for multi-modality imaging. (2007).

[57] Chen, D. et al. Cone beam x-ray luminescence computed tomography: A feasibility study. *Med. Phys.* 40, 031111 (2013).

[58] Liu, X., Liao, Q. & Wang, H. In vivo x-ray luminescence tomographic imaging with single-view data. *Optics letters* 38, 4530-4533 (2013).

[59] Naczynski, D. J., Tan, M. C., Riman, R. E. & Moghe, P. V. Rare earth nanoprobes for functional biomolecular imaging and theranostics. *Journal of Materials Chemistry B* 2, 2958-2973 (2014).

[60] Sokolov, K. et al. Real-time vital optical imaging of precancer using anti-epidermal growth factor receptor antibodies conjugated to gold nanoparticles. *Cancer Res.* 63, 1999-2004 (2003).

7

Species-independent Pipeline for Quantitative Analysis of Electroencephalogram with Application in Classification of Traumatic Brain Injury in Mice and Humans

Manoj Vishwanath[1], Carolyn Jones[2,3], Miranda M. Lim[2,3], and Hung Cao[1,4,5]

[1]Department of Computer Science, University of California, Irvine, CA 92697
[2]VA Portland Health Care System, Portland, OR 97239, USA
[3]Oregon Health Science University, Portland, OR, 97239
[4]Department of Electrical Engineering and Computer Science, University of California, Irvine, CA 92607, USA
[5]Department of Biomedical Engineering, University of California, Irvine, CA 92697, USA

Abstract

Electroencephalogram (EEG) is an electrophysiological recording of the brain potentials whose analysis has been crucial in the study of various neurological conditions. However, due to the non-stationarity and noisy nature of EEG, it is extremely difficult to extract meaningful information. In this chapter, we introduce a standard pipeline consisting of an amalgamation of advanced signal processing and machine learning techniques that can be used to analyze EEG signals irrespective of the organism it was collected from. The working of the proposed standard EEG analysis pipeline is demonstrated in the detection of traumatic brain injury (TBI) in mouse and human sleep EEG.

Keywords: Traumatic Brain Injury, Electroencephalogram, Machine Learning, Sleep, Standard Pipeline, Mice

7.1 Introduction

The electrical potentials generated by billions of neurons in the brain reach the scalp through conduction across multiple layers which are recorded as EEG at the scalp. Therefore, EEG reflects a cumulative synchronous activity of a population of neurons at high temporal resolution [47]. However, it is extremely noisy and poses low spatial resolution which becomes a hindrance in extracting meaningful information. EEG has been used to investigate a range of neurological conditions such as epilepsy, stroke, Alzheimer's disease, and topics of interest in cognitive sciences such as sensory and auditory pathways, memory and motor processes due to the advantages it poses over other methods of study such as computerized tomography (CT) or magnetic resonance imaging (MRI) such as lower cost, lesser need for highly trained clinicians, and bulky equipment. Along with the challenges mentioned above, high variability across subjects makes it difficult to build generalized models for EEG analysis. Signal processing pipelines with domain-specific knowledge are frequently employed to solve the aforementioned issues. The use of advanced machine learning (ML) techniques has led to the development of more automated methods with accurate generalization [54].

ML [55] is the study of computer algorithms that focus on the use of statistical methods and data to automatically improve through experience. Rule-based ML is characterized by identifying a set of rules that represent the knowledge learned by the algorithm. It is dependent on the user to determine

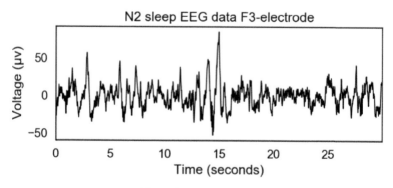

Figure 7.1   EEG signal.

the set of input features based on domain knowledge. ML models can be categorized into three categories: supervised learning, where the model has access to labeled data; unsupervised learning, where the dataset being used is unlabeled; and semisupervised learning, where a small subset of the labeled data is used to guide classification and feature extraction from a larger, unlabeled dataset. In this chapter, we introduce relevant EEG features and various supervised rule-based ML algorithms formulating an efficient standard pipeline to analyze EEG from different organisms. The results obtained using the proposed standard analysis pipeline on the classification of mice and human EEG to detect mild TBI (mTBI) are presented.

The subsequent sections are organized as follows: Section 7.2 discusses the need for preprocessing EEG signals and various methods to do so, Section 7.3 deals with extracting relevant QEEG features, Sections 7.4 and 7.5 outline feature normalization and selection procedures, respectively. Section 7.6 outlines the importance of splitting the data into train, validation, and held-out test datasets before diving into different rule-based ML models in Section 7.7. Finally, we review the evaluation metrics used to evaluate ML models and examine the results obtained on an application of classifying mTBI in mice and humans in Sections 7.8 and 7.9, respectively.

7.2 Preprocessing

In this section, we introduce some of the crucial and most commonly used preprocessing techniques in the EEG analysis pipeline. Preprocessing is a necessary step in EEG analysis due to the noisy nature of the signal. Some of the well-known preprocessing pipelines are PREP [3], ASR_10* [40] (artifact subspace reconstruction-based algorithm), and Statistical Control of Artifacts in Dense Arrays Studies (SCADS) [36]. Other techniques such as dimensionality reduction and downsampling are also often performed as a preprocessing step so as to facilitate the efficient handling of huge EEG datasets. Prior to their implementation, a careful understanding of the effects of these processes on the EEG signal is necessary to make an informed choice [46].

7.2.1 Independent component analysis

EEG is a composite signal consisting of electrical potentials from different parts of the body including artifacts from the eye-electrooculogram (EOG), heart-electrocardiogram (ECG), and muscle activity. First introduced by J. Herault, C. Jutten, and B. Ans in the 1980s [25], independent component analysis (ICA) is a signal processing technique used to separate underlying independent components from observed multivariate non-gaussian data. Multi-channel EEG signals can be modeled as n random variables $x_1, \ldots, x_n$, that are linear combinations of n random variables $s_1, \ldots, s_n$. In vector notation, it is represented as

$$\mathbf{X} = \mathbf{AS},\qquad(7.1)$$

where $\mathbf{X}$ and $\mathbf{S}$ are column vectors of the observed variables and the latent variables respectively and $\mathbf{A}$ denotes the mixing matrix. In the above equation, both, the mixing coefficients a_{ij} which constitutes $\mathbf{A}$ and the independent components (ICs) s_i which constitutes $\mathbf{S}$ are unknown and the goal of ICA is to estimate the mixing coefficients with minimum assumptions [30] and thereby determining the ICs using

$$\mathbf{S} = \mathbf{WX},\qquad(7.2)$$

where $\mathbf{W}$ is inverse of $\mathbf{A}$.

Figure 7.2 shows the result of applying the ICA method used in MNE [23] -an open-source Python package for analyzing human neurophysiological data. The top panel shows a raw EEG recording typically consisting of ECG components. This can be confirmed by observing the ECG recorded simultaneously shown in the middle panel. The last panel overlays the

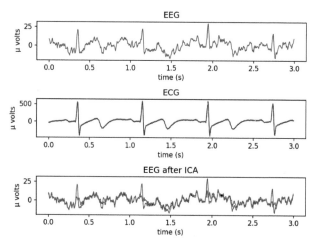

Figure 7.2   Removing ECG component in EEG using ICA.

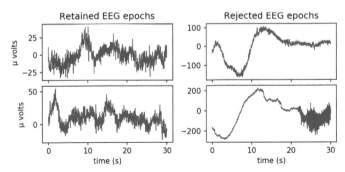

Figure 7.3   Examples of retained (left) and rejected (right) epochs.

ECG removed-preprocessed EEG signal over the original unprocessed EEG signal.

7.2.2 Thresholding

Muscle artifacts, typically in the order of milli-volts (mV) are usually identified by careful visual inspection which is a very tedious task. Bad channels in the recording may manifest as being much noisier when compared to other channels due to an increase in the scalp-electrode impedance. This increase may be attributed to the relative displacement of the electrode on the scalp or to the prolonged use of certain EEG electrodes at a stretch. In order to filter out bad epochs based on amplitude range, variance, and channel deviation it is more reasonable to calculate the thresholds based on the Z-score rather than having an absolute threshold value. A Z-score of ±3 is used as a threshold to identify contaminated data. However, a suitable value can be chosen based on the application. Figure 7.3 shows two of the retained as well as rejected epochs based on thresholding.

7.2.3 Filtering

Much of the information in EEG is contained between 0 and 60 Hz. Various studies have shown that numerous tasks are mapped to certain specific frequency bands in the EEG signals. As a result, filtering is one of the most commonly performed preprocessing tasks in neural signal processing. Filters are used to remove slow frequency drifts, power noise, or to extract various frequency band information which results in an increased signal-to-noise ratio. More often than not, the effect of filters on the phase of the signal is neglected and the temporal shift caused due to change in phase is not taken into account which can lead to misinterpretation of results regarding the underlying process [49], [70].

The impact of filtering on the phase of the signal can be understood as follows. Consider an input signal $x(t) = A\cos(\omega_0 t)$ filtered using a filter of frequency response

$$H(j\omega) = M(\omega)e^{j\phi(\omega)}, \tag{7.3}$$

where $M(\omega)$ and $\phi(\omega)$ are the magnitude and phase response of the filter, respectively. The filtered signal is given by

$$y(t) = AM(\omega_0)\cos(\omega_0 t + \phi(\omega_0)), \tag{7.4}$$

which can be rewritten as

$$y(t) = AM(\omega_0)\cos\left(\omega_0\left(t + \frac{\phi(\omega_0)}{\omega_0}\right)\right) = AM(\omega_0)\cos(\omega_0(t - \tau_p(\omega_0))). \tag{7.5}$$

From eqn (7.5), we see that a sinusoidal signal of frequency ω_0 experiences a delay of $\tau_p(\omega_0)$. To overcome the effect of phase shift due to filtering, the filtering operation is performed in the frequency domain as follows. The Fourier transform (FT) [12] of the signal is obtained which is multiplied by a window defining the frequency of interest. The resulting frequency signal is reconstructed back to the time domain by taking inverse fourier transform (IFT).

The above formulation is demonstrated using a simple experiment shown in Fig.7.4.

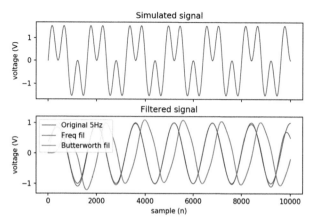

Figure 7.4   Effect of filtering on the signal.

Consider a signal which is a mixture of 2 sine waves (5 and 15 Hz sinusoids) as shown in the top panel of Figure 7.4. The goal here is to filter this signal in the time domain using sixth order Butterworth filter and also in the frequency domain as described earlier in the passband of 0.5Hz-10 Hz to understand the effect these filtering techniques on the phase of the filtered signal. As seen from the bottom panel of Figure 7.4, filtering in the time domain induces a slight phase shift whereas, frequency filtering does not. Also, due to the availability of faster algorithms to calculate FT and IFT, filtering in the frequency domain might be more efficient when compared to the implementation of convolution steps involved in time-domain filtering [72].

7.3 Feature Extraction

In this section, we discuss some of the prominent features extracted from EEG signals by grouping them into four main categories: spectral features, connectivity features, time-domain features, and non-linear features. While using rule-based classification algorithms, the choice of features from a plethora of features becomes highly significant. The performance of rule-based models will significantly depend on the features chosen and hence, appropriate and relevant features have to be carefully chosen depending on the task being performed. Hence, domain knowledge becomes crucial in the development of such models. Since the application discussed in this chapter is mTBI classification, we will be focusing on features that are relevant to this task [52], [68]. However, most of these features contain significant information which can be used in other EEG-related tasks as well.

7.3.1 Spectral features

The significance of the study of frequency domain features in EEG analysis is showcased in numerous studies that report statistically significant alteration in at least one frequency band when target groups in the study are compared. Some studies show these changes return to normal over a period of time in the target subjects suggesting that care must be taken when selecting these features for longitudinal studies [10]. The following subsections outlay prominent spectral features used in QEEG discriminant models.

7.3.1.1 Average and relative power

Welch's periodogram [69] calculates power spectral density (PSD) by averaging multiple FT obtained from consecutive sections of small segments of the signal. This is necessary as the spectral content obtained from the signal is non-stationary and has to be averaged over multiple windows to obtain an accurate estimate. The optimal window size is often considered to be twice the length of the cycle of the lowest frequency of interest. Figure 7.5 shows the periodogram of a typical resting state EEG epoch along with prominent EEG frequency bands-delta (0.5-4 Hz), theta (4-8 Hz), alpha (8-12 Hz), sigma (12-16 Hz), and beta (12-35 Hz) shaded in different colors. The average power in the specific frequency band is then equal to the area of the shaded region corresponding to that frequency band.

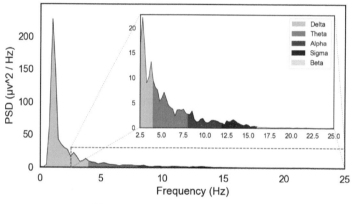

Figure 7.5   EEG frequency bands.

The ratio of power in the frequency band of interest to the total power of the signal is referred to as relative power:

$$\text{Relative Power} = \frac{\text{Power in frequency band}}{\text{Total Power}}. \tag{7.6}$$

In EEG analysis, change in power ratios corresponding to $theta : alpha$, and $alpha1 : alpha2$ are also considered as prominent QEEG features. These are referred to as slow:fast power ratio.

7.3.1.2 Frequency amplitude asymmetry

The study of relative change in power across different regions of the brain becomes imperative in most EEG studies. To study these relations, frequency amplitude asymmetry calculated as differences in absolute power between pairs of electrodes is considered to be an efficient indicator [62]. For inter-hemisphere comparison, it is calculated as

$$\text{Frequency Amplitude Asymmetry} = \frac{\text{Left} - \text{Right}}{\text{Left} + \text{Right}}. \tag{7.7}$$

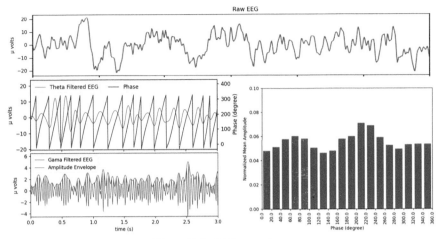

Figure 7.6 PAC procedure.

7.3.1.3 Phase amplitude coupling

One of the better ways to study cross-frequency coupling in EEG signals is through the observation of modulation of the amplitude of high-frequency oscillations by the phase of the low-frequency component of the signal. This is characterized by phase-amplitude coupling (PAC) [45] quantified by a modulation index put forth by Tort et al. [65]. For an observed EEG signal $x(t)$, phase $\phi_{x_l}(t)$ of the signal filtered at lower frequency range f_l and amplitude envelope $A_{x_h}(t)$ of the signal filtered at higher frequency f_h is calculated using Hilbert transform which is shown in Figure 7.6. Then the normalized mean amplitude is calculated for each frequency bin denoted as

$$P(j) = \frac{\langle A_{x_h} \rangle_{\phi_{x_l}}(j)}{\sum_{k=1}^{N} \langle A_{x_h} \rangle_{\phi_{x_l}}(k)}, \tag{7.8}$$

where $\langle \, \rangle$ denotes mean operation. The modulation index is calculated using Kullback-Leibler (KL) distance [41] as the deviation of the phase-amplitude plot from a uniform distribution.

7.3.2 Connectivity features

Connectivity features give insights into adirectional relationships between concurrently recorded EEG signals. Commonly used frequency domain connectivity features include coherence, phase synchronization, phase-locking value, and phase lag index [59], some of which are detailed in this section. Prior to performing computational analysis, care must be taken to correctly reference the electrodes as this plays a significant role in the conclusions drawn from the connection metrics [1].

7.3.2.1 Coherence

One of the prominent ways to analyze spatial properties of the brain is through coherence [48]. The strength of synchrony between two brain regions is directly proportional to the coherence value between the corresponding pairs of electrodes. It is a normalized quantity ranging between 0 and 1 given by the squared correlation of two spectral density functions over trials

$$\text{Coh}(f, t) = \frac{|\sum_n S_{1n} \cdot S'_{2n}|^2}{\sum_n |S_{1,}|^2 \cdot \sum_n |S_{2n}|^2}, \qquad (7.9)$$

where the spectral–temporal density function of a signal is expressed as

$$S(f, t) = A(f, t) \cdot e^{i\phi(f,t)}, \qquad (7.10)$$

where A and ϕ are the amplitude and phase of the signal at a certain frequency f and time t.

As the coherence analysis takes into account the amplitude of the signal, it is not the best parameter to study the phase synchrony unless the signals considered in the analysis are normalized or a different parameter called phase locking value (PLV) discussed in 7.3.2.3 is used.

7.3.2.2 Phase difference

The phase of a signal is calculated as the angle of its analytical signal obtained by taking the Hilbert transform of the signal [11],[20].

$$\phi(t) = \arg\left[s(t) + j\hat{s}(t)\right]. \qquad (7.11)$$

Hilbert transform ($\hat{s}(t)$) of signal $s(t)$ is given by

$$hats(t) = \left[\frac{1}{\pi t} * s(t)\right], \qquad (7.12)$$

where $*$ conveys convolution operation.

A panel in Figure 7.6 illustrates the phase of a section of filtered EEG signal in the theta frequency band plotted between 0 and 360 degrees.

7.3.2.3 Phase locking value

The extent of phase synchrony between brain different regions is measured using phase locking value (PLV). This was introduced by Lachaux et al. [42] to overcome the shortcoming in the interpretation of phase synchrony by coherence calculation [7]:

$$\text{PLV}(f, t) = \left| \frac{1}{N} \sum_{n} e^{i(\phi_{1n} - \phi_{2n})} \right|, \tag{7.13}$$

where ϕ_1 and ϕ_2 are phases of the two EEG signals, respectively.

7.3.3 Time-domain features

Some of the commonly used time-domain QEEG features include mean, peak-to-peak amplitude, maximum and minimum value, variance, kurtosis, and zero-crossing rate to name a few. They are simple EEG characteristics that are computationally very easy to calculate.

Hjorth parameters consist of three measures–activity, mobility, and complexity which can also be derived from the first five statistical moments of the power spectrum [26]. The power of EEG signal $x(t)$ can be characterized using activity, which is calculated as the variance of the EEG signal:

$$\text{Activity}\,(x(t)) = \text{var}(x(t)). \tag{7.14}$$

Mobility is defined as the square root of the ratio between the variances of the first derivative and the amplitude of the EEG signal expressed as a ratio per time unit.

$$\text{Mobility}(x(t)) = \sqrt{\frac{\text{var}\left(\frac{dx(t)}{dt}\right)}{\text{var}(x(t))}}. \tag{7.15}$$

Complexity is a dimensionless quantity evaluated as the ratio between the mobility of the first derivative of the signal and the mobility of the signal itself.

$$\text{Complexity}\,(x(t)) = \frac{\text{Mobility}\left(\frac{dx(t)}{dt}\right)}{\text{Mobility}\,(x(t))}. \tag{7.16}$$

7.3.4 Non-linear features

Spectral entropy quantifies the uniformity of energy distribution in frequency-domain [32][58]. Spectral entropy is calculated using the standard entropy

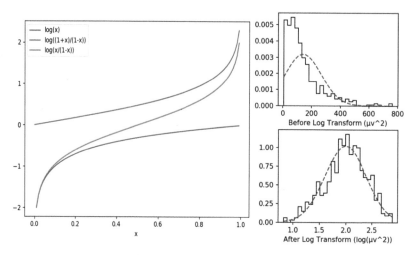

Figure 7.7   Log transformation.

formula

$$H(x, sf) = -\sum_{f=0}^{f_s/2} P(f) \log_2[P(f)], \qquad (7.17)$$

where $P(f)$ is normalized PSD and f_s is the sampling frequency.

7.4 Feature Normalization

7.4.1 Log transformation

In general, the majority of EEG characteristics often conform to skewed distribution. Depending on the class of log transformation being used, it results in the compression or expansion of specific ranges of values. Figure 7.7 plots the range of input versus output values for different transformations.

Generally, relative band power R is transformed using $\log(R/(1 - R))$, magnitude squared coherence C is transformed with $\log(C/(1 - C))$, amplitude asymmetry X is transformed with $\log((2 + X)/(2 - X))$ and spectral entropy SpEn using $-\log(1 - \text{SpEn})$ [34][22]. Figure 7.7 demonstrates the significance of log transforms as it transforms the skewed input feature to a Gaussian distribution. However, the characteristics will no longer have the same significance as the original QEEG features when the log transformation is complete, and they cannot be directly interpreted as before.

7.4.2 Age regression

The dominant frequency of EEG increases with age, thereby suggesting that age is reflected in EEG [35]. The two ways to minimize the effect of age on QEEG features are age stratification, which is the process of dividing subjects into groups based on their age [63], and age regression, which regresses the features with respect to the age of subjects assuming a linear relationship between the calculated QEEG features and $\log_{10}$ of subject's age expressed in years. The intercept and the coefficients obtained from fitting a straight line to QEEG features in the age regression method are then used to regress the same feature using the below equation:

$$y_i = x_i - \log 10(\text{ SubjectAge }) \cdot m_i, \tag{7.18}$$

where x_i and y_i represent original and age-regressed features respectively and m_i represents the age-regression parameter.

An example of age regression is shown in Figure 7.8. The delta power from the F3 electrode is displayed in the left pane prior to age regression and an obvious correlation between this variable and age can be observed. This is removed once the age regression is performed, as shown in the right pane. The regression parameters are calculated for the control group only and are then used to age-regress TBI subjects [51]. If the changes in the

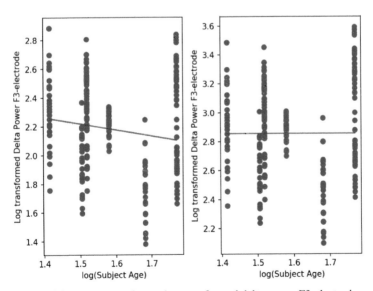

Figure 7.8 Age regression on log transformed delta power F3-electrode.

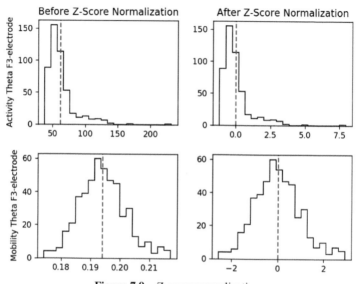

Figure 7.9   Z-score normalization.

QEEG features with age are more rapid than a simple linear regression, a higher-order polynomial may account for a better fit which has to be carefully investigated.

7.4.3 Z-score standardization

Feature scaling is a transform that brings all features to a similar scale of zero mean and unit standard deviation which aids to minimize the cost function quickly, thereby facilitating a faster training process. This technique helps gradient descent optimization-based ML models the most however, such standardization has little effect on tree-based ML models. Z-score standardization is performed by subtracting the mean and then dividing it by the feature's standard deviation. Each QEEG feature in the training dataset has its mean and standard deviation determined, which are subsequently utilized to standardize the corresponding features of the test individuals [21]. The effect of Z-score normalization is shown in Figure 7.9.

All the plots display histograms of the respective features. One has to take note of the range of values taken by the features before and after standardization. Once the standardization is performed both the features attain zero mean and unit standard deviation and transform to a similar scale giving equal importance to both features in the ML algorithm.

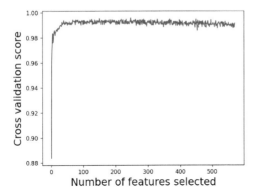

Figure 7.10 Relationship between the number of features and CV accuracy.

7.5 Feature Selection

In this section, we introduce some well-known feature selection techniques. Feature selection becomes essential to eliminate overfitting and increase model performance by reducing the computational cost and training time [8]. The distinction between dimensionality reduction and feature selection should be well understood as dimensionality reduction refers to projecting the existing features onto a new space obtaining new input features with lower dimensionality, whereas feature selection methods only remove less relevant features.

The feature selection procedure may be broadly divided into the wrapper, filter, and embedded methods. In order to pick the best features, filter techniques rank the features according to scores produced based on statistical quantities in relation to their correlation to the target variable. Pearson's correlation [24], mutual information [2], and analysis of variance (ANOVA) are some examples of filter methods. Wrapper methods [39] are computationally demanding techniques that assess a given ML model's performance on a subset of features using metrics like accuracy to decide whether to add or remove a specific feature. This procedure is repeated until the required number of features is obtained. Forward feature selection and recursive feature elimination (RFE) are the best examples of this technique. Embedded feature selection procedures are faster than wrapper methods as they incorporate feature selection as a part of the training process of the model. Regularization methods are the most common type of embedded feature selection. Decision trees [43], least absolute shrinkage and selection operator (LASSO) [64], and ridge regression [28] are some of the examples of embedded feature selection methods.

A plot of the number of best features used versus the cross-validation accuracy obtained on the human EEG dataset discussed in Section 7.9.1.2 using the RFE method with the random forest as the base estimator is shown in Figure 7.10. The decision on the optimal number of features for a particular dataset is still ongoing research [29]. The authors in [33] propose that the optimal number of features for a collection of uncorrelated features is $N - 1$ whereas it is $\sqrt{N}$ for highly correlated features where N denotes the total number of features.

7.6 Train/Test Split

Data snooping occurs when the model anticipates the kind of data it receives and overfits the data which would result in a much higher accuracy on the existing dataset and a failure to classify new, previously unseen data [21]. K-Fold cross validation and individual validation are two different train/test data splits that will be discussed in this section.

In k-Fold cross validation, the data are shuffled and evenly divided into "k" number of folds. While one fold is chosen as the validation set in each iteration, the model is trained on the rest of the data [38]. The final accuracy metric is calculated as the mean of accuracies obtained for models in each iteration. This is shown in Figure 7.11. However, this is not appropriate for datasets with significant imbalances. It should be noted here that the impact of subject-specific variations is not taken into consideration in k-fold validation as a part of the data from each individual is present in the training set due to initial shuffling. Since this style of data arrangement uses historical data from the same individual, rather than detection problems, it allows one to develop models that may be utilized for daily patient monitoring.

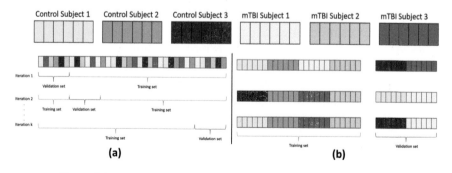

Figure 7.11   (a) k-Fold cross-validation and (b) individual validation.

In contrast, independent-validation emphasizes how various patients' EEGs differ from one another as the model is evaluated on completely held-out subjects. To maintain similar composition between the training and validation sets, the validation data comprises data from a subject of each class and the training set includes data from the rest of the subjects. This is done for all conceivable combinations of training and validation sets and the mean accuracy of all iterations is calculated as in the case of k-Fold cross-validation. Since the model is evaluated on held-out data, this approach is used to study the generalizability of the trained model. This arrangement mostly reflects a more practical scenario of TBI detection.

7.7 Rule-based ML

Classical rule-based ML models take manually specified domain-specific features as inputs and are best suited for small to medium-structured datasets. This section introduces the working principle of a number of classical ML algorithms.

7.7.1 Decision tree and random forest

Constructed as a tree of nodes, decision trees (DTs) are categorized as non-parametric supervised learning techniques. It successively divides the root node consisting of the source data based on certain metrics such as Gini impurity, information gain, and variance reduction. Classification and regression tree (CART) which is one of the most used algorithms use Gini impurity given by

$$I_G(p) = \sum_{i=1}^{J} \left(p_i \sum_{k \neq i} p_k \right), \tag{7.19}$$

where p_i is the probability of an instance with label i, and J is the number of classes to decide on the splits.

During the training process, the decision on the best split is made based on optimization which involves maximizing the Gini gain. This process is repeated recursively until all data in a node has the same label or when splitting no longer adds value to the predictions. DTs are easy to interpret but are prone to overfit. Figure 7.12 shows a DT built on wake sleep stage of mice data with gamma and beta power as features.

The overfitting problem of DTs can be overcome by the use of another supervised ML algorithm-random forest (RF) [27] that introduces

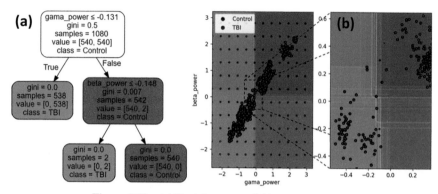

Figure 7.12   (a) Decision tree and (b) random forest.

randomization by developing an ensemble of DTs fit on different sub-samples of the dataset and features with replacement. This technique is known as bagging [5]. The output of the random forest algorithm is the averaged prediction of individual classifiers. Figure 7.12 shows the overlapping RF decision boundaries of 30 different DTs.

7.7.2 Support vector machine

The main goal of support vector machines (SVM) [13] used in classification tasks is to produce a hyperplane in an *N*-dimensional space that provides the greatest separation between various classes and clearly categorizes the data points. The margin is determined by the separation between the hyper-plane and the closest subset of data points called support vectors [57]. Linear, polynomial, and radial basis function (RBF) kernels are often used to transform the input data onto a space in which they are linearly separable. Figure 7.13 shows the SVM decision boundary and the support vectors obtained using a linear kernel.

7.7.3 k-Nearest neighbor

k-Nearest neighbor (kNN) is a supervised, instance-based learning algorithm developed by Evelyn Fix and Joseph Hodges [17]. It assumes that the data points from the same class exist in close proximity in the feature space. The labels of the test data points are inferred by a majority vote of the 'k' (a predefined number) closest instances to the unknown data point that is stored during the training phase. Any scaling operations performed on the data

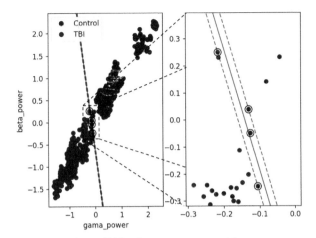

Figure 7.13 Support vector machine.

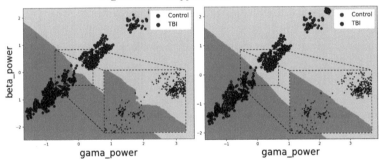

Figure 7.14 k-Nearest neighbor (Left: $K = 5$, Right: $K = 11$).

results in a change in the distance between the data points and thereby may impact the assignment of the label. Figure 7.14 shows the decision boundary obtained for two different values of "k", 5, and 11. As can be observed, a higher value of "k" reduces overfitting. The major hindrance one experiences in using kNN is the running time as it uses the brute force method. It should also be kept in mind that if the dataset is highly skewed, then the class with more samples tends to dominate the prediction of the label of new data points. This is overcome by using a weighted nearest neighbor algorithm.

7.7.4 Extreme gradient boosting

XGBoost (XGB) [9] is an ensemble learning technique that uses a gradient-boosting decision tree algorithm to enable more rapid learning through parallel and distributed computing [14]. In the sequential process known as "boosting," the model's output is the weighted average of earlier models. In each iteration, the residual error is updated and the current stump is constructed by assigning higher weights to misclassified points from earlier stumps. It improves its performance through algorithmic enhancement techniques like regularization, sparsity awareness, and weighted quantile sketch, as well as system optimization approaches like parallelization, tree trimming, and hardware optimization.

7.8 Evaluation Metrics

This section introduces the most frequently used evaluation metrics in the biomedical field that plays a significant role in evaluating model performance and provide better insights on the results obtained [50],[61]. Understanding confusion matrix is fundamental to understanding evaluation metrics as they can be derived from one another. Confusion matrix for binary classification summarizes the prediction results into four cells as shown in Table 7.1. True positive and true negative refers to the number of correctly predicted positive and negative labels, respectively. Similarly, false positive and false negative refers to the number of incorrectly predicted positive and negative labels, respectively.

Accuracy is the ratio of the total number of correct predictions to the total number of cases:

$$\text{Accuracy} = \frac{\text{TP} + \text{TN}}{\text{TP} + \text{TN} + \text{FP} + \text{FN}}. \tag{7.20}$$

Sensitivity (Sen) also known as recall or true positive rate is the ratio of true positive to the sum of true positive and false negative:

$$\text{Sensitivity} = \frac{\text{TP}}{\text{TP} + \text{FN}}. \tag{7.21}$$

Table 7.1   Confusion matrix.

| | | Actual values | | |
		Positive	Negative	
Predicted values	Positive	True positive (TP)	False positive (FP)	Row entries for determining positive predictive value
	Negative	False negative (FN)	True negative (TN)	Row entries for determining negative predictive value
		Column entries for determining sensitivity	Column entries for determining specificity	

Specificity (Spec) is the ratio of true negatives to the sum of true negatives and false positives:

$$\text{Specificity} = \frac{\text{TN}}{\text{TN} + \text{FP}}. \tag{7.22}$$

Sensitivity and specificity convey the ability of the ML model to detect positive and negative cases relative to the known reference standard, respectively, and are not concerned with the correctness of the actual true labels of the subjects [66].

Performance comparisons between different models are accomplished using the receiver operating characteristic (ROC) curve which is a two-dimensional plot representing true positive rate against false positive rate obtained using different thresholds [16]. The area under ROC (AUC) is defined as the ratio of the area under the ROC curve to the area of the unit square [4].

7.9 Application: mTBI Detection and Classification in Mice and Humans

TBI is defined as a change in brain functioning or brain pathology evoked by external impacts [37] causing a wide range of functional changes affecting thinking, emotions, and sleep [18], [56]. The effects of TBI are considered to last from a few days upto an entire lifetime depending on the severity of the injury. This is currently assessed clinically based on Glasgow coma scale (GCS). GCS is a qualitative measure that is highly observer-dependent [60]. The subject is assessed based on their ability to perform certain actions which are outlined in Table 7.2.

TBI is highly prevalent among US service members and sports personnel. More than 400,000 TBI cases have been reported among US service members between 2000 and 2019 [15]. Concussions and mild TBI (mTBI) constitute about 75% of TBIs that occur each year [19] and the lack of consensus regarding what constitutes mTBI adds to the complication of the under-diagnosis of the disease. Studies indicate that EEG returns to baseline after a few days of the incident [31], but some mTBI cases result in lasting disability. There are currently no prognostic signs to identify those who are most at risk. Therefore, innovative methods for the accurate identification and prognostication of mTBI are crucial.

Table 7.2 Glasgow coma scale.

Response	Scale	Score/points
Eye opening response	Open spontaneously	4
	Open to sound	3
	Open to pressure	2
	No eye opening	1
Verbal response	Oriented	5
	Confused conversation	4
	Words discernible	3
	Incomprehensible sounds	2
	No verbal response	1
Motor response	Obeyes command	6
	Movement to stimulus	5
	withdraws from pain	4
	Abnormal flexion	3
	Extensor response	2
	No response	1
Mild TBI = 13–15 points	Moderate TBI = 9–12 points	Severe TBI = 3–8 points

7.9.1 Data acquisition

7.9.1.1 Mice data

EEG and EMG signals were recorded from 4 to 6 months old Male C57BL6 mice (Jackson Laboratories, Bar Harbor, ME) for the experiments. After a period of two weeks left craniotomy and electromagnetic controlled cortical impact (CCI) or left craniotomy only (sham) was performed ipsilateral to the implanted probes [71], [6]. SleepSign (Kissei Comtec Co., Ltd., Nagano, Japan)- an automatic sleep scoring software was used to categorize 4-second epochs into the wake, REM, and non-REM [53]. After a thorough visual inspection, it was later manually over-scored and corrected by a single investigator blinded to the intervention. On day 5, the animals were sacrificed.

7.9.1.2 Human data

The initial study consisted of 337 subjects who underwent overnight polysomnography using Polysmith (NihonKohden, Japan), 38 of whom met the criteria for PTSD and were age-matched to controls. According to the 10–20 system for EEG placement, six scalp electrodes were positioned at F3, F4, C3, C4, O1, and O2 as shown in Figure 7.15. The recordings were epoched into 30-second chunks and manually classified into different sleep stages (Awake, Rapid Eye Movement, non-REM stages N1, N2,

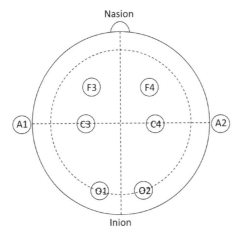

Figure 7.15   Electrode placement.

and N3) by an American Academy of Sleep Medicine (AASM)-accredited polysomnographic technician. The details of the procedure can be found in [44].

7.9.2 Pipeline and tools used

Although the analysis pipeline consisting of the steps outlined in the earlier sections is identical in case of mice and human dataset, there exists minor differences because of the inherent differences in the dataset itself. Age regression is carried out for human data since the human cohort belongs to

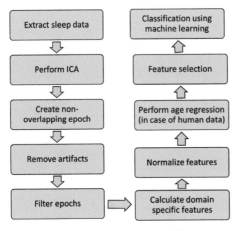

Figure 7.16   Standard pipeline.

Table 7.3   Accuracy (%) obtained for mice data.

Sleep stage	DT	RF	kNN			MLP	SVM	XGB
			k = 5	k = 11	k = 19			
				CV				
Wake	99.54	99.63	99.54	99.63	99.63	99.44	98.8	99.63
NR	99.72	99.81	99.81	99.72	99.81	99.07	99.63	99.91
REM	100	100	100	94.55	96.36	85.27	97.27	100
				IV				
Wake	**98.26**	**99.16**	**96.76**	**97.89**	**97.94**	**98.53**	**93.23**	**98.65**
NR	97.54	96.44	91.27	92.03	92.85	95.02	90.23	98.38
REM	95.67	94.48	88.58	91.63	93.83	89.69	87.14	95.27

Table 7.4   Accuracy (%) obtained for human data.

Sleep stage	DT	RF	kNN			MLP	SVM	XGB
			k = 5	k = 11	k = 19			
				CV				
W	89.29	97.86	98.75	97.86	95.36	97.32	94.11	96.43
N1	88.04	97.41	99.11	98.57	97.77	94.64	92.77	97.41
N2	**93.19**	**99.43**	**99.62**	**99.33**	**99.29**	**99.43**	**97.71**	**99.24**
N3	76.39	82.08	66.94	69.31	70.42	73.61	67.08	78.61
REM	91.79	99.14	98.79	98.21	97.71	93.86	82.29	98.86
				IV				
W	55.78	65.45	67.87	68.03	67.06	71.81	70.64	64.37
N1	57.58	64.72	69.48	69.81	70.29	68.84	70.63	62.86
N2	**70.14**	**75.25**	**75.55**	**75.27**	**75.35**	**80.55**	**80.89**	**73.9**
N3	62.8	64.44	70.91	66.81	69.81	69.35	73.28	63.79
REM	64.91	69.79	72.73	72.82	72.62	77.77	76.85	69.53

a wide range of age demographics and since mice data consists of single electrode recording, connectivity features are not estimated. Figure 7.16 shows the common pipeline.

7.9.3 Result

This section discusses the results obtained using the above pipeline. Tables 7.3 and 7.4 show the accuracy obtained for multiple algorithms across different sleep stages on the mice and human datasets, respectively, for cross-validation (CV) and IV data arrangement. Detailed analysis of the model performance is present in [67].

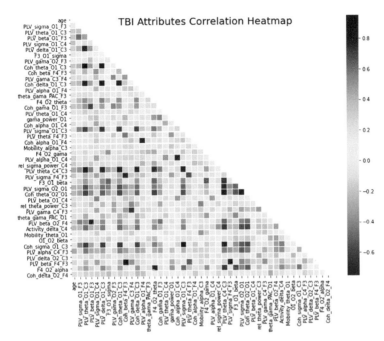

Figure 7.17 Correlation matrix of frequently chosen features in N2 sleep stage of human data.

The variance in accuracy obtained in different sleep stages can be attributed to the amount of data present in the respective sleep stage. While the mice are recorded over a 24-hour period through a process of controlled experiment, human data consists of overnight PSG recording of subjects whose location of impact and time since impact varies greatly. As a result, the distribution of data across the sleep stages is not similar and plays a huge role in the results obtained. We expect an increase in accuracy obtained in IV data arrangement when more data are added to the training since this adds heterogeneity to the training dataset thereby aiding the model to learn more general parameters. The top panel of Figure 7.17 shows the correlation matrix obtained for top features selected using RFE for the human N2 stage. The pool of the selected features predominantly consists of connectivity features such as PLV and coherence and the features are not highly correlated. As seen from the above discussion, the effective working of the standard EEG processing pipeline is validated for both mice and human datasets. This has to be further evaluated on other neurological conditions as well.

References

[1] André M Bastos and Jan-Mathijs Schoffelen. A tutorial review of functional connectivity analysis methods and their interpretational pitfalls. *Frontiers in systems neuroscience*, 9:175, 2016.

[2] Roberto Battiti. Using mutual information for selecting features in supervised neural net learning. *IEEE Transactions on neural networks*, 5(4):537–550, 1994.

[3] Nima Bigdely-Shamlo, Tim Mullen, Christian Kothe, Kyung-Min Su, and Kay A Robbins. The prep pipeline: standardized preprocessing for large-scale eeg analysis. *Frontiers in neuroinformatics*, 9:16, 2015.

[4] Andrew P Bradley. The use of the area under the roc curve in the evaluation of machine learning algorithms. *Pattern recognition*, 30(7):1145–1159, 1997.

[5] Leo Breiman. Bagging predictors. *Machine learning*, 24(2):123–140, 1996.

[6] David L Brody, Christine Mac Donald, Chad C Kessens, Carla Yuede, Maia Parsadanian, Mike Spinner, Eddie Kim, Katherine E Schwetye, David M Holtzman, and Philip V Bayly. Electromagnetic controlled cortical impact device for precise, graded experimental traumatic brain injury. *Journal of neurotrauma*, 24(4):657–673, 2007.

[7] Ricardo Bruña, Fernando Maestú, and Ernesto Pereda. Phase locking value revisited: teaching new tricks to an old dog. *Journal of neural engineering*, 15(5):056011, 2018.

[8] Girish Chandrashekar and Ferat Sahin. A survey on feature selection methods. *Computers & Electrical Engineering*, 40(1):16–28, 2014.

[9] Tianqi Chen and Carlos Guestrin. Xgboost: A scalable tree boosting system. In *Proceedings of the 22nd acm sigkdd international conference on knowledge discovery and data mining*, pages 785–794, 2016.

[10] Xi-Ping Chen, Lu-Yang Tao, and A Cn Chen. Electroencephalogram and evoked potential parameters examined in chinese mild head injury patients for forensic medicine. *Neuroscience bulletin*, 22(3):165, 2006.

[11] Leon Cohen. *Time-frequency analysis*, volume 778. Prentice Hall PTR Englewood Cliffs, NJ, 1995.

[12] James W Cooley and John W Tukey. An algorithm for the machine calculation of complex fourier series. *Mathematics of computation*, 19(90):297–301, 1965.

[13] Corinna Cortes and Vladimir Vapnik. Support-vector networks. *Machine learning*, 20(3):273–297, 1995.

[14] Raphael Couronné, Philipp Probst, and Anne-Laure Boulesteix. Random forest versus logistic regression: a large-scale benchmark experiment. *BMC bioinformatics*, 19(1):1–14, 2018.

[15] Center DaVBI. Department of defense numbers for traumatic brain injury. 2011.

[16] Tom Fawcett. An introduction to roc analysis. *Pattern recognition letters*, 27(8):861–874, 2006.

[17] E Fix and J Hodges. An important contribution to nonparametric discriminant analysis and density estimation. *International Statistical Review*, 3(57):233–238, 1951.

[18] Centers for Disease Control, Prevention, et al. Report to congress on traumatic brain injury in the united states: epidemiology and rehabilitation. *National Center for Injury Prevention and Control*, pages 1–72, 2015.

[19] National Center for Injury Prevention and Control (US). *Report to Congress on mild traumatic brain injury in the United States: steps to prevent a serious public health problem*. Centers for Disease Control and Prevention, 2003.

[20] Walter J Freeman. Hilbert transform for brain waves. *Scholarpedia*, 2(1):1338, 2007.

[21] Jerome Friedman, Trevor Hastie, Robert Tibshirani, et al. *The elements of statistical learning*, volume 1. Springer series in statistics New York, 2001.

[22] Theo Gasser, Petra Bächer, and Joachim Möcks. Transformations towards the normal distribution of broad band spectral parameters of the eeg. *Electroencephalography and clinical neurophysiology*, 53(1):119–124, 1982.

[23] Alexandre Gramfort, Martin Luessi, Eric Larson, Denis A Engemann, Daniel Strohmeier, Christian Brodbeck, Roman Goj, Mainak Jas, Teon Brooks, Lauri Parkkonen, et al. Meg and eeg data analysis with mne-python. *Frontiers in neuroscience*, 7:267, 2013.

[24] Mark A Hall. Correlation-based feature selection of discrete and numeric class machine learning. 2000.

[25] Jeanny Hérault, Christian Jutten, and Bernard Ans. Détection de grandeurs primitives dans un message composite par une architecture de calcul neuromimétique en apprentissage non supervisé. In *10 Colloque sur le traitement du signal et des images, FRA, 1985*. GRETSI, Groupe d'Etudes du Traitement du Signal et des Images, 1985.

[26] Bo Hjorth. Eeg analysis based on time domain properties. *Electroencephalography and clinical neurophysiology*, 29(3):306–310, 1970.

[27] Tin Kam Ho. Random decision forests. In *Proceedings of 3rd international conference on document analysis and recognition*, volume 1, pages 278–282. IEEE, 1995.

[28] Arthur E Hoerl and Robert W Kennard. Ridge regression: Biased estimation for nonorthogonal problems. *Technometrics*, 12(1):55–67, 1970.

[29] Jianping Hua, Zixiang Xiong, James Lowey, Edward Suh, and Edward R Dougherty. Optimal number of features as a function of sample size for various classification rules. *Bioinformatics*, 21(8):1509–1515, 2005.

[30] Aapo Hyvtirinen, Juha Karhunen, and Erkki Oja. Independent component analysis. 2001.

[31] Jéssica Natuline Ianof and Renato Anghinah. Traumatic brain injury: An eeg point of view. *Dementia & neuropsychologia*, 11(1):3–5, 2017.

[32] Tsuyoshi Inouye, Kazuhiro Shinosaki, H Sakamoto, Seigo Toi, Satoshi Ukai, Akinori Iyama, Y Katsuda, and M Hirano. Quantification of eeg irregularity by use of the entropy of the power spectrum. *Electroencephalography and clinical neurophysiology*, 79(3):204–210, 1991.

[33] Anil K Jain and William G Waller. On the optimal number of features in the classification of multivariate gaussian data. *Pattern recognition*, 10(5-6):365–374, 1978.

[34] ER John. Normative data bank and neurometrics. basic concepts, methods and results of norm constructions. *Methods od analysis of brain electrical and magnetic signals. EEG handbook*, 1:449–498, 1987.

[35] ER John, H Ahn, L Prichep, M Trepetin, D Brown, and H Kaye. Developmental equations for the electroencephalogram. *Science*, 210(4475):1255–1258, 1980.

[36] Markus Junghöfer, Thomas Elbert, Don M Tucker, and Brigitte Rockstroh. Statistical control of artifacts in dense array eeg/meg studies. *Psychophysiology*, 37(4):523–532, 2000.

[37] Erin S Kenzie, Elle L Parks, Erin D Bigler, Miranda M Lim, James C Chesnutt, and Wayne Wakeland. Concussion as a multi-scale complex system: an interdisciplinary synthesis of current knowledge. *Frontiers in neurology*, 8:513, 2017.

[38] Ron Kohavi et al. A study of cross-validation and bootstrap for accuracy estimation and model selection. In *Ijcai*, volume 14, pages 1137–1145. Montreal, Canada, 1995.

[39] Ron Kohavi and George H John. Wrappers for feature subset selection. *Artificial intelligence*, 97(1-2):273–324, 1997.

[40] Christian Andreas Edgar Kothe and Tzyy-ping Jung. Artifact removal techniques with signal reconstruction, April 28 2016. US Patent App. 14/895,440.

[41] Solomon Kullback and Richard A Leibler. On information and sufficiency. *The annals of mathematical statistics*, 22(1):79–86, 1951.

[42] Jean-Philippe Lachaux, Eugenio Rodriguez, Jacques Martinerie, and Francisco J Varela. Measuring phase synchrony in brain signals. *Human brain mapping*, 8(4):194–208, 1999.

[43] Wei-Yin Loh. Classification and regression trees. *Wiley interdisciplinary reviews: data mining and knowledge discovery*, 1(1):14–23, 2011.

[44] Mo H Modarres, Ryan A Opel, Kristianna B Weymann, and Miranda M Lim. Strong correlation of novel sleep electroencephalography coherence markers with diagnosis and severity of posttraumatic stress disorder. *Scientific reports*, 9(1):1–10, 2019.

[45] Tamanna TK Munia and Selin Aviyente. Time-frequency based phase-amplitude coupling measure for neuronal oscillations. *Scientific reports*, 9(1):1–15, 2019.

[46] Hugh Nolan, Robert Whelan, and Richard B Reilly. Faster: fully automated statistical thresholding for eeg artifact rejection. *Journal of neuroscience methods*, 192(1):152–162, 2010.

[47] Paul L Nunez, Ramesh Srinivasan, et al. *Electric fields of the brain: the neurophysics of EEG*. Oxford University Press, USA, 2006.

[48] Paul L Nunez, Ramesh Srinivasan, Andrew F Westdorp, Ranjith S Wijesinghe, Don M Tucker, Richard B Silberstein, and Peter J Cadusch. Eeg coherency: I: statistics, reference electrode, volume conduction, laplacians, cortical imaging, and interpretation at multiple scales. *Electroencephalography and clinical neurophysiology*, 103(5):499–515, 1997.

[49] Alan V Oppenheim. *Discrete-time signal processing*. Pearson Education India, 1999.

[50] David MW Powers. Evaluation: from precision, recall and f-measure to roc, informedness, markedness and correlation. *arXiv preprint arXiv:2010.16061*, 2020.

[51] Leslie S Prichep, Arnaud Jacquin, Julie Filipenko, Samanwoy Ghosh Dastidar, Stephen Zabele, Asmir Vodencarevic, and Neil S Rothman. Classification of traumatic brain injury severity using informed data

reduction in a series of binary classifier algorithms. *IEEE transactions on neural systems and rehabilitation engineering*, 20(6):806–822, 2012.

[52] Paul E Rapp, David O Keyser, Alfonso Albano, Rene Hernandez, Douglas B Gibson, Robert A Zambon, W David Hairston, John D Hughes, Andrew Krystal, and Andrew S Nichols. Traumatic brain injury detection using electrophysiological methods. *Frontiers in human neuroscience*, 9:11, 2015.

[53] John J Renger, Susan L Dunn, Sherri L Motzel, Colena Johnson, and Kenneth S Koblan. Sub-chronic administration of zolpidem affects modifications to rat sleep architecture. *Brain research*, 1010(1-2):45–54, 2004.

[54] Yannick Roy, Hubert Banville, Isabela Albuquerque, Alexandre Gramfort, Tiago H Falk, and Jocelyn Faubert. Deep learning-based electroencephalography analysis: a systematic review. *Journal of neural engineering*, 16(5):051001, 2019.

[55] Arthur L Samuel. Some studies in machine learning using the game of checkers. *IBM Journal of research and development*, 3(3):210–229, 1959.

[56] Danielle K Sandsmark, Jonathan E Elliott, and Miranda M Lim. Sleep-wake disturbances after traumatic brain injury: synthesis of human and animal studies. *Sleep*, 40(5), 2017.

[57] Hinrich Schütze, Christopher D Manning, and Prabhakar Raghavan. *Introduction to information retrieval*, volume 39. Cambridge University Press Cambridge, 2008.

[58] Claude Elwood Shannon. A mathematical theory of communication. *ACM SIGMOBILE mobile computing and communications review*, 5(1):3–55, 2001.

[59] Cornelis J Stam, BF Jones, G Nolte, M Breakspear, and Ph Scheltens. Small-world networks and functional connectivity in alzheimer's disease. *Cerebral cortex*, 17(1):92–99, 2007.

[60] Graham Teasdale and Bryan Jennett. Assessment of coma and impaired consciousness: a practical scale. *The Lancet*, 304(7872):81–84, 1974.

[61] Alaa Tharwat. Classification assessment methods. *Applied Computing and Informatics*, 2020.

[62] Robert W Thatcher, RA Walker, I Gerson, and FH Geisler. Eeg discriminant analyses of mild head trauma. *Electroencephalography and clinical neurophysiology*, 73(2):94–106, 1989.

[63] RW Thatcher, RA Walker, and S Giudice. Human cerebral hemispheres develop at different rates and ages. *Science*, 236(4805):1110–1113, 1987.

[64] Robert Tibshirani. Regression shrinkage and selection via the lasso. *Journal of the Royal Statistical Society: Series B (Methodological)*, 58(1):267–288, 1996.

[65] Adriano BL Tort, Robert Komorowski, Howard Eichenbaum, and Nancy Kopell. Measuring phase-amplitude coupling between neuronal oscillations of different frequencies. *Journal of neurophysiology*, 104(2):1195–1210, 2010.

[66] Robert Trevethan. Sensitivity, specificity, and predictive values: foundations, pliabilities, and pitfalls in research and practice. *Frontiers in public health*, 5:307, 2017.

[67] Manoj Vishwanath. *Detection of Traumatic Brain Injury Using a Standard Machine Learning Pipeline in Mouse and Human Sleep Electroencephalogram*. University of California, Irvine, 2021.

[68] Brian E Wallace, Amy K Wagner, Eugene P Wagner, and James T McDeavitt. A history and review of quantitative electroencephalography in traumatic brain injury. *The Journal of head trauma rehabilitation*, 16(2):165–190, 2001.

[69] Peter Welch. The use of fast fourier transform for the estimation of power spectra: a method based on time averaging over short, modified periodograms. *IEEE Transactions on audio and electroacoustics*, 15(2):70–73, 1967.

[70] Andreas Widmann, Erich Schröger, and Burkhard Maess. Digital filter design for electrophysiological data–a practical approach. *Journal of neuroscience methods*, 250:34–46, 2015.

[71] Jon T Willie, Miranda M Lim, Rachel E Bennett, Allan A Azarion, Katherine E Schwetye, and David L Brody. Controlled cortical impact traumatic brain injury acutely disrupts wakefulness and extracellular orexin dynamics as determined by intracerebral microdialysis in mice. *Journal of neurotrauma*, 29(10):1908–1921, 2012.

[72] Dorin Yael, Jacob J Vecht, and Izhar Bar-Gad. Filter-based phase shifts distort neuronal timing information. *Eneuro*, 5(2), 2018.

8

Novel Image Processing to Restore Scattered Light-sheet Microscopic Imaging Technique and its Application for Quantifying Biomechanics

Cynthia Dominguez, Tanveer Teranikar, and Juhyun Lee

Department of Bioengineering, UT Arlington, Texas, USA

Abstract

Research of cellular and molecular processes by way of histological methods allows for some insight but comes with a fundamental set of constraints that are challenging to overcome. Traditional histological methods are laborious, as well as severely limiting for in-depth study of developmental processes or disease processes in vivo. In traditional histology, fixing and sectioning tissue necessarily eliminates its dynamic function, while tissue section thickness limits the scope of investigation with conventional imaging tools. Noninvasive in vivo study of tissues and biomarkers is therefore paramount in gaining a fuller understanding of the pathophysiology surrounding conditions like congenital heart disorders. Light-sheet fluorescence microscopy (LSFM) is a powerful and noninvasive optical microscopy tool that can image in vivo tissue function in 4D (3D + time). LSFM boasts benefits such as short pixel dwell time (and therefore minimal photobleaching) while maintaining the ability to image a high dynamic range, as well as deep-tissue optical sectioning. Researchers have been seeking to overcome this problem by developing tissue-clearing techniques to attempt to homogenize the refractive index across the tissue via the removal of light-scattering pigments and lipids.

Keywords: Light-sheet fluorescent microscope, optical microscope, 4D imaging, tissue-clearing techniques, light-scattering pigments

Even so, anisotropy and light scatter are pervasive effects stemming from tissue thickness and refractive index mismatching of mounting media, making optical sectioning with perfectly tuned acquisition parameters difficult to achieve. Therefore, pre- and post-processing techniques are critical for yielding images suitable for biomedical research. Two such novel techniques are presented here.

8.1 Correcting Anisotropic Intensity in Light Sheet Images using Dehazing and Image Morphology

In biomedical research, there has long been a demand for specimen images that are free of aberrations in the region of interest (ROI) [7]. However, in

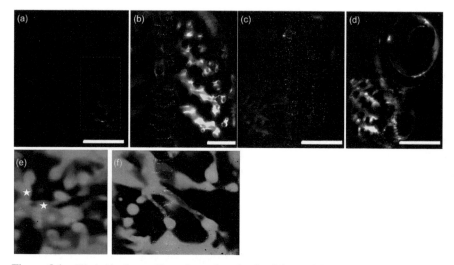

Figure 8.1 [7] Achieving uniform intensity of zebrafish cranial vasculature with a combination of dehazing and background subtraction. (A) Raw image acquired with oblique scanning super-resolution stage on a light sheet fluorescence microscope. Scale bar: 200 μm. (B) Magnified view of ROI in red box from (A). Intensity was restored by dehazing to fix attenuation, and background subtracting to remove diffuse light. Scale bar: 100 μm. (C) Raw image acquired using the same methods as (A). Scale bar: 200 μm. (D) Intensity-corrected version of (C). Scale bar: 200 μm. (E) Z-stack images acquired using light sheet fluorescence microscope with dual-sided illumination. White stars: poor separation of objects results in indistinguishable local morphology. This is due to light scatter in the FOV due to refractive index mismatch, which saturates the ROI and yields poorly delineated overlapping structures. (F) Processed 3D image. Intensity is uniform and edges of overlapping objects in the z-stack are enhanced, resulting in spatial integrity in the 3D image.

LSFM, unfocused fluorescence in the axial direction while scanning deeper into tissue creates a challenge in distinguishing foreground from background [7]. Additionally, light scatters due to refractive index heterogeneity throughout the tissue and in the mounting medium [10] results in axial and lateral blurriness [11], making isotropic resolution of volumetric reconstruction challenging, especially in high-density tissue samples.

To correct anisotropic intensity in a raw image (Figure 8.1 (A), (C), 8.2 (A), 8.3 (B), (E), and 8.4 (A)), the dark channel prior (DCP) illumination correction algorithm was used in conjunction with a background subtraction method. First, dehazing was performed with the DCP algorithm, which was used under the assumption that light is traveling underwater rather than through air[12–15] and is highly sensitive to noise while correcting medium-dependent attenuation along the line of sight [16]. Dehazing takes atmospheric light and transmission distance into account, resulting in the restoration of object radiance at zero viewing distance without loss of information or addition of noise [7] (Figure 8.3 (C)).

The DCP algorithm also tends to restore autofluorescence experiencing forward scatter (Figures 8.1 (E) and 8.3 (C)), so background subtraction

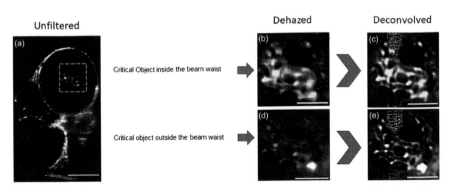

Figure 8.2 [7] Dehazing and deconvolution. (A) Raw image acquired with light sheet fluorescence microscope using oblique scanning super-resolution stage. Scale bar: 200 μm. (B) Magnification of blue box ROI from image (A). Image is background subtracted and dehazed. Scale bar: 100 μm. (C) Blurred overlapping depth structures in image (B), due to axial gradient refractive index, are now visible after deconvolution. Scale bar: 100 μm. (D) The same procedure is performed on an attenuated image with the light sheet passing through the critical object to compare against saturated image (b) where the light sheet is perpendicular to the critical object. Scale bar: 100 μm. (E) The contour in the ROI is defined after deconvolution. An edge-preserving bilateral smoothening filter is applied to images (c) and (e) to remove noise introduced by deconvolution. Scale bar: 100 μm.

with a rolling ball averaging algorithm is used to eliminate the scatter [7] (Figures 8.1 (B), (D), 8.2 (B), (D), and 8.3 (D)). The background subtraction isolates fluorescence-tagged structures in the field of view (FOV) and subtracts out-of-focus objects [7].

Anisotropy of the fluorescent bead manifests as skew in the axial and lateral resolution point spread function (PSF) graphs [7]. After dehazing by DCP, axial and lateral resolution improvements were seen in the PSF [7]. Random noise introduced by deconvolution was then eliminated with an

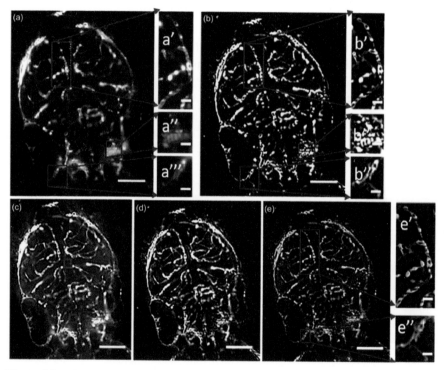

Figure 8.3 [7] Image processing to resolve fine details. (A) Unprocessed image. Scale bar: 200 μm. (A') Magnification of middle mesencephalic artery. (A") Magnification of primordial hindbrain. (A''') Magnification of posterior cerebral vein. Scale bar for (A')–(A"'): 50 μm. (b) Super-resolved image. Scale bar: 200 μm. Scale bar (B')–(B'''): 50 μm. (c) Autofluorescence or tissue scattering is amplified along with the fluorescent signal after dehazing. Scale bar: 200 μm. Blue squares: local pixel patch with autofluorescence. (D) The fluorescent signal-to-noise ratio of zebrafish vasculature is enhanced using background subtraction. Scale bar: 200 μm. (e) Fine vasculature is resolved after deconvolution of dehazed PSF. Scale bar: 200 μm. Scale bar (e') and (e"): 50 μm.

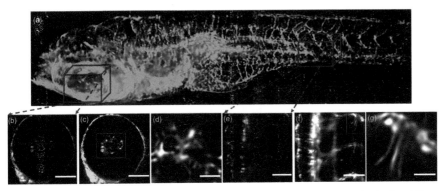

Figure 8.4 [7] Application of technique to super-resolution image acquired via oblique scanning method. (A) 3D reconstruction of the vasculature of a 4 days post fertilization (dpf) zebrafish. (B) High resolution image taken from the large FOV. Scale bar: 100 μm. (C) Intensity of image (B) is corrected and edges between features are resolved through dehazing and background subtraction. Scale bar: 100 μm. (D) Magnification of red box area from image (C) showing detail of super-resolution scanning. Scale bar: 30 μm. (E) High-resolution image is taken from the large FOV. Scale bar: 100 μm. (F) Intensity correction and resolution of edges between features of image (E) are achieved through dehazing and background subtraction. Scale bar: 100 μm. (G) Super-resolution details from the red box area in image (f). Scale bar: 30 μm.

edge-preserving bilateral smoothening filter [7]. Performing deconvolution on background-subtracted images using dehazed PSF has shown to be a promising method for symmetric restoration of objects in the focal plane[7] (Figures 8.2 (C), 8.2 (e), and 8.3 (e)).

Next, a top-hat morphological transform was used to 1) remove vague connections between ROI leftover from the background subtraction step, and 2) further resolve image boundaries [7]. Finally, a bottom-hat transform improves blurred depth details caused by autofluorescence impacts on the axial resolution [7].

This method was successfully applied to images acquired with various LSFM modalities including single/dual selective plane illumination microscopy (SPIM), multiview SPIM with dual illumination, and the voxel super-resolution technique using an oblique scanning stage with dual illumination [7]. In all instances, regardless of image dimensions, anisotropy was resolved, and isotropic structural integrity was achieved [7] (Figures 8.1 (F), 8.3 (B), (B')-3(B'''), and 8.4).

8.2 Feature Detection to Segment Cardiomyocytes for Investigating Cardiac Contractility

Based on recovering the anisotropic fluorescent intensity, further application to in vivo zebrafish heart for biomechanical quantification. Myocardial contractility is an important factor in healthy cardiac function [17]. Contractility is regulated by cardiomyocytes, which comprise a significant portion of the myocardium [18]. Proper investigation of cardiomyocyte physiology and pathophysiology depends on the ability to visualize individual cardiomyocytes with respect to the surrounding tissue [19]. For instance, precise counting of cardiomyocytes to understand cell proliferation during cardiogenesis [20, 21] can only be achieved if separate and distinct cardiomyocytes are observable. Limitations of invasive histological investigation, such as small sampling size and low cell viability, make it difficult to demonstrate statistical significance [4].

Imaging transgenic zebrafish with LSFM enables in vivo 4D optical sectioning for the study of cardiac architecture [6, 7, 23, 24]. Even so, specimen movement and sampling artifacts negatively affect the focus, making it a challenge to quantify biomarker data [5].

On the whole, optical microscopy offers a vast spectrum of complex image attributes available for feature selection [1], which presents an obstacle in research that relies on confidence in biomarker data. With a large range of target attributes available across image datasets, manual analysis becomes impractical and unreliable [2, 3]. Feature detection methods can be employed to discard irrelevant attributes and reduce data dimensionality [1, 2]. This is useful in the volumetric reconstruction of images, which requires high sensitivity in feature detection. This applies to aberrations like tissue protrusions, illumination changes, scaling differences, and motion [27], which can all cause feature redundancy in images [26].

Manual boundary delineation of fused cardiomyocyte nuclei volumes is time-consuming because boundaries are poorly defined or entirely undiscernible [5] (Figure 8.5 (A)). Additionally, because of background fluorescence present in light sheet fluorescence microscopy [5], poor contrast between neighboring cardiomyocytes is a pervasive issue [1] (Figure 8.6 (A) and (D)). Further contributing to the problem are low sampling rates, autofluorescence, and dynamic motion of the heart convoluting the lateral and axial imaging planes and interfering with optimal image quality [5] (Figure 8.5 (B)–(D)).

Intensity-based separation techniques like Otsu's method, iso data thresholding, entropy-based thresholding, and adaptive thresholding can be used for automated cell tracking, but these methods are known to perform poorly when noise is present, and consequently do not allow separation of clustered objects into distinctly individual objects [28]. These techniques are further hindered by optical aberrations, short exposure times, movement of cells in and out of the field of view, and poor contrast between cells [3, 27]. The watershed algorithm is another popular strategy, but it is highly prone to over-segmentation, and factors like noise or complex cell morphology can result in the false detection of features [3, 28]. Owing to these limitations, an alternate method is needed for separating the target biomarker from its immediate surroundings and creating distinct and meaningful biological regions.

The pre-processing technique that addresses this need involves first the difference of Gaussian (DoG) scale-space bandpass operation to reduce noise

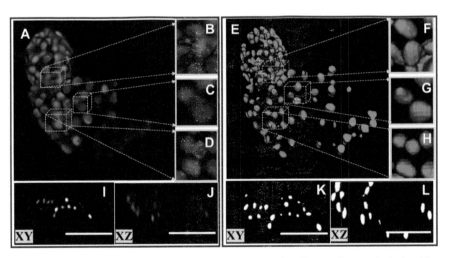

Figure 8.5 [5] Using the Difference between the Gaussian filter and watershed algorithm to individualize cardiomyocyte nuclei. (A) 2 dpf volumetric reconstruction of zebrafish cardiomyocytes, acquired using light sheet fluorescence microscopy and used to visualize the time-dependent motion of cardiomyocytes. (B-D) Magnifications of image (A) showing tight clusters of cardiomyocytes. Tracking and counting cardiomyocytes is hampered by the density of clusters. (E) The difference of the Gaussian (DoG) method is applied along with the watershed algorithm to individualize clustered cardiomyocytes. (F–H) Magnifications of image (E) demonstrate the distinctness of each cardiomyocyte compared to those seen in images (B–D). (I–J) 2D lateral and axial views showing object overlap due to increased noise from complex tissue morphology. Scale bar: 50 μm. (K–L) Lateral and axial segments as binary data are used to verify the curvature of the marker cardiomyocytes. Scale bar: 50 μm.

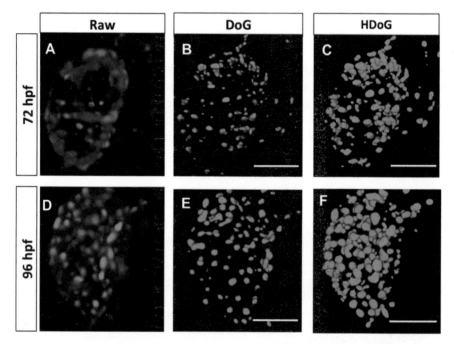

Figure 8.6 [5] Segmenting overlapping cardiomyocyte nuclei using the watershed algorithm and Hessian Difference of Gaussian. (A, D) Unprocessed volumetric reconstructions of (A) 3 dpf and (D) 4 dpf zebrafish cardiomyocytes. Higher density of cardiomyocytes results in more light scattering, yielding blurred images. (B, E) Watershed algorithm and Difference of Gaussian (DoG) detector applied to (B) 3 dpf and (E) 4 dpf volumes result in under-segmentation and inaccurate cardiomyocyte tracking. Scale bar: 50 μm. (C, F) Watershed algorithm and Hessian Difference of Gaussian (HDoG) are applied to the (C) 3 dpf and (F) 4 dpf cardiomyocyte volumes, resulting in more sensitive and accurate blob detection and segmentation. Scale bar: 50 μm.

and minimize false positives [29]. Rather than relying on brightness variation to identify features as binary images, this detection technique instead focuses on sensitivity to edges through the human visual perception (HVP) model [26, 30, 31]. Next, the watershed algorithm is used, yielding segmented cardiomyocytes even in a relatively dense 48 hours post fertilization (hpf) cluster of cells [5] (Figure 8.5 (E)–(H)). Since the processed images were binary, the watershed algorithm does not over-segment by detecting background or autofluorescent noise.

Because the DoG method filters out high-frequency noise, under-detection of features is a possiblility [25]. In comparatively dense fields of

72 and 96 hpf cardiomyocytes, low pixel intensities produced by the DoG edge detector results in under-segmentation and incorrect cell tracking [5] (Figure 8.6 (B) and (E)). Such under-detection is counteracted through the application of local contour detection with the Hessian matrix [22]. The HPV-based Hessian difference of Gaussian (HDoG) strategy locates saddle points [32], which are points in a function that represents neither an intensity maximum nor minimum and can indicate merged nuclei borders. [5] allows for precise identification of cardiomyocyte nuclei boundaries and individual volumes even in dense cell environments [5] (Figure 8.6 (C) and (F)).

This edge detection technique can be coupled with post-processing methods like top-hat and bottom-hat transform, to remove redundant binary

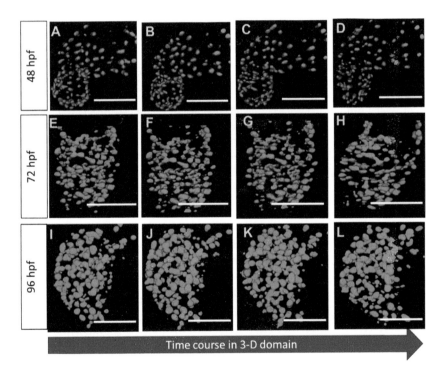

Figure 8.7 [5] Separating cardiomyocytes in differing tissue morphologies of multiple developmental stages. (A–D) Visualizations of dynamic 2 dpf zebrafish cardiomyocytes. Scale bar: 50 μm. (E–H) 3 dpf dynamic cardiomyocytes individualized through the use of Hessian Difference of Gaussian detector. Scale bar: 50 μm. (I–L) 4 dpf cardiomyocytes are detected and isolated using the image processing methods, despite the fast frame rates needed to sample contractility. Scale bar: 50 μm.

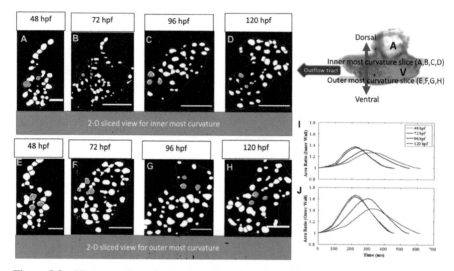

Figure 8.8   [5] Area ratio analysis using selected markers. (A–D) 2D slice views of zebrafish dorsal area of ventricle highlighting the innermost curvature for (A) 2 dpf, (B) 3 dpf, (C) 4 dpf, and (D) 5 dpf zebrafish. Scale bar for (A): 30 μm. Scale bar for (B–D): 50 μm. (E–H) 2D slices of ventral part of zebrafish heart highlighting the outermost ventricle curvature for (E) 2 dpf, (F) 3 dpf, (G) 4 dpf, and (H) 5 dpf zebrafish. Scale bar for (E): 30 μm. Scale bar for (F–H): 50 μm. (I) Three cardiomyocytes highlighted in (A-D) are tracked in area ratio for innermost curvature and demonstrate increasing contractility of the developing zebrafish heart. (J) Three cardiomyocytes (E–H) are tracked in the area ratio of the outermost curvature and show that the outermost curvature has higher contractility than the innermost curvature.

features [22, 26], and applied to cell studies in developmental biology [3, 34]. It has been used to identify and track cardiomyocyte nuclei, thereby allowing quantification of in vivo contractility in the zebrafish heart during distinct phases of development [5]. Using this technique, dynamic cardiomyocyte movement through the cardiac cycle was visualized at 48 hpf (Figure 8.7 (A)–(D)), at 72 hpf (Figure 8.7 (E)–(H)), and at 96 hpf (Figure 8.7 (I)–(L)). The HDoG edge detector was able to segment the cardiomyocyte nuclei regardless of anisotropic Gaussian luminance, fast frame rates required for imaging, and the increasingly dense cell environments of more mature developmental stages [5]. The HDoG edge detector can be used not only for segmenting and counting cardiomyocytes but for any other type of cell, no matter the orientation or heterogeneity of sizes [5].

Application of this biomarker edge detection method is proving to be a useful tool in cell morphology studies, cell proliferation studies, and developmental signaling mechanotransduction [5]. For instance, this method was used to demonstrate cardiac maturation through quantification of the outermost curvature having a higher area ratio than the innermost curvature [5]. Cardiomyocyte nuclei were tracked across developmental stages ranging from 48 hpf to 120 hpf (Figure 8.8 (A)–(H)). Stretch level changes in the developing zebrafish heart were investigated along with area ratio comparisons between innermost and outermost curvature areas [5]. After analyzing the time course of area ratio using three cardiomyocytes as markers, it was found that the area ratio of the outermost curvature area (the opposite side of the atrioventricular canal receiving blood pumped in from the atrium) displays a higher area ratio than the innermost curvature of the ventricle, though the area ratio for both regions increases consistently [5] (Figure 8.8 (I) and (J)).

8.3 Summary

Novel image processing for LSFM adopted for biomechanical quantification would be able to solve important biological questions including developmental biology, molecular biology, and genetics in a Mechanobiology manner. Specifically, dynamic sample images, such as heart or rapid cell movement as well as calcium transient of tissue, previously remained challenging problems although other image modalities have been developed. Light-sheet microscope with a high-end camera has overcome the aforementioned problems, still diminishing fluorescent intensity or tissue scattering exacerbated original image qualities to biomechanical quantification or cell signaling analysis. These recently published papers will be beneficial in various research societies.

References

[1] Bolot'n-Canedo V, Remeseiro B. Feature selection in image analysis: a survey. *Artif Intell Rev*. 2020;53(4):2905-2931. doi:10.1007/s10462-019-09750-3

[2] Torres R, Judson-Torres RL. Research Techniques Made Simple: Feature Selection for Biomarker Discovery. *J Invest Dermatol*. 2019;139(10):2068-2074.e1. doi:https://doi.org/10.1016/j.jid.2019.07.682

[3] Meijering E, Dzyubachyk O, Smal I, van Cappellen WA. Tracking in cell and developmental biology. *Semin Cell Dev Biol.* 2009;20(8):894-902. doi:https://doi.org/10.1016/j.semcdb.2009.07.004

[4] Wright PT, Tsui SF, Francis AJ, MacLeod KT, Marston SB. Approaches to High- Throughput Analysis of Cardiomyocyte Contractility. *Front Physiol.* 2020;11:612. doi:10.3389/fphys.2020.00612

[5] Teranikar, T., Villarreal, C., Salehin, N., Lim, J., Ijaseun, T., Cao, H., . . . Lee, J. (2021). Feature Detection to Segment Cardiomyocyte Nuclei for Investigating Cardiac Contractility. bioRxiv, 2021.2003.2003.433810. https://doi.org/10.1101/2021.03.03.433810

[6] Olarte, O. E., Andilla, J., Gualda, E. J., & Loza-Alvarez, P. (2018). Light-sheet microscopy: a tutorial. Advances in Optics and Photonics, 10(1), 111-179. https://doi.org/10.1364/AOP.10.000111

[7] Teranikar, T., Messerschmidt, V., Lim, J., Bailey, Z., Chiao, J.-C., Cao, H., . . . Lee, J. (2020). Correcting anisotropic intensity in light sheet images using dehazing and image morphology. APL Bioengineering, 4(3), 036103. https://doi.org/10.1063/1.5144613

[8] Johnsen, S. (2014). Hide and Seek in the Open Sea: Pelagic Camouflage and Visual Countermeasures. Annual Review of Marine Science, 6(1), 369-392. https://doi.org/10.1146/annurev-marine-010213-135018

[9] Jing, D., Zhang, S., Luo, W., Gao, X., Men, Y., Ma, C., . . . Zhao, H. (2018). Tissue clearing of both hard and soft tissue organs with the PEGASOS method. Cell Research, 28(8), 803-818. https://doi.org/10.1038/s41422-018-0049-z

[10] Crosignani, V., Dvornikov, A., Aguilar, J., Stringari, C., Edwards, R., Mantulin, W., & Gratton, E. (2012). Deep tissue fluorescence imaging and in vivo biological applications. Journal of Biomedical Optics, 17(11), 116023.

[11] Bagge, L. E., Kinsey, S. T., Gladman, J., & Johnsen, S. (2017). Transparent anemone shrimp (Ancylomenes pedersoni) become opaque after exercise and physiological stress in correlation with increased hemolymph perfusion. Journal of Experimental Biology, 220(22), 4225-4233. https://doi.org/10.1242/jeb.162362

[12] Žuži M, Čejka J, Bruno F, Skarlatos D and Liarokapis F (2018) Impact of Dehazing on Underwater Marker Detection for Augmented Reality. Front. Robot. AI 5:92. doi: 10.3389/frobt.2018.00092M. A. Malathi V, Int. J. Eng. Adv. Technol. (IJEAT) 9(2), 280 (2019).

[13] Lu, H., Li, Y., Zhang, L., & Serikawa, S. (2015). Contrast enhancement for images in turbid water. Journal of the Optical Society of America A, 32(5), 886-893. https://doi.org/10.1364/JOSAA.32.000886

[14] T. M. Nimisha, K. Seemakurthy, A. N. Rajagopalan, N. Vedachalam, and, and R. Raju, in Proceedings of the Tenth Indian Conference on Computer Vision, Graphics and Image Processing (Association for Computing Machinery, Guwahati, Assam, 2016), Article No. 26.

[15] He, K., Sun, J., & Tang, X. (2011). Single Image Haze Removal Using Dark Channel Prior. IEEE Transactions on Pattern Analysis and Machine Intelligence, 33(12), 2341-2353. https://doi.org/10.1109/TPAMI.2010.168

[16] Nollet EE, Manders EM, Goebel M, Jansen V, Brockmann C, Osinga J, van der Velden J, Helmes M and Kuster DWD (2020) Large-Scale Contractility Measurements Reveal Large Atrioventricular and Subtle Interventricular Differences in Cultured Unloaded Rat Cardiomyocytes. Front. Physiol. 11:815. doi: 10.3389/fphys.2020.00815

[17] Morrissy, S., & Chen, Q. M. (2010, July 12). Oxidative stress and heart failure. Comprehensive Toxicology (Second Edition). Retrieved from https://www.sciencedirect.com/science/article/pii/B97800804688 46007119

[18] Legrice I, Pope A, Smaill B. The Architecture of the Heart: Myocyte Organization and the Cardiac Extracellular Matrix. In: Vol 253. ; 2005:3-21. doi:10.1007/0-387-22825-X_1

[19] de Pater E, Clijsters L, Marques SR, et al. Distinct phases of cardiomyocyte differentiation regulate growth of the zebrafish heart. *Development*. 2009;136(10):1633 LP - 1641. doi:10.1242/dev.030924

[20] Bensley JG, De Matteo R, Harding R, Black MJ. Three-dimensional direct measurement of cardiomyocyte volume, nuclearity, and ploidy in thick histological sections. *Sci Rep*. 2016;6(1):23756. doi:10.1038/srep23756

[21] Fei P, Lee J, Packard RR, et al. Cardiac Light-Sheet Fluorescent Microscopy for Multi- Scale and Rapid Imaging of Architecture and Function. *Sci Rep*. 2016;6:22489. doi:10.1038/srep22489

[22] Beucher S, Mathmatique C. The Watershed Transformation Applied To Image Segmentation. *Scanning Microsc*. 2000;6.

[23] Bakkers J. Zebrafish as a model to study cardiac development and human cardiac disease. *Cardiovasc Res*. 2011;91(2):279-288. doi:10.1093/cvr/cvr098

[24] Major, R. J., & Poss, K. D. (2007). Zebrafish Heart Regeneration as a Model for Cardiac Tissue Repair. Drug discovery today. Disease models, 4(4), 219–225. https://doi.org/10.1016/j.ddmod.2007.09.002

[25] Yin X, Ng BW-H, He J, Zhang Y, Abbott D. Accurate Image Analysis of the Retina Using Hessian Matrix and Binarisation of Thresholded Entropy with Application of Texture Mapping. *PLoS One.* 2014;9(4):1-17. doi:10.1371/journal.pone.0095943

[26] El-gayar MM, Soliman H, meky N. A comparative study of image low level feature extraction algorithms. *Egypt Informatics J.* 2013;14(2):175-181. doi:https://doi.org/10.1016/j.eij.2013.06.003

[27] Xu Y, Wu T, Gao F, Charlton JR, Bennett KM. Improved small blob detection in 3D images using jointly constrained deep learning and Hessian analysis. *Sci Rep.* 2020;10(1):326. doi:10.1038/s41598-019-57223-y

[28] Bharodiya AK, Gonsai AM. An improved edge detection algorithm for X-Ray images based on the statistical range. *Heliyon.* 2019;5(10):e02743-e02743. doi:10.1016/j.heliyon.2019.e02743

[29] Zhang, H., Li, Y., Chen, H., Yuan, D., & Sun, M. (2013). Perceptual Contrast Enhancement with Dynamic Range Adjustment. Optik, 124(23), 10.1016/j.ijleo.2013.04.046. https://doi.org/10.1016/j.ijleo.2013.04.046

[30] So Í S, Johnsen S. *Lifting the Cloak of Invisibility: The Effects of Changing Optical Conditions on Pelagic Crypsis 1.* Vol 43.; 2003.

[31] Kulikov V, Guo S-M, Stone M, et al. DoGNet: A deep architecture for synapse detection in multiplexed fluorescence images. *PLOS Comput Biol.* 2019;15(5):1-20. doi:10.1371/journal.pcbi.1007012

[32] Marsh BP, Chada N, Sanganna Gari RR, Sigdel KP, King GM. The Hessian Blob Algorithm: Precise Particle Detection in Atomic Force Microscopy Imagery. Sci Rep. 2018;8(1):978. doi:10.1038/s41598-018-19379-x

9

Assessment of Cardiac Functions in Developing Zebrafish using Imaging Techniques

Amir Mohammad Naderi[1], Daniel Jilani[1], and Hung Cao[1,2,3]

[1]Department of Electrical Engineering and Computer Science, UC Irvine, USA
[2]Department of Biomedical Engineering, UC Irvine, USA
[3]Department of Computer Science, UC Irvine, USA

Abstract

In this chapter, the application of imaging in the assessment of zebrafish (zf) cardiology is discussed. Medical imaging is used to expose internal structures and establish a database of normal anatomy and physiology. Through the analysis of these images, diseases and abnormalities are diagnosed and treated. Considering that the embryonic zebrafish is transparent, bright field microscopic videos could reveal the heart mechanism and could be useful for quantification of it; although microscopic imaging can be useful for adult zebrafish as well. However, alternative imaging methods used in other works will be discussed first. Later, the cardiovascular parameters that can be measured using imaging will be defined. We will then compare different digital image processing and deep learning algorithms that have been employed to process or segment images from zebrafish. At the end of the chapter, challenges in mutant variants of zf will be investigated.

Keywords: Digital image processing, deep learning algorithms, image segmentation, medical imaging, embryonic zebrafish

9.1 Introduction

By looking at a bright field microscopic image from a zf embryo, a significant amount of information such as heart rate, ejection fraction (EF), and dimensions of different organs can be ascertained. In Figure 9.1, a microscopic image from a 3 days post fertilization (dpf) zf is shown.

For zf embryos, blood flow velocities are measured to determine cardiovascular function which can be assessed using microscopic videos[1]. This can be accomplished simply by following the motions of red blood cells (RBCs) throughout the embryo's body, which are clearly visible due to the embryo's transparent skin. RBC motions (and thus blood flow) can have their acceleration, deceleration, and peak velocity quantified for analysis. RBC motions in the dorsal aorta and the cardinal vein, two major blood arteries in the body, can be observed for this purpose. Individual cell locations are obtained using consecutive frames, and RBC velocity is computed using the coordinates of the cell's location and the time interval between frames as follows:

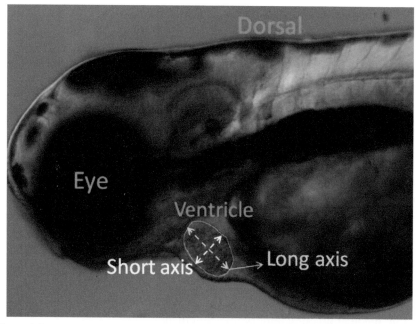

Figure 9.1 Microscopic image from a 3-dpf zebrafish where the ventricle border, long, and short axes are highlighted. Figure from [13].

$$\text{RBC velocity} = \frac{\sqrt{(x_2 - x_1)^2 + (y_2 - y_1)^2}}{\Delta t} \tag{9.1}$$

where x_1, y_1, and x_2, y_2 are the position change of an RBC in the time of Δt. The embryonic zf (up to –3 dpf) are transparent and have good visibility of their internal organs, including the heart and blood circulation. At this stage, bright-field microscopic videos can be used to quantify heart mechanism and morphology. Typically, two-dimensional (2D) movies are recorded for cardiovascular analysis. Then, during the cardiac cycle, continual changes in ventricular wall position would be tracked by first selecting a linear region of interest for the ventricle's borders.

Measurement of myocardial thickness, for example, is critical for determining the magnitude of an induced defect in hypertrophic cardiomyopathy. In zf embryos, fractional area change (FAC) is a well-established ventricular function metric for evaluating contractility. It can be estimated using 2D still frames of the ventricle at end-diastole (ED) and end-systole (ES). The fully dilated ventricle is designated as ED, whereas the fully contracted ventricle is designated as ES. In Figure 9.2 , a time series progression of the heartbeat process for a 3-dpf zf embryo is illustrated and the manual segmentation of the ventricle and its area is also provided.

The goal is to track the ongoing changes in the position of the ventricular wall during the cardiac cycle. You can accomplish this by first deciding on a linear region of interest. Either the short axis or the long axis of the ventricle would correspond to this area. In literature, the shape of the ventricle is assumed to have a spheroidal shape. This assumption is not completely accurate, however, with 2-D imaging that is the most simple and accurate model for the ventricle. For a spheroidal shape, having long and short axis can result in volume and other dimensions. At these two positions, the ventricular areas (EDA and ESA) are determined, and the fractional area change (FAC) is derived as follows:

$$\text{FAC} = \frac{(\text{EDV} - \text{ESV})}{\text{EDV}} \times 100. \tag{9.2}$$

Another measure of ventricular contractility is fractional shortening (FS), which can be calculated using the ventricular diameters at ED and ES (Dd and Ds) as follows:

$$\text{FS} = \frac{(D_\mathrm{d} - D_\mathrm{s})}{D_\mathrm{d}}. \tag{9.3}$$

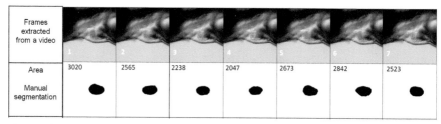

Frames extracted from a video							
Area	3020	2565	2238	2047	2673	2842	2523
Manual segmentation							

Figure 9.2   A series of eight frames extracted from a bright field microscopy video of a 3-dpf zf. Each frame on the top has its corresponding manually segmented ventricle mask and the measured area using the software ImageJ (https://imagej.nih.gov/ij/).

To determine stroke volume, ejection fraction, and cardiac output, ventricular volumes must be computed. The long- and short-axis diameters (DL and DS) are initially measured from 2D still images. The following volume formula can be employed if the ventricle has a prolate spheroidal shape:

$$\text{Volume} = \frac{1}{6} \times \pi \times D_{\text{L}} \times D_{\text{S}}^2. \tag{9.4}$$

The blood volume pumped from the ventricle for each beat is called stroke volume (SV), and it is easily determined using the ventricle volumes at ED (EDV) and ES (ESV):

$$\text{SV} = (\text{EDV} - \text{ESV}). \tag{9.5}$$

The fraction of blood evacuated from the ventricle with each heartbeat is known as ejection fraction (EF), and it may be computed using the formula:

$$\text{EF}\,(\%) = \frac{(\text{EDV} - \text{ESV})}{\text{EDV}} \times 100 = \frac{\text{SV}}{\text{EDV}} \times 100. \tag{9.6}$$

The following formula can be used to compute cardiac output (CO) from SV and heart rate (HR):

$$\text{CO}\left(\frac{\text{nanoliter}}{\text{min}}\right) = \text{SV} \times \text{HR}. \tag{9.7}$$

The time between two identical subsequent points (i.e., ED or ES) in the captured images is used to calculate HR.

9.2 Image Processing Methods

For calculating the HR of the zf from videos, there are many differ-ent methods that have been discussed in the literature. To name a few, frequency transforms such as fast Fourier transform, filtering, and pixel intensity changes are all employed for the quantification of HR and heart rate variability. These methods can be grouped into three categories: Time domain, frequency domain, and blind source separation. However, most of the methods for HR measurement cannot measure the other cardiovascular like heart contractibility measures like EF and FS. In contrast, most methods used for the quantification of heart contractibility can also measure HR. Hence, the focus of this chapter will be on these methods. For a more comprehensive review, Ling et al. discuss quantitative measurements of zf heart rate and heart rate variability at greater length [2]

The general idea for evaluating contractility via video is semantic seg-mentation of the ventricle. By segmenting the ventricle into a series of consecutive frames, the ES and ED frames can be found, and thereby the ven-tricular contractility. In colored microscopic videos, segmentation of the heart can be done much easier by filtering red color as red has more intensity in the heart region. In black-and-white recordings, more complicated approaches are needed. Two of the most important features in the segmentation of the ventricle in a beating heart video are the motion of the ventricle and the edges in the border.

9.2.1 Background subtraction

Background subtraction is a common technique for dividing the moving parts of a scene taken by a static camera by segmenting it into background and foreground. Continuous frames from a video are subtracted from each other to locate moving objects. The static pixels can be eliminated because the majority of the fish body is motionless and the only pixels moving in the video are blood cells and the heart. Frame difference, Gaussian mixture model, kernel density estimation, and fuzzy model are just a few examples of background subtraction approaches. All of the methods have varying degrees of accuracy for different applications [3]. In a video in which the fish is completely static and minimal noise is present this method could be helpful in the segmentation of the ventricle. Evidently, other dynamic features such as blood cells and vessels, gill movement due to respiration, and noisy pixels will be detected in the output of this method and require varying approaches to remove. For example, large moving objects detected that are not a part of the ventricle can be removed using specifying a region of interest (ROI) and thresholding the size of the object. Small detected particles and noise can be removed by filtering. For example, an arithmetic mean filter can be used for smoothening and a geometric mean filter can be used for removing salt and paper noise.

9.2.2 Morphological image processing

Morphology is a broad range of image processing techniques that manipulate images depending on their shapes[4]. An input image is given a structural element by morphological procedures, which results in an output image of the same size. In a morphological operation, the value of each pixel in the output image is determined by comparing it to its neighbors in the corresponding pixel in the input image. Morphological image processing and spatial filtering are fundamentally comparable. Every pixel in the original image has the structural element moved across it to create a pixel in the newly processed image. The morphological process used determines the value of this new pixel. Erosion and dilation are the two most frequently used operations. Erosion shrinks the image pixels and removes pixels on object boundaries. Dilation expands the image pixels, or it adds pixels to object boundaries. Binary graphics could have a lot of flaws. Noise and texture in particular alter the binary regions created by straightforward thresholding. By considering the shape and structure of the image, morphological image processing aims to achieve the objective of erasing these flaws. Greyscale photos can be used with these methods as well. Morphological filters are useful for smoothing binary images, especially for removing small structures and border detection. The morphological filter's concept is a shrink-and-let-grow procedure. The term "shrink" refers to the use of a median filter to round off large structures and remove small structures, with the surviving structures being grown back by the same amount during the growth process[5]. In zf videos, employing the background subtraction oftentimes leads to detecting the ventricle as a region containing a group of multiple objects that form the shape which represents ROI. Applying morphological filters can be helpful because we need to have a single object to represent the ventricle. Nevertheless, background subtraction can result in inaccurate results that cannot guarantee.

Additionally, if we want to detect borders there are a number of algorithms that can enhance or detect the edges. Enhancing the image in a way that the ventricular border can be more visible can be beneficial to researchers for manual and automatic segmentation.

9.2.3 High-pass filter

High-pass (sharpening) filters are a form of taking derivative from the time domain signal. Taking the derivative of an image will amplify rapid changes like edges. Laplacian and Sobel filters are some examples of high-pass filters used to sharpen an image. Additionally, Gaussian and Butterworth filters can be designed to be high-pass filters.

9.2.4 Thresholding

One of the most basic approaches to image segmentation is Histogram-based thresholding. Thresholding can be used to create binary pictures from a grayscale image. In the most basic form of this method, each pixel's intensity will be changed to black if it is less than the specified constant threshold or white if it is more than it. There are algorithms like Otsu that find the best threshold automatically[6]. Otsu finds a threshold that segments the background and foreground classes of the histogram by minimizing intra-class intensity variance.

9.2.5 Histogram equalization

The effectiveness of histogram thresholding is diminished when the range of pixel intensity values is narrow. By employing histogram equalization, the range of intensities represented by the histogram is spread evenly, resulting in an image with higher global contrast. In essence, local areas of low contrast in the image become more distinguishable after histogram equalization is applied. In zf videos, it is very common for the region of interest to be too dark due to poor placement of the fish on the microscope, thickness and transparency of some tissues compared with the surrounding tissues, etc. Having a dynamic range of contrast will help with manual and automated segmentation of the heart. Histogram equalization works well when the distribution of pixel values is similar all throughout the image. However, the contrast in certain sections will not be appropriately improved if the image contains regions that are much lighter or darker than the rest of the image. In zf videos, this is especially troublesome since the background of the light sheet microscopy has the highest intensity in the image and there are different sections of the fish with different transparencies. Hence, Adaptive Histogram Equalization

(AHE) solves the problem by transforming each pixel with a transformation function consequent to a neighborhood region. Finally, Contrast Limited AHE (CLAHE) is a variant of adaptive histogram equalization, which does not have the issue of over-amplification of noise in regular AHE [7].

9.2.6 Edge detection

Edge detection refers to a set of mathematical techniques for detecting edges, or curves, in a digital image when the brightness of the image abruptly changes or, more formally, has discontinuities. The Canny algorithm, which is one of the most prominent edge detection methods, is a multi-stage algorithm to detect a wide range of edges in images [8]. For fully automated heart segmentation in zf microscopic videos, edge detection algorithms like Canny are usually not robust. The most important problem is that in the zf videos have numerous edges and tissues therefore the canny algorithm detects many different edges next to each other. This makes it impossible to tell which edge belongs to the heart. However, edge detection can be used as preprocessing or as one of the steps in an automatic segmentation framework.

9.2.7 Color filtering

In colored microscopic videos, the red coloration associated with the cardiac system can be exploited to improve image segmentation. By thresholding the intensity of redness of each pixel, a binary image can be produced. In this image, the heart will be distinctly separate from all other features except the blood vessels and other noise.

In literature transgenic animals expressing the myocardial-specific fluorescent reporter have been frequently used to improve image segmentation. However, for fully automated quantification of cardiovascular metrics like EF, segmentation of the ventricle is needed. Using a simple color filtering of the color that the heart is highlighted in is going to segment the whole heart. Hence, further segmentation of the chambers of the heart and identifying the ventricle is necessary. A convolutional neural network (CNN) architecture that automatically identifies the chambers from the videos and calculates the EF was proposed by Akerberg et al. [9]. In this paper, a CNN was used to segment the ventricle and atrium individually in a video that the heart is annotated. In conclusion, color filtering alone can only be employed as a feature selection method to increase the accuracy of segmentation. Machine learning's application will be discussed more in the next sections.

9.2.8 Machine learning

Aforementioned methods, namely edge detection, color filtering, and background subtraction, are not robust, as the ventricle edges might have multiple scales of gray and have different textures all around. Additionally, the border could be partially obstructed by other tissues. Thus, researchers tried to apply machine learning approaches to build a fully automated framework. First, unsupervised learning segmentation methods like the Gaussian mixture model (GMM) and K-means will be discussed. Secondly, supervised deep learning methods are brought.

9.2.8.1 K-means

Clustering algorithms are unsupervised algorithms that are similar to classification algorithms, but differ in their underlying principles. When you employ clustering algorithms on your dataset, unexpected features like structures, clusters, and groupings can appear that you would not have imagined. The unsupervised K-means clustering algorithm is utilized to separate the interest area from the background. Based on the K-centroids, it clusters or partitions the given data into K-clusters or sections. When you have unlabeled data, the algorithm is used (i.e., data without annotations or groups). The purpose is to locate specific groups based on some form of data similarity, with K being the number of groups. In clustering an image-based, K different color ranges get clustered. In the black-and-white videos, the same concept could be applied to ranges of gray levels. The K-means algorithm is divided into two parts. In the first step, the k centroid is determined, and in the second phase, each point is moved to the cluster with the centroid that is closest to the data point. The Euclidean distance is one of the most widely used methods for determining the distance to the nearest centroid. In the second phase, after grouping, the algorithm recalculates the new centroid of each cluster. Then it calculates a new Euclidean distance between each center and each data point based on that centroid and allocates the points in the cluster with the shortest Euclidean distance. The centroid of each cluster is the place at which the sum of the distances between all of the items in the cluster is the smallest. So, overall, K-means is an iterative algorithm that minimizes the sum of distances between each item and its cluster centroid. To generate a noise-free image, median filtering is utilized as a noise-removal technique. The segmented image may still have some undesired regions or noise after it has been segmented. As a result, the median filter is applied to the segmented image to improve its quality [10].

9.2.8.2 Gaussian mixture model

Histogram thresholding, one of the most used methods for segmenting images, is the foundation of the Gaussian-mixture-based segmentation technique. In histogram thresholding, a picture is divided into two classes or regions: target and background, each with its own uni-modal gray level distribution. As a result, the segmentation problem requires selecting a desirable threshold for partitioning the image into target and background regions. In this algorithm, it is then assumed that there are two probability density functions (PDF) of the gray levels in the image with given the means, standard deviations, and proportions. If we know the means, standard deviations, and proportions for both the background and target area, the required threshold can be calculated. The accuracy of model parameter estimations and how closely the histogram of an image approximates a Gaussian mixture determine the effectiveness of Gaussian-mixed-based segmentation techniques [11].

The unsupervised methods can be used for preprocessing for other methods like deep learning models. All these are shown in Figure 9.3, A–D panels. As seen in the figure, mentioned methods have all enhanced the image and made the borders of the ventricle more visible compared to the original image. Hence, they could be used as preprocessing for a deep learning framework to increase its accuracy.

None of the methods mentioned above can be used as a stand-alone robust system for fully automated segmentation. Additionally, manual segmentation is extremely tedious work and in most practical research scenarios there are numerous videos recorded, and manual segmentation can take be time-consuming as well. In conclusion, for a fully automated framework, a more robust method is required. For achieving this goal, a few recent papers proposed using deep learning methods.

9.2.8.3 Semantic image segmentation

Grouping portions of an image that belong to the same object class together through the classification of individual pixels is known as semantic segmentation. The method is a supervised machine-learning technique that learns mask patterns for an image using techniques such as fuzzy measures, decision

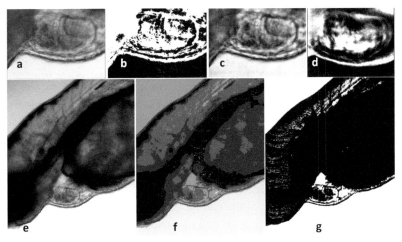

Figure 9.3 Ventricle segmentation using different methods. Panels A–D: A frame from the video of a 3 dpf zebrafish with 40× zoom undergoing different HBS algorithms. (A) Original frame. (B) Manual histogram thresholding. (C) CLAHE. (D) Otsu thresholding. Panels E–G: A frame from the video of a 3 dpf zebrafish with 10× zoom undergoing GMM and K-means approaches. (E) Original frame. (F) GMM. (G) K-means [13].

trees, support vector machines, and artificial neural network-based methods. Artificial neural network models have been showing great performance and accuracy, particularly in biomedical images. The goal of semantic-level image classification is to assign a distinct semantic class to each scene image. The large-scale remote sensing image is manually processed to obtain the scene images. For example, here the object would be the ventricle of the zebrafish. A mask will have two classes: the ventricle and the background.

9.2.8.4 Semantic image segmentation validation metrics
We often divide our predictions into four groups when assessing a conventional machine learning model: true positives, false positives, true negatives, and false negatives. However, it is not immediately evident what constitutes a "true positive" and, more broadly, how we might evaluate our predictions for the complex prediction task of picture segmentation. The most used metrics for semantic picture segmentation include pixel-wise accuracy, dice coefficient, and intersection over union (IoU). In quantification of cardiovascular metrics from videos using deep learning methods, our goal is to accurately predict the geometrical shape of the ventricle based on the actual ventricle's position, size, and shape. We would anticipate that the predicted mask and

the ground truth overlap with an acceptable ratio. We use these ratios as loss functions and metrics to optimize and validate the segmentation task.

9.2.8.4.1 Pixel-wise accuracy

In segmentation of the ventricle, the ground truth is either the target or background represented by black or white in the mask. Thus, it is a binary classification. The pixel wise accuracy can be defined as bellow:

$$pixel - wise\ accuracy = \frac{pixels\ classified\ correctly}{All\ pixels} \times 100. \qquad (9.8)$$

In these videos, the ventricle has a much smaller area compared with the rest of the frame, so this metric alone can be misleading. The background is usually the dominant pixel distribution and in most pixels the system predicts background resulting in high accuracy. In most architecture, a combination of binary accuracy and the Dice or Jacard coefficient are used.

   a. *Dice coefficient*

      The dice coefficient is a widely utilized metric for determining the pixel-wise agreement of a segmented image and its ground truth mask. The measure has a scale of 0 to 1, with 1 indicating perfect match or complete overlap. For the binary case, the coefficient is calculated as:

$$Dice = \frac{2\,|(A \cap B)|}{|A| + |B|} \qquad (9.9)$$

      where A is the predicted image, and B is the ground truth.

   b. *Intersection over union*

      The IoU (or Jaccard index) measures the area of overlap between the predicted segmentation and the ground truth divided by the area of union between both. This measure ranges from 0 to 1, with 0 indicating no overlap and 1 indicating complete overlap. For the binary case, it can be calculated as:

$$J = \frac{|A \cap B|}{|A \cup B|}. \qquad (9.10)$$

9.2.8.5 Semantic segmentation using deep learning

Convolutional neural networks (CNNs, or ConvNets) are a type of artificial neural network (ANN) used most frequently in deep learning to interpret image data. The algorithms for convolutional neural networks (CNN) advanced quickly with the emergence of artificial intelligence. Medical imaging classification, object detection, and semantic segmentation all benefit from the use of CNN and its extension approaches. Comparatively speaking to other image classification algorithms, CNNs employ a minimal amount of pre-processing. This means that, unlike traditional methods where these filters are hand-engineered, the network learns to optimize the filters (or kernels) through automatic learning [12]. This feature extraction's independence from prior information and human interaction is a significant benefit.

Akerberg et al. proposed a CNN with encoder-decoder SegNet architecture for automatically segmenting and calculating the EF from mice videos using MATLAB environment. A ground-truth reference dataset was created by manually segmenting systole and diastole for both chambers, across four animals [9]. Nonetheless, unique transgenic mice expressing the myocardial-specific fluorescent reporter and high-end fluorescence microscopes were used, which are not universally applicable to the scientific community, particularly those who do not have access to transgenic lines or fluorescence microscopes.

For a more inclusive and available example of a fully automatic cardiovascular segmentation for zf, Naderi et al. proposed a framework using U-net

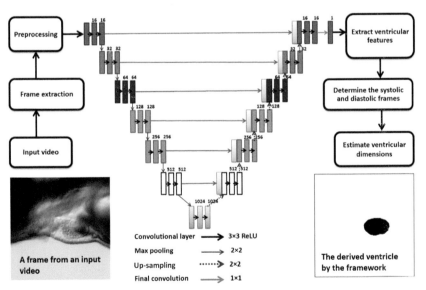

Figure 9.4   The process flow and the U-net architecture. Each rectangle represents a layer, and the number above it shows the number of the neurons inside. A trained model can estimate a mask of the ventricle from all the extracted frame of the input video. When all the frames have a predicted mask, by determination of ES and ED frames, important cardiac indices like EF, FS, and stroke volume can be automatically calculated and saved in a desired format [13].

to segment monochromic light sheet microscopy videos [13]. The process for the proposed framework can be seen in Figure 9.4. In this framework, after preprocessing using sharpening filter and CLAHE, 50 videos of wild-type and mutant type zf have been manually segmented to be used for the training dataset.

The U-net was then trained and validated using dataset. The deep learning model showed 99.1% pixel-wise accuracy, 95.04% for the Dice coefficient, and 91.24% for IoU. A graphical user interface was created to provide an end-to-end platform so researchers can use it conveniently. The framework inputs raw videos and segments the ventricle in each frame. The output for each frame is a binary mask of the ventricle. From there, diameters of the ventricle in each frame can be calculated. The frames with the largest and smallest area represent ED and ES, respectively. Having the ED and ES frames, important cardiovascular parameters, namely EF, FS, CO, and SV, can be quantified. The EF quantification was validated using 8 videos that were not included in the training set. The averages of absolute errors and standard deviations for the automatically calculated EF of the 8 wild-type test videos compared to the expert's manual calculation were 6.13% and 3.68%, respectively.

Figure 9.5 illustrates a series of frame from a ventricle masks that are manually and automatically segmented.

9.3 Estimations in 2D Videos

To estimate the volume of the ventricle from a 2D zf image, the short and long axes of the ventricle are used to model the ventricle three-dimensionally as a perfect ellipsoid. Given the ventricle takes on a complex shape, this estimation is inaccurate; this is especially significant for mutant variants of zf where the ventricle is not shaped like an ellipsoid. However, the only solution to this problem is 3D imaging. In literature, there are several studies that have extensively used 3D imaging techniques like Z-stack imaging [14]. In 3D segmentation of the chambers using deep learning, the process is fairly similar and can be easily adopted with enhancement of the architecture.

9.4 Consistency of Measurement

Looking at the two frameworks that used deep learning for automatic segmentation of the zf heart, it can be seen that this method is promising. The fully automated frameworks do the manual quantification of the cardiovascular metrics in a fraction of the time that it takes to do it manually. Performing the segmentation task manually is challenging due to the small size, partial obstruction, and unclear edges of the heart. These challenges, especially the unclearness of the edges will result in inconsistent manual measurements. In [13], this inconsistency in measurement was quantified. Two experts were asked to segment the ventricle in single frames of 12 sample videos. They were told to manually do the measurement twice for each frame, pausing briefly in between each attempt. Twelve frames with four measurements each were the outcome. The average standard deviation of the measurements in these 12 frames was roughly 150 pixels, with a standard deviation of 50 pixels. This represents about 8% of the area of an average ventricular region in our scenario. This demonstrates how inconsistent the manual segmentation was. With mutant embryos, whose EF is often quite low, this might be especially important. However, deep learning models like ZACAF are consistent and inputting the same image to the system will always result in the same measurement.

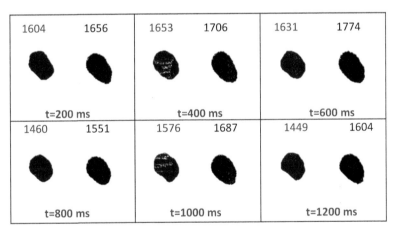

Figure 9.5   Validation of U-net image segmentation framework. The sequential frames from a wild type zebrafish recorded video with FPS of 5 are extracted. The respective ventricle mask of each frame is shown in each panel via manual and automatic segmentation. The area of each ventricle is measured and written above its own box.

9.5 Frame Rate Issue

The ground truth is created using the frames extracted from the videos. Therefore, effect of the frame rate of the videos cannot be evaluated using the metrics like Dice and accuracy. When it comes to quantifying parameters like HR, EF, and FS, the ES and ED frames are the most crucial ones. The camera shutter takes a sequence of images with a certain frames per second (fps). Higher fps will result in a higher probability of capturing ES and ED frames.

9.5.1 Complications of measurements in mutant types

Heart failure is a chronic, inherited condition known as dilated cardiomyopathy (DCM) [19]. Therefore, it is crucial to assess the early cardiac functions connected to DCM. In genetic studies of cardiomyopathy, dozens of harmful genes have been discovered, and the incidence rate of DCM is approximately 1/250 [20]. Titin truncated variants (TTNtv), which cause 25% of DCM cases, are the most prevalent genetic component [21]. Heart functions must be

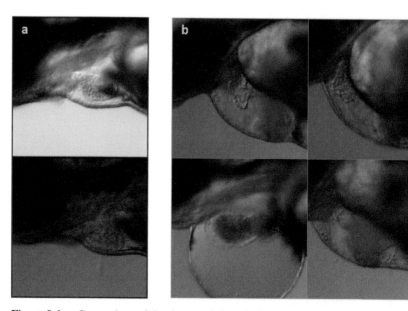

Figure 9.6 . Comparison of the shape and size of wildtype (A) and TTNtv mutant zebrafish (B). Besides the abnormal shape of the heart with the swollen ventricular wall, the smaller size of the ventricle is also found with TTNtv mutants. Further, the swollen chest can also be noticed.

accurately assessed in order to examine the mechanobiology of the generated abnormalities of these disease models [18]. There are two notable differences between the mutant and wild-type fish from the perspective of segmentation. In numerous mutant types, the ventricle and the heart as a whole have abnormal shape. Second, in mutant types like TTNtv the hear contractibility is highly effected. As a result, in TTNtv mutants, the ventricle area difference between ES and ED frames is quite small. Comparing wild with TTNtv zf using examples is shown in Figure 9.6. The area difference between the ED and ES frames in some cases is less than the segmentation error because the ventricle is hardly beating. To put it another way, the ventricle area hardly ever varies to the point where the nominator of the EF formula occasionally is in the scale of the estimation error. This is the main reason why EF measurements of TTNtv mutants are inaccurate, and future preprocessing enhancements or framework optimization would not have a substantial impact on the outcome. Low-resolution videos were used in [13] to show the capabilities of our system. Higher resolution would assist alleviate this problem, boosting the robustness and accuracy for TTNtv mutant and wild-type fish generally. Although this is advantageous for researchers in that required storage space is reduced. Moreover, properly zooming on the zf larva would greatly help with this issue by making the segmentation error negligible compared to the ventricle area. Generally, a $20\times$ zoom shows good results.

References

[1] Benslimane, F., et al., Cardiac function and blood flow hemodynamics assessment of zebrafish (Danio rerio) using high-speed video microscopy. Micron, 2020. 136: pp. 102876.

[2] Ling, D., et al., Quantitative measurements of zebrafish heartrate and heart rate variability: A survey between 1990—2020. Computers in Biology and Medicine, 2021: pp. 105045.

[3] Sobral, A. and A. Vacavant, A comprehensive review of background subtraction algorithms evaluated with synthetic and real videos. Computer Vision and Image Understanding, 2014. 122: pp. 4-21.

[4] Soille, P., Morphological image analysis: principles and applications. Vol. 2. 1999: Springer.

[5] Maragos, P., Chapter 13 - Morphological Filtering, in The Essential Guide to Image Processing, A. Bovik, Editor. 2009, Academic Press: Boston. pp. 293-321.

[6] Bangare, S. L., et al., Reviewing Otsu's method for image thresholding. International Journal of Applied Engineering Research, 2015. 10(9): pp. 21777-21783.

[7] Pizer, S. M., et al., Adaptive histogram equalization and its variations. Computer Vision, Graphics, and Image Processing, 1987. 39(3): pp. 355-368.

[8] Canny, J., A Computational Approach to Edge Detection. IEEE Transactions on Pattern Analysis and Machine Intelligence, 1986. PAMI-8(6): pp. 679-698.

[9] Akerberg, A. A., et al., Deep learning enables automated volumetric assessments of cardiac function in zebrafish. Disease models & mechanisms, 2019. 12(10): pp. dmm040188.

[10] Dhanachandra, N., K. Manglem, and Y. J. Chanu, Image Segmentation Using K -means Clustering Algorithm and Subtractive Clustering Algorithm. Procedia Computer Science, 2015. 54: pp. 764-771.

[11] Gupta, L. and T. Sortrakul, A gaussian-mixture-based image segmentation algorithm. Pattern Recognition, 1998. 31(3): pp. 315-325.

[12] Albawi, S., T. A. Mohammed, and S. Al-Zawi. Understanding of a convolutional neural network. in 2017 International Conference on Engineering and Technology (ICET). 2017.

[13] Naderi, A. M., et al., Deep learning-based framework for cardiac function assessment in embryonic zebrafish from heart beating videos. Computers in biology and medicine, 2021. 135: pp. 104565.

[14] Mickoleit, M., et al., High-resolution reconstruction of the beating zebrafish heart. Nature methods, 2014. 11(9): pp. 919-922.

[15] Merlo, M., et al., Evolving concepts in dilated cardiomyopathy. Eur J Heart Fail, 2018. 20(2): pp. 228-239.

[16] Hershberger, R. E., D. J. Hedges, and A. Morales, Dilated cardiomyopathy: the complexity of a diverse genetic architecture. Nat Rev Cardiol, 2013. 10(9): pp. 531-47.

[17] Wheeler, F. C., et al., QTL mapping in a mouse model of cardiomyopathy reveals an ancestral modifier allele affecting heart function and survival. Mamm Genome, 2005. 16(6): pp. 414-23.

[18] Hoage, T., Y. Ding, and X. Xu, Quantifying cardiac functions in embryonic and adult zebrafish. Methods Mol Biol, 2012. 843: p. 11-20.

[19] Hershberger, R. E., D. J. Hedges, and A. Morales, *Dilated cardiomyopathy: the complexity of a diverse genetic architecture.* Nat Rev Cardiol, 2013. **10**(9): pp. 531-47.

[20] Wheeler, F. C., et al., *QTL mapping in a mouse model of cardiomyopathy reveals an ancestral modifier allele affecting heart function and survival.* Mamm Genome, 2005. **16**(6): pp. 414-23.

[21] Hoage, T., Y. Ding, and X. Xu, *Quantifying cardiac functions in embryonic and adult zebrafish.* Methods Mol Biol, 2012. **843**: p. 11-20.

Index

4D imaging 271

About the Editors

Anh Hung Nguyen is a research scientist for The HERO Laboratory in University of California, Irvine. He received his Ph.D. degree in Chemical and Biological Engineering under the supervision of Dr. Sang Jun Sim. He joined The HERO Lab in 2019 and was promoted to his current position in 2020, working in cardiac regenerations and biosensor engineering. He has published articles in biosensors, molecular biology, and protein engineering.

Sang Jun Sim is a Professor of Chemical and Biological Engineering, Korea University, South Korea. He is a renowned leader in biosensors.

Hung Cao is an Associate Professor of Electrical Engineering, Biomedical Engineering, and Computer Science at UC Irvine. He is one of the pioneers in developing systems to assess the biomedical signals in zebrafish and rodent models. He directs the HERO Laboratory at UC Irvine.